Bio
105,- €
'21

METHODS IN MOLECULAR BIOLOGY

Series Editor
John M. Walker
School of Life and Medical Sciences
University of Hertfordshire
Hatfield, Hertfordshire, AL10 9AB, UK

For further volumes:
http://www.springer.com/series/7651

Protein Misfolding Diseases

Methods and Protocols

Editor

Cláudio M. Gomes

BioISI – Biosystems and Integrative Sciences Institute, Faculty of Sciences
University of Lisbon, Lisbon, Portugal;
Department of Chemistry and Biochemistry, Faculty of Sciences
University of Lisbon, Lisbon, Portugal

Humana Press

Editor
Cláudio M. Gomes
BioISI – Biosystems and Integrative Sciences Institute
Faculty of Sciences University of Lisbon
Lisbon, Portugal

Department of Chemistry and Biochemistry
Faculty of Sciences University of Lisbon
Lisbon, Portugal

ISSN 1064-3745 ISSN 1940-6029 (electronic)
Methods in Molecular Biology
ISBN 978-1-4939-8819-8 ISBN 978-1-4939-8820-4 (eBook)
https://doi.org/10.1007/978-1-4939-8820-4

Library of Congress Control Number: 2018958479

Cover illustration: SOD1 structure and a cartoon representing a fibril superimposed on a transmission electron microscopy image of amyloid fibrils.

This Humana Press imprint is published by the registered company Springer Science+Business Media, LLC, part of Springer Nature.
The registered company address is: 233 Spring Street, New York, NY 10013, U.S.A.

Preface

During their lifetimes, proteins are involved in multiple folding and unfolding processes in the cell. In many instances, in spite of tight regulatory processes for proteostasis maintenance and protein quality control, these events lead to misfolded protein conformers. The deregulated accumulation of misfolded or aggregated proteins in the cellular environment perturbs the biological function of the altered protein or of other proteins within its interactome that participate in common biochemical processes. Such perturbations may be due to genetic defects that impair protein folding, trafficking, or stability or, as it occurs more frequently, be caused by altered biochemical and physicochemical conditions in the cellular environment. The consequence of these perturbations in protein folding is frequently the emergence of pathophysiological states, and therefore such proteinopathies are known as protein misfolding diseases (also frequently referred to as protein folding diseases or conformational disorders). Collectively, protein misfolding diseases comprise a wide group of pathologies that can be broadly grouped as amyloid-forming diseases (e.g., Alzheimer's disease, light chain amyloidosis, or cataracts), chaperonopathies affecting molecular chaperones (e.g., hereditary spastic paraplegia), and non-amyloid misfolding diseases (e.g., cystic fibrosis or phenylketonuria). Such diversity of pathologies with distinct underlying molecular processes that involve defects in different proteins and biological functions calls for a multitude of experimental approaches and specialized methods.

This volume of the *Methods in Molecular Biology* series on protein misfolding diseases gathers a broad collection of experimental approaches to assist researchers in setting up different methods to investigate protein conformational disorders. The 21 chapters composing the volume are organized in three parts. Part I presents assays focusing on biophysical approaches to study protein (mis)folding, as the in vitro structural and mechanistic investigation of misfolding and aggregation is quintessential to understand disease processes. The section outlines assays to monitor aggregation kinetics, conformational dynamics, toxicity, and identification of aggregation biomarkers. Part II focuses on cellular and proteostasis assays which allow understanding of the misfolding of a given protein in the broader context of the cell, through implementing screening assays, imaging of aggregates in cells, cellular models, and implications in clearance and protein quality control machineries. Part III overviews assays for protein folding correction and recovery, combining methods such as thermal shift assays, in silico improvement of protein solubility, and compound screening, an important area of research as it may open avenues for therapeutic strategies.

Editorial projects require stamina, patience, and plenty of goodwill. In this respect, I have to firstly thank the nearly 80 authors from 14 different nationalities that have kindly agreed to participate in this volume, for their hard and rigorous work as well as for their tolerance in respect to the inevitable delays one faces when putting a book together. I would also like to acknowledge the professional and expedited assistance from Springer US staff David C. Casey, who, along with Patrick Marton and Anna Rakovsky, provided help and assured the smoothness of online chapter submission procedures. Finally, I would like to express my gratitude to the series editor, Dr. John M. Walker, for the opportunity to edit this book and for his counseling and guidance throughout the process.

I dedicate this book to all enthusiastic young researchers whose research will advance knowledge in protein folding diseases.

Lisbon, Portugal *Cláudio M. Gomes*

Contents

Contributors

ALIREZA ABDOLVAHABI • *Department of Chemical Biology and Therapeutics, St. Jude Children's Research Hospital, Memphis, TN, USA*
YOUSSRA K. AL-HILALY • *School of Life Sciences, University of Sussex, East Sussex, UK; Department of Chemistry, College of Sciences, Al-Mustansiriyah University, Baghdad, Iraq*
OSCAR AUBI • *Department of Biomedicine, University of Bergen, Bergen, Norway*
MICHAEL R. BALDWIN • *Department of Molecular Microbiology and Immunology, University of Missouri, Columbia, MO, USA*
LUIS M. BLANCAS-MEJIA • *Department of Biochemistry and Molecular Biology, Mayo Clinic, Rochester, MN, USA*
PETER BROSS • *Research Unit for Molecular Medicine, Department of Clinical Medicine, HEALTH, Aarhus University, and Department of Clinical Biochemistry Aarhus University Hospital, Aarhus, Denmark; Department of Biochemistry, Aarhus University Hospital, Aarhus, Denmark*
DONATELLA BULONE • *Institute of Biophysics, SL Palermo, National Research Council, Palermo, Italy*
NEIL R. CASHMAN • *Djavad Mowafaghian Centre for Brain Health, University of British Columbia, Vancouver, BC, Canada*
BYRON CAUGHEY • *Rocky Mountain Laboratories, NIAID, NIH, Hamilton, MT, USA*
CAGLA CÖMERT • *Research Unit for Molecular Medicine, Department of Clinical Medicine, HEALTH, Aarhus University, and Department of Clinical Biochemistry Aarhus University Hospital, Aarhus, Denmark; Department of Biochemistry, Aarhus University Hospital, Aarhus, Denmark*
SHAWNA A. COOPER • *Department of Biochemistry and Molecular Biology, Mayo Clinic, Rochester, MN, USA*
JOANA S. CRISTÓVÃO • *BioISI – Biosystems and Integrative Sciences Institute, Faculty of Sciences University of Lisbon, Lisbon, Portugal; Department of Chemistry and Biochemistry, Faculty of Sciences University of Lisbon, Lisbon, Portugal*
CORBIN M. CROOM • *Department of Chemistry and Biochemistry, Baylor University, Waco, TX, USA*
DANIELLY C. FERRAZ DA COSTA • *Instituto Nacional de Ciência Tecnologia de Biologia Estrutural e Bioimagem, UFRJ, Rio de Janeiro, Brazil; Instituto de Nutrição, UERJ, Rio de Janeiro, Brazil*
GIULIA DINIZ DA SILVA FERRETTI • *Instituto de Bioquímica Médica Leopoldo de Meis, UFRJ, Rio de Janeiro, Brazil; Instituto Nacional de Ciência Tecnologia de Biologia Estrutural e Bioimagem, UFRJ, Rio de Janeiro, Brazil*
EVERLY CONWAY DE MACARIO • *Department of Microbiology and Immunology, School of Medicine, University of Maryland at Baltimore-Institute of Marine and Environmental Technology (IMET), Baltimore, MD, USA; Euro-Mediterranean Institute of Science and Technology (IEMEST), Palermo, Italy*
CHRISTOPHER J. DICK • *Department of Biochemistry and Molecular Biology, Mayo Clinic, Rochester, MN, USA*

Evelien Eenjes • *Department of Neurology, Columbia University, New York, NY, USA; Department of Pediatric Surgery, Erasmus Medical Center-Sophia Children's Hospital, Rotterdam, The Netherlands*

Matthew Fares • *Department of Chemistry, The Pennsylvania State University, University Park, PA, USA*

Paula Fernandez-Guerra • *Research Unit for Molecular Medicine, Department of Clinical Medicine, HEALTH, Aarhus University, and Department of Clinical Biochemistry Aarhus University Hospital, Aarhus, Denmark; Department of Biochemistry, Aarhus University Hospital, Aarhus, Denmark*

Mark T. Fisher • *Department of Biochemistry and Molecular Biology, University of Kansas Medical Center, Kansas, KS, USA*

Cláudio M. Gomes • *BioISI – Biosystems and Integrative Sciences Institute, Faculty of Sciences University of Lisbon, Lisbon, Portugal; Department of Chemistry and Biochemistry, Faculty of Sciences University of Lisbon, Lisbon, Portugal*

Bradley R. Groveman • *Rocky Mountain Laboratories, NIAID, NIH, Hamilton, MT, USA*

Bárbara J. Henriques • *BioISI – Biosystems and Integrative Sciences Institute, Faculty of Sciences University of Lisbon, Lisbon, Portugal; Department of Chemistry and Biochemistry, Faculty of Sciences University of Lisbon, Lisbon, Portugal*

Quyen Q. Hoang • *Department of Biochemistry and Molecular Biology, Indiana University School of Medicine, Indianapolis, IN, USA*

Andrew G. Hughson • *Rocky Mountain Laboratories, NIAID, NIH, Hamilton, MT, USA*

Raz Jelinek • *Department of Chemistry, Ben Gurion University of the Negev, Beer Sheva, Israel; Ilse Katz Institute for Nanotechnology, Ben Gurion University of the Negev, Beer Sheva, Israel*

John Karanicolas • *Molecular Therapeutics Program, Fox Chase Cancer Center, Philadelphia, PA, USA*

Karen R. Khar • *Molecular Therapeutics Program, Fox Chase Cancer Center, Philadelphia, PA, USA*

Per M. Knappskog • *Center for Medical Genetics and Molecular Medicine, Haukeland University Hospital, Bergen, Norway; K. G. Jebsen Centre for Neuropsychiatric Disorders, Bergen, Norway*

Sofiya Kolusheva • *Ilse Katz Institute for Nanotechnology, Ben Gurion University of the Negev, Beer Sheva, Israel*

Allison Kraus • *Rocky Mountain Laboratories, NIAID, NIH, Hamilton, MT, USA*

Igor K. Lednev • *Department of Chemistry, University at Albany, SUNY, Albany, NY, USA*

Yu Liu • *Department of Chemistry, The Pennsylvania State University, University Park, PA, USA*

Tânia G. Lucas • *BioISI – Biosystems and Integrative Sciences Institute, Faculty of Sciences University of Lisbon, Lisbon, Portugal; Department of Chemistry and Biochemistry, Faculty of Sciences University of Lisbon, Lisbon, Portugal*

Gergely L. Lukacs • *Department of Physiology, McGill University, Montreal, QC, Canada; Department of Biochemistry, McGill University, Montréal, QC, Canada*

Alberto J. L. Macario • *Department of Microbiology and Immunology, School of Medicine, University of Maryland at Baltimore-Institute of Marine and Environmental Technology (IMET), Baltimore, MD, USA; Euro-Mediterranean Institute of Science and Technology (IEMEST), Palermo, Italy*
Alexandra J. Machen • *Department of Biochemistry and Molecular Biology, University of Kansas Medical Center, Kansas, KS, USA*
Ravit Malishev • *Department of Chemistry, Ben Gurion University of the Negev, Beer Sheva, Israel*
Marta Marin-Argany • *Institute for Research in Biomedicine (IRB Barcelona), Barcelona, Spain*
Aurora Martinez • *Department of Biomedicine, University of Bergen, Bergen, Norway; K. G. Jebsen Centre for Neuropsychiatric Disorders, Bergen, Norway*
Michael Metrick • *Rocky Mountain Laboratories, NIAID, NIH, Hamilton, MT, USA*
Pinaki Misra • *Department of Biochemistry and Molecular Biology, Mayo Clinic, Rochester, MN, USA*
Jeremy Nan • *Djavad Mowafaghian Centre for Brain Health, University of British Columbia, Vancouver, BC, Canada*
Susanna Navarro • *Institut de Biotecnologia i de Biomedicina, Universitat Autònoma de Barcelona, Barcelona, Spain; Departament de Bioquímica i Biologia Molecular, Universitat Autònoma de Barcelona, Barcelona, Spain*
Pierce T. O'Neil • *Department of Biochemistry and Molecular Biology, University of Kansas Medical Center, Kansas, KS, USA*
Christina D. Orrù • *Rocky Mountain Laboratories, NIAID, NIH, Hamilton, MT, USA*
Devon L. Plewman • *Department of Chemistry and Biochemistry, Baylor University, Waco, TX, USA*
Edward Pokrishevsky • *Djavad Mowafaghian Centre for Brain Health, University of British Columbia, Vancouver, BC, Canada*
Vibha Prasad • *Department of Neurology, University Medical Center, RWTH Aachen University, Aachen, Germany*
Tatiana Quiñones-Ruiz • *Department of Chemistry, University at Albany, SUNY, Albany, NY, USA*
Marina Ramirez-Alvarado • *Department of Biochemistry and Molecular Biology, Mayo Clinic, Rochester, MN, USA; Department of Immunology, Mayo Clinic, Rochester, MN, USA*
Luciana Pereira Rangel • *Instituto Nacional de Ciência Tecnologia de Biologia Estrutural e Bioimagem, UFRJ, Rio de Janeiro, Brazil; Faculdade de Farmácia, UFRJ, Rio de Janeiro, Brazil*
Sanaz Rasouli • *Department of Chemistry and Biochemistry, Baylor University, Waco, TX, USA; Institute of Biomedical Studies, Baylor University, Waco, TX, USA*
Keely R. Redhage • *Department of Biochemistry and Molecular Biology, Mayo Clinic, Rochester, MN, USA*
Frank T. Robb • *Department of Microbiology and Immunology, School of Medicine, University of Maryland at Baltimore-Institute of Marine and Environmental Technology (IMET), Baltimore, MD, USA; Institute for Bioscience and Biotechnology Research (IBBR), Rockville, MD, USA*
Ariel Roldan • *Department of Physiology, McGill University, Montreal, QC, Canada*

Manuel Rosario-Alomar • *Department of Chemistry, University at Albany, SUNY, Albany, NY, USA*
Frederic Rousseau • *Switch Laboratory, VIB Center for Brain and Disease Research, Leuven, Belgium; Switch Laboratory, Department of Cellular and Molecular Medicine, KU Leuven, Leuven, Belgium; VIB-KU Leuven Center for Brain and Disease Research, KU Leuven, Leuven, Belgium*
Steffen J. Sahl • *Max Planck Institute for Biophysical Chemistry, Göttingen, Germany*
Eri Saijo • *Rocky Mountain Laboratories, NIAID, NIH, Hamilton, MT, USA*
Pier Luigi San Biagio • *Institute of Biophysics, SL Palermo, National Research Council, Palermo, Italy*
Joost Schymkowitz • *Switch Laboratory, VIB Center for Brain and Disease Research, Leuven, Belgium; Switch Laboratory, Department of Cellular and Molecular Medicine, KU Leuven, Leuven, Belgium; VIB-KU Leuven Center for Brain and Disease Research, KU Leuven, Leuven, Belgium*
Louise C. Serpell • *School of Life Sciences, University of Sussex, East Sussex, UK*
Jerson L. Silva • *Instituto de Bioquímica Médica Leopoldo de Meis, UFRJ, Rio de Janeiro, Brazil; Instituto Nacional de Ciência Tecnologia de Biologia Estrutural e Bioimagem, UFRJ, Rio de Janeiro, Brazil*
Naoto Soya • *Department of Physiology, McGill University, Montreal, QC, Canada*
Jackie A. Thompson • *Department of Biochemistry and Molecular Biology, University of Kansas Medical Center, Kansas, KS, USA*
Devkee M. Vadukul • *School of Life Sciences, University of Sussex, East Sussex, UK*
Rob van der Kant • *Switch Laboratory, VIB Center for Brain and Disease Research, Leuven, Belgium; Switch Laboratory, Department of Cellular and Molecular Medicine, KU Leuven, Leuven, Belgium; VIB-KU Leuven Center for Brain and Disease Research, KU Leuven, Leuven, Belgium*
Joost van Durme • *Switch Laboratory, VIB Center for Brain and Disease Research, Leuven, Belgium; Switch Laboratory, Department of Cellular and Molecular Medicine, KU Leuven, Leuven, Belgium; VIB-KU Leuven Center for Brain and Disease Research, KU Leuven, Leuven, Belgium*
Salvador Ventura • *Institut de Biotecnologia i de Biomedicina, Universitat Autònoma de Barcelona, Barcelona, Spain; Departament de Bioquímica i Biologia Molecular, Universitat Autònoma de Barcelona, Barcelona, Spain*
Tuane Cristine R. G. Vieira • *Instituto Federal do Rio de Janeiro, IFRJ, Rio de Janeiro, Brazil; Instituto Nacional de Ciência Tecnologia de Biologia Estrutural e Bioimagem, UFRJ, Rio de Janeiro, Brazil*
Aaron Voigt • *Department of Neurology, University Medical Center, RWTH Aachen University, Aachen, Germany; JARA-BRAIN Institute Molecular Neuroscience and Neuroimaging, Forschungszentrum Jülich GmbH and RWTH Aachen University, Aachen, Germany*
Willianne I. M. Vonk • *Princess Máxima Center for Pediatric Oncology, Utrecht, The Netherlands*
Wei Wang • *Department of Biochemistry and Molecular Biology, Indiana University School of Medicine, Indianapolis, IN, USA*
Ai Yamamoto • *Department of Neurology, Columbia University, New York, NY, USA; Department of Pathology and Cell Biology, Columbia University, New York, NY, USA*

YOUNG JOO YANG-KLINGLER • *Department of Pathology and Cell Biology, Columbia University, New York, NY, USA*

XIN ZHANG • *Department of Chemistry, The Pennsylvania State University, University Park, PA, USA; Department of Biochemistry and Molecular Biology, The Pennsylvania State University, University Park, PA, USA; The Huck Institutes of the Life Sciences, The Pennsylvania State University, University Park, PA, USA*

Part I

Protein Biophysics Assays

Chapter 1

Biophysical and Spectroscopic Methods for Monitoring Protein Misfolding and Amyloid Aggregation

Joana S. Cristóvão, Bárbara J. Henriques, and Cláudio M. Gomes

Abstract

Proteins exhibit a remarkable structural plasticity and may undergo conformational changes resulting in protein misfolding both in a biological context and upon perturbing physiopathological conditions. Such nonfunctional protein conformers, including misfolded states and aggregates, are often associated to protein folding diseases. Understanding the biology of protein folding diseases thus requires tools that allow the structural characterization of nonnative conformations of proteins and their interconversions. Here we present detailed procedures to monitor protein conformational changes and aggregation based on spectroscopic and biophysical methods that include circular dichroism, ATR-Fourier-transformed infrared spectroscopy, fluorescence spectroscopy and dynamic light scattering. To illustrate the application of these methods we report to our previous studies on misfolding, aggregation and amyloid fibril formation by superoxide dismutase 1 (SOD1), a protein whose toxic deposition is implicated in the neurodegenerative disease amyotrophic lateral sclerosis (ALS).

Key words Circular dichroism, Fourier-transformed infrared spectroscopy, Fluorescence, Dynamic light scattering, Protein misfolding, Amyloid, Protein aggregation, Thioflavin T, SOD1, ALS

Abbreviations

CD	Circular dichroism
FTIR	Fourier-transformed infrared spectroscopy
DLS	Dynamic light scattering
SOD1	Superoxide dismutase 1
1,8-ANS	8-Anilinonaphthalene-1-sulfonic acid
ThT	Thioflavin T

Joana S. Cristóvão and Bárbara J. Henriques contributed equally to this work.

Cláudio M. Gomes (ed.), *Protein Misfolding Diseases: Methods and Protocols*, Methods in Molecular Biology, vol. 1873,
https://doi.org/10.1007/978-1-4939-8820-4_1, © Springer Science+Business Media, LLC, part of Springer Nature 2019

1 Introduction

1.1 Protein Misfolding and Aggregation in Disease

Proteins fold into a well-defined three-dimensional structure whose maintenance is, in most cases, critical to assure biological function. However, in the biological context and during its life time, as dynamic entities, protein molecules can assume a variety of conformations and interconvert between different conformational states. These conformational changes in proteins may be triggered by functional interactions (e.g., with ligands, substrates, and other proteins), as a response to biochemical changes in the cellular environment (e.g., pH, crowding, redox state, metal ion binding) or result from mutations that alter the folding landscape of the protein. While some of these conformations and transitions are functional, others are potentially deleterious for the cell as they result in persistent nonfunctional protein states. The latter are frequently associated to protein folding diseases and range from misfolded states to aggregates and amyloid fibrils. The understanding of the biology of protein folding diseases thus requires tools that allow the structural characterization of nonnative conformations of proteins and their interconversions. This is particularly relevant in what concerns conformational changes and structural transitions such as those taking place during protein fibrillization processes. Typically, in these cases, a given protein undergoes some type of conformational change that results in populating aggregation-prone conformers whose self-assembly results in the formation of precursor aggregate species, usually, β-structured, which further evolve into larger assemblies and amyloid fibrils. These processes involve a plethora of conformational changes: interconversion between secondary structure elements (exposure of aggregation-prone regions), modifications in tertiary interactions (in early misfolded conformers), variations in particle shape and size (during self-assembly of misfolded monomers into fibrils), emergence of species with a particular structural signature (cross-β patterns in amyloid fibrils). Therefore, the structural and mechanistic understanding of the structural biology of protein misfolding in disease calls for the integrated use of different spectroscopic methods and biophysical approaches that include circular dichroism, ATR-Fourier-transformed infrared spectroscopy, fluorescence spectroscopy and dynamic light scattering to monitor protein conformational changes and aggregation. In this chapter we present methods employing a combination of biophysical and spectroscopic methodologies that we have implemented to study misfolding and amyloid formation by superoxide dismutase 1 (SOD1), a protein whose aggregation is implicated in the neurodegenerative disease amyotrophic lateral sclerosis (ALS) [1]. Natively folded SOD1 is a highly stable β-sheeted protein containing a binuclear copper–zinc site. However, in its apo state, which is populated during SOD1 biosynthesis or upon a destabilizing condition, SOD1 decreased stability and misfolding results in exposure of aggregation-prone segments, making it prone to aggregation [2]. Like in other amyloid diseases [3, 4], metal ion binding to SOD1 influences its aggregation [5, 6]. Indeed, SOD1-enriched protein inclusions and calcium overload are hallmarks in ALS-affected motor neurons [7]. Interestingly, we observed that calcium binding to SOD1 induces conformational changes that influence its aggregation pathway from fibrillar to amorphous aggregates [5, 8].

1.2 Circular Dichroism

Circular dichroism (CD), is a very useful method to rapidly study the folded state of a purified protein [9], evaluate the effect of mutations on protein conformation or stability [10], study protein interactions [11], or follow conformational changes during protein (un)folding processes and aggregation [12]. CD results from differentially absorbed right and left circularly polarized light from chiral molecules in solution [13]. To study protein structure, CD is informative essentially in the far-UV region (180–260 nm) for secondary structure analysis and in the near-UV region (250–350 nm) for information on tertiary structure [12]. The peptide bond is the most important chromophore responsible for the absorption in the far-UV region due to its $n \rightarrow \pi^*$ (≈190 nm) and $\pi \rightarrow \pi^*$ (≈222 nm) electronic transitions. Depending whether the protein backbone is folded as an α-helix, β-sheet or as a random coil, this wi II result in a distinct overlap of the involved molecular orbitals and their energy levels, thus resulting in rather distinct spectroscopic signatures for the different types of secondary structure. Therefore, a characteristic CD spectral fingerprint is obtained for the different types of regular secondary structure [13]. Typically, folded proteins with a high degree of order present large distinctive CD signals, while in contrast, unfolded proteins present low or null signals. Thus, changes in CD intensity can be used to follow protein conformational alterations during thermal or chemical denaturation or upon interaction with a ligand. Amyloid processes can also be monitored by this technique due to conversion of the native state of the protein into a β-sheet rich structure monitored by the formation of a negative band centered at 220 nm, characteristic of amyloid fibrils [14]. Absorption in the near-UV region arises mostly from side chains of aromatic residues and disulfide bonds, being therefore a valuable tool to gain insights into changes in tertiary interactions. Conformational changes in proteins containing absorbing cofactors in chiral environments such as heam, flavins or pyridoxal-5′-phosphate can also be probed via changes in the CD signal of the cofactors [13]. In summary, CD is a highly powerful method to monitor the folded state of a protein in solution requiring low amounts of materials (<50 μg protein for far-UV CD characterization) and is thus the spectroscopic method of choice to characterize protein structure and conformational changes. A number of excellent reviews on protein CD are available [9, 11–13].

1.3 Fourier-Transform Infrared Spectroscopy

Fourier-transform infrared (FTIR) spectroscopy is a very powerful spectroscopic method for the structural analysis of biomolecules that detects the absorption bands associated with molecular bond vibrational frequencies [15]. Proteins have a prominent and sensitive vibrational band in the 1700–1600 cm^{-1} region (the amide I band), which results from the C=O stretching vibrations of the peptide bond, which account for most of the vibrational absorption. Regretfully the protein amide I band overlaps with the strong water bending absorption band, which masks the protein signal. This limitation can be however overcome by carrying out measurements in buffers containing D_2O and/or highe protein concentrations. Proteins also exhibit peaks centered at 1550 cm^{-1} (the amide II band), which correspond to the out-of-phase N-H bending and C-N stretching vibrations, and at 1400–1200 cm^{-1} (the amide III band) resulting from in-phase C-N stretching and N-H bending vibrations. The amide I band of proteins is sensitive to the type of secondary structure. In general, folded conformations exhibit a structured amide I band that can be deconvoluted into multiple bands associated different types of secondary structures. In contrast, unfolded proteins are characterized by a broad amide I band centered around 1650 cm^{-1}, which is characteristic of disordered structures. FTIR excels at identifying different types of β-structures, being for this reason particularly useful to characterize amyloidogenic conformers. Fibrils and β-oligomers exhibit an absorption band around 1620–1625 cm^{-1} which corresponds to inter-chained antiparallel β-sheet structure [16]. These differences permit to use this technique to follow protein unfolding and aggregation and to discriminate the conformational changes occurring during ligand binding or environmental changes (pH, salt concentration, crowding, metal ion binding). The use of an ATR (attenuated total reflectance) accessory further enhances the analytical capability of FTIR as in this case protein samples may be analyzed as a thin film deposited over the ATR crystal. In spite of the limitation imposed by interference of water with the amide I band, FTIR is a powerful method for protein structural analysis as it usually requires short measuring times and low amount of sample (typically 10–100 μg).

1.4 Fluorescence Spectroscopy

Fluorescence spectroscopy of proteins is extremely valuable to characterize protein structure and dynamics and may involve the use of intrinsic or extrinsic fluorophores. The most relevant intrinsic fluorophores in proteins are the aromatic residues, notably tryptophan which has the highest quantum yield among those.

Tryptophan (Trp) fluorescence is very useful to monitor conformational changes in proteins as emission of the indole group is highly sensitive to solvent polarity. Tryptophan residues are frequently found buried in the protein hydrophobic core or at the interface between domains or subunits. When in such apolar environments, a blue-shifted emission (310–340 nm) is observed. However, as Trp becomes hydrogen bonded or solvent exposed, the emission red shifts to longer wavelengths (down to 340 nm). Therefore, the position of the emitting band of this residue is an excellent reporter of polarity changes and tertiary structure. Therefore, Trp emission is highly sensitive to the surrounding environment, to small differences in conformation for example due to subunit association or substrate binding, and to misfolding and conformational destabilization. Additionally, some protein cofactors found in proteins such as flavins or NADH also have fluorescence properties that will be suited to monitor folding changes [17]. The group of extrinsic fluorophores that can be employed to monitor protein conformational changes and aggregation is very wide, and comprises probes that bind noncovalently with proteins via hydrophobic or electrostatic interactions, as well as covalently linked probes, for example via the ε-amino group of lysines, the α-amino group at the *N*-terminus, or cysteine thiols [18]. 1-Anilinonaphthalene-8-sulfonic acid (ANS) is a particularly useful fluorophore to analyze protein nonnative states. This molecule is barely fluorescent in pure water but its fluorescence is strongly enhanced in hydrophobic environments, as it happens when it interacts with hydrophobic patches exposed on proteins. ANS is therefore very useful to follow protein unfolding, detect molten globule states, monitor ligand-induced hydrophobic exposure as well as protein aggregation [19–21]. In regard to fluorophores specific to amyloid aggregates, Thioflavin T (ThT) is among the most widely used probe to monitor in vitro protein aggregation. In solution, ThT is barely emitting. However, when it intercalates with the β-sheet structures formed during aggregation this dye displays enhanced fluorescence and a characteristic red shift of its emission spectrum [22]. The emission is extensively employed to monitor the kinetics of amyloid aggregation making it possible to study reaction mechanisms and effects of small molecules, metal ions, or other proteins in the fibril formation process [5, 6, 23–25]. To characterize and monitor the formation of early amyloid aggregates and prefibrillar oligomers, other fluorophores such as oligothiophenes, ANS and Bis-ANS can also be used [26, 27]. Overall, protein fluorescence spectroscopy, which requires relatively small amounts of material [28], is a highly sensitive spectroscopic method to analyze protein tertiary structure, conformational changes and to characterize nonnative conformers.

1.5 Dynamic Light Scattering

Dynamic light scattering (DLS) measures the translational diffusion coefficient of a macromolecule undergoing Brownian motion in solution [29]. It is commonly used to determine the size of biomolecules, to monitor ligand binding and to detect protein aggregates in solution [30]. Briefly, moving particles in solution will scatter monochromatic light and the intensity of the scattered light fluctuates over time due to the Brownian motion of the particles. The Brownian motions depend on particle diffusion rate which is a function of particle size, viscosity and temperature of solution [31]. An autocorrelation function of the light scattering signal, that is a measure of the time-dependence of intensity fluctuations, will provide information on particle behavior in solution. Using DLS one can detect size and size distribution of proteins with diameter in the range of 0.001–1 nm [32], which allows effective monitoring of protein self-assembly and aggregation reactions in solution. We have previously employed DLS to monitor the effect of small molecules and metal ion binding during SOD1 aggregation [5, 6, 23]. Recent instruments implement a technique called nanoparticle tracking analysis (NTA) to measure the size and distribution of particles in very heterogeneous formulations, such as aggregated protein solutions [33]. Current DLS instruments implement user-friendly software for the analysis of sample polydispersity and for measuring particle size distributions as a function of intensity, volume, number, or mass [30]. One of the main advantages of DLS in what concerns the analysis of protein misfolding processes is its ability to accurately determine particle size, being extremely sensitive to aggregation, and to detect small fractions of larger particles [30, 31]. Other advantages of the technique relate to the low volumes and protein concentration required (0.1–50 mg ml^{-1}), fast data acquisition, high sensitivity and reproducibility [31, 32].

2 Materials

2.1 SOD1 Expression and Purification

1. pACA(pMa5-8)-Human SOD1 plasmid.
2. *E. coli* BL21(DE3) competent cells.
3. LB agar ampicillin plates.
4. LB medium and 100 μg.ml^{-1} ampicillin.
5. 1 M IPTG; 50 mM $ZnSO_4$; 300 mM $CuSO_4$.
6. Cooled incubator.
7. Resuspension buffer: 50 mM Tris–HCl, pH 7.4, 0.3 mM PMSF, 1 μg.ml^{-1} DNaseI.
8. Ammonium sulfate.

9. Size exclusion chromatography Buffer: 50 mM Tris–HCl, pH 7.4, 150 mM NaCl.
10. Anion-exchange chromatography Buffer A: 10 mM Tris–HCl, pH 7.4.
11. Anion-exchange chromatography Buffer B: 10 mM Tris–HCl pH 7.4, 1 mM NaCl.
12. Centrifuge.
13. HiPrep Sephacryl S100 16/60 column (GE Healthcare).
14. Q-Sepharose FF column (GE Healthcare).
15. Amicon Ultra-Centrifugal filters, cut-off 10 kDa.
16. UV–visible spectrophotometer.
17. Bradford assay kit.

2.2 Demetallation of SOD1

1. Dialysis membrane with a cutoff lower than 15 kDa.
2. Chelex 100 resin.
3. 50 mM acetate buffer, pH 3.8, 10 mM EDTA.
4. 50 mM acetate buffer, pH 3.8, 100 mM NaCl.

2.3 CD

1. CD spectropolarimeter with a peltier-controlled thermostated cell support (e.g., Jasco J-1500).
2. Quartz cell 1 mm (e.g., Hellma quartz SUPRASIL precision cells).

2.4 ATR-FTIR

1. FTIR spectrometer (e.g., Bruker TENSOR II).
2. Thermostatized attenuated total reflectance (ATR) cell suitable for liquid samples (e.g., Harrick BioATR II cell).

2.5 1,8-ANS Fluorescence

1. Spectrofluorometer (e.g., FP-8200 Jasco).
2. Quartz cell 10 nm (e.g., Hellma quartz SUPRASIL precision cells).
3. 1 mM 8-Anilinonaphthalene-1-sulfonic acid (ANS) stock solution (Sigma).

2.6 ThT Fluorescence Kinetics

1. 5 mM Thioflavin T (ThT) stock solution (Millipore).
2. 5 mM $CaCl_2$.
3. 1.5 ml low binding microtubes (e.g., Axygen, Corning, MCT-150-L-C).
4. 96 well assay plate (e.g., Nunc, catalog no. 732-2701).
5. Polyethylene sealing tape (e.g., Costar 6524).
6. Fluorescence microplate reader (e.g., FLUOstar OPTIMA, BMG Labtech).
7. Teflon beads with 1/8-inch diameter.

2.7 DLS

1. Malvern Zetasizer Nano ZS Instrument.
2. Quartz cell 45 μl (e.g., Hellma quartz SUPRASIL precision cells).
3. Thermoshaker (e.g., Grant-Bio Thermo-shaker).

3 Methods

3.1 SOD1 Expression and Purification

SOD1 is produced using *E. coli* as expression host.

1. Transform *E. coli* competent BL21(DE3) cells with pACA(pMa5-8) plasmid according to the instructions of the manufacturer (*see* **Note 1**).
2. Spread transformed cells on LB agar plates containing ampicillin (100 μg.ml^{-1}) and incubate overnight at 37 °C.
3. Inoculate a single colony in 50 ml LB medium containing ampicillin (100 μg.ml^{-1}) and incubate overnight at 30 °C shaking at 200 rpm.
4. The overnight culture is used to inoculate the expression culture and incubated at 30 °C and shaking at 200 rpm until $OD_{600\ nm}$ = 0.5–0.6 is reached.
5. Induce SOD1expression by addition of 0.5 mM IPTG, 30 μM $ZnSO_4$, and 3 mM $CuSO_4$ and let the cells grow overnight at 24 °C shaking at 200 rpm (*see* **Note 2**). The culture is harvested by centrifugation on the day after, at 20,000 × *g* for 10 min. Cell pellets can be frozen at −80 °C.

SOD1 purification

6. Thaw frozen cell pellets on ice and suspend the cells in 50 mM Tris pH 7.4 with 1 μg.ml^{-1} DNAse I and 0.3 mM PMSF. Break the cells with two passages through a French pressure cell set at 320 Mpa (*see* **Note 3**).
7. Centrifuge the cell lysate at 15,000 × *g* for 25 min.
8. Incubate the clear cell lysate for 30 min at 65 °C. Centrifuge the resulting suspension for 45 min at 15,000 × *g* and collect the supernatant.
9. Perform an ammonium sulfate precipitation adding 50% of ammonium sulfate and incubate for 2 h at 4 °C with slow agitation. Centrifuge at 15,000 × *g* for 20 min and discard the pellet. Add 60% of ammonium sulfate and incubate for 1 h at 4 °C. Centrifuge at 15,000 × *g* for 20 min and discard the pellet. Add 90% of ammonium sulfate and incubate for 2 h at 4 °C. Centrifuge at 15,000 × *g* for 40 min and collect the pellet. Resuspend the pellet in 10 ml of 50 mM Tris, pH 7.4.

10. Load the previously resuspended pellet into a size exclusion chromatography (ex. Sephacryl S100 or S200, GE Healthcare) (*see* **Note 4**).
11. Load the fractions containing SOD1 in an anion-exchange chromatography (ex. Q-Sepharose FF, GE Healthcare) with 10 mM Tris, pH 7.4 and elute the protein with a linear gradient from 0 to 1 M NaCl, in the same buffer (*see* **Note 5**). Collect the fractions containing SOD1 and concentrate the solution by ultrafiltration with 10 kDa Amicon filters. Store at −80 °C until further use (*see* **Note 6**).

3.2 Demetallation of SOD1

In order to understand the conformational changes of metal ions on protein conformation, SOD1 must be demetallated. All buffers should be passed through a Chelex column to insure removal of trace metal ions.

1. Dialyze 2 $mg.ml^{-1}$ of SOD1 in 50 mM acetate buffer, pH 3.8, 10 mM EDTA at 4 °C for 24 h to release all the metal ions of SOD1 (*see* **Note** 7).
2. Exchange the buffer to 50 mM acetate buffer, 100 mM NaCl, pH 3.8 to remove EDTA from the solution.
3. Exchange the buffer to the desired buffer.
4. Quantitate SOD1 spectrophotometrically using the molar extinction coefficient $\varepsilon_{280\,nm} = 10{,}800\ M^{-1}.cm^{-1}$ or the Bradford assay (*see* **Note 8**).

3.3 Characterization of SOD1 Conformational Changes in Presence of Calcium

3.3.1 Far-UV CD Analysis to Evaluate Secondary Structure Changes

1. Prepare 200 μl of 30 μM apoSOD1 in 50 mM Tris, pH 7.4 and incubate overnight with increasing concentration of $CaCl_2$.
2. Set acquisition parameters: from 260 to 190 nm, data collecting 0.1 nm and 8 scans.
3. Record the baseline spectrum using 200 μl of buffer solution, containing $CaCl_2$ (*see* **Note 9**).
4. Remove the blank solution and replace it by 200 μl SOD1 solution and record spectra, repeat for all $CaCl_2$ concentrations (*see* **Notes 10** and **11**).
5. The results can be presented as mean residue molar ellipticity [θ] with units of $deg.cm^2/dmol$, as calculated by the equation, $[\theta] = ([\theta]_{obs} \times MRW)/(10 \times c \times l)$, where $[\theta]_{obs}$ is the ellipticity in milidegrees, MRW is the mean residue molecular weight, c is the protein concentration in $mg.ml^{-1}$ and l is the optical path of the cell in cm. SOD1 shows a typical spectrum of a β-sheet structure and addition of calcium leads to an increase in the negative signal at 218 nm (Fig. 1a).

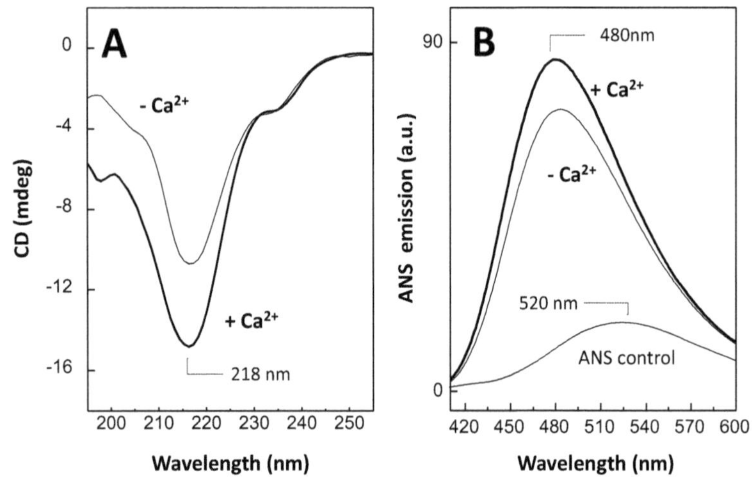

Fig. 1 Far-UV circular dichroism and ANS fluorescence analysis of SOD1 conformational changes in the presence of calcium. (**a**) Far-UV CD secondary structure analysis of 30 μM apo-SOD1 in the absence ($-Ca^{2+}$) and in the presence of threefold excess of $CaCl_2$ ($+Ca^{2+}$). (**b**) 1,8-ANS fluorescence emission spectra of apo-SOD1 ($-Ca^{2+}$) and upon incubation with calcium ($+Ca^{2+}$), overlaid with the emission spectrum of unbound ANS. Figure adapted from Leal et al., with permission [5]

3.3.2 1,8-ANS Fluorescence to Evaluate Hydrophobic Exposure

1. Prepare 400 μl of 15 μM apoSOD1 in 50 mM Tris, pH 7.4 with 30 μM 1,8-ANS and incubate for 30 min at RT with and without $CaCl_2$.
2. Set acquisition parameters: emission spectra from 400–600 nm, excitation wavelength at 370 nm, 8 scans.
3. Record the reference spectrum using 400 μl of buffer solution with 30 μM 1,8-ANS.
4. Remove the control solution and replace it by 400 μl SOD1 solution and record its spectrum; repeat procedure for SOD1 in the presence of $CaCl_2$.
5. Changes in SOD1 hydrophobicity induced by calcium binding are noticeable from the observed increase in fluorescence emission and from the blue shift of the maximum emission wavelength from 520 nm when ANS is in free state to 480 nm when ANS is bound to SOD1 hydrophobic patches (Fig. 1b).

3.3.3 ATR-FTIR Secondary Structure Analysis and Intermolecular β-Strand Formation in Amyloid Aggregates

1. Prepare 30 μl of 150 μM apoSOD1 with twofold of $CaCl_2$ in 50 mM Tris, pH 7.4 and incubate overnight.
2. Set acquisition parameters: 1760–1600 cm^{-1} for spectral range, 2 cm^{-1} resolution, each spectrum comprising a mean of 150 scans, 20 kHz scanner velocity, 12 mm aperture. Set the acquisition mode to continuous spectra acquisition.
3. Record the baseline spectrum: transfer 20 μl of buffer solution, containing $CaCl_2$ to the ATR cell surface. Close carefully the ATR to avoid air bubble formation. Record the spectrum.

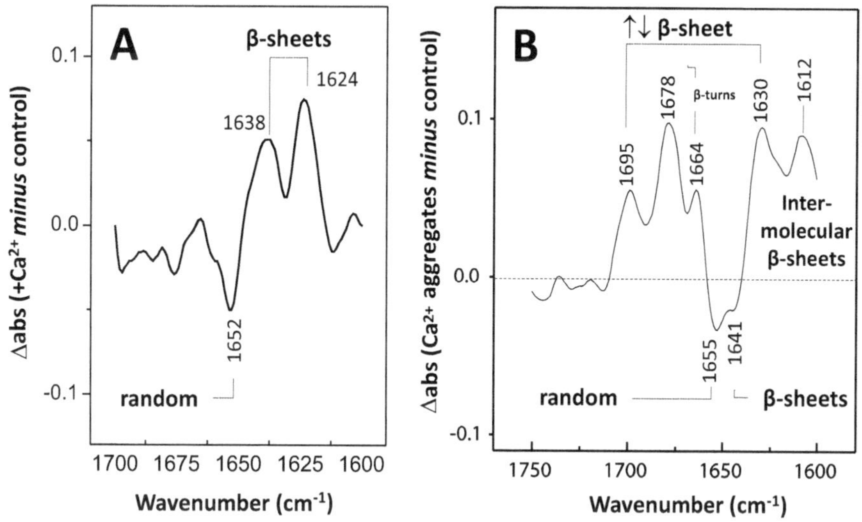

Fig. 2 ATR-FTIR structural analysis of native and aggregated SOD1 in the presence of calcium. (**a**) ATR-FTIR difference spectrum in the Amide I region between SOD1 incubated with $CaCl_2$ and metal-free SOD1. Positive peaks correspond to appearance of structural interconversion arising from calcium binding. (**b**) ATR-FTIR difference spectrum in the Amide I region of SOD1 aggregates formed in the presence of $CaCl_2$ minus SOD1 alone. Positive peaks correspond to features increased with the presence of $CaCl_2$

4. Remove the blank solution and replace it by 20 μl SOD1 solution and record spectra of all conditions. Contribution of water and buffer in the absorption spectra should be subtracted from the spectra.
5. Changes in secondary structure of SOD1 induced by calcium are perceived in a difference spectra which denotes an increase in β-sheet content (positive band between 1620 and 1630 cm^{-1}, suggesting a conformational rearrangement of SOD1 (Fig. 2a).
6. Incubate 150 μM apoSOD1 with and without twofold excess of $CaCl_2$ in 50 mM Tris, pH 7.4 at 37 °C under constant agitation at 600 rpm with a Teflon bead for 150 h.
7. Record the baseline spectrum as previously indicated and measure SOD1 samples with the same acquisition parameters described above.
8. SOD1 aggregates formed in the presence of calcium display β-sheet content evidenced by absorption bands at 1630 and 1695 cm^{-1}, coherent with intermolecular antiparallel arrangement of the β-strand (Fig. 2b) (*see* **Note 12**).

3.4 Monitoring SOD1 Aggregation

3.4.1 ThT Fluorescence

1. Prepare 600 μl of each condition to distribute in the 96 well-plate in triplicates. Dilute SOD1 to 60 μM in 50 mM Tris, pH 7.4. $CaCl_2$ may be added as twofold molar equivalents over SOD1. 120 μM ThT solution is added in the final solution (*see* **Note 13**).

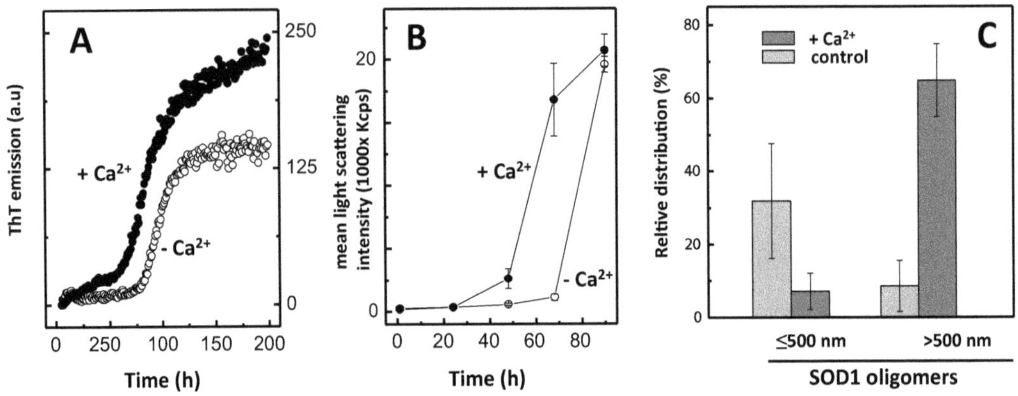

Fig. 3 SOD1 aggregation kinetics and size distribution of aggregates in the presence of calcium. (**a**) Aggregation kinetic of SOD1 monitored by ThT fluorescence emission, in the absence (open circles) and in the presence (full circles) of twofold $CaCl_2$. (**b**) Aggregation kinetic of SOD1 monitored by mean light scattering intensity in the absence (open circles) and in the presence (full circles) of two-fold $CaCl_2$. (**c**) Relative distribution of total light scattering intensities arising from aggregated SOD1 species with a hydrodynamic diameter under and above 500 nm, for the Ca^{2+}-incubated SOD1 and control. Figure adapted from Leal et al., with permission [5]

2. Homogenize the solution gently and add 200 μl to each well of the 96-well plate. Add also a Teflon bead to promote faster aggregation.
3. Measure at 37 °C ThT fluorescence emission at 480 nm, exciting at 440 nm. Readings should be taken each 7 min, with plate agitation at 600 rpm, 5 min before each measurement.
4. Visualizing an increase in the ThT fluorescence is a suggestion of the formation of amyloid-like species (Fig. 3a). This assay provides a simple, inexpensive and quantitative way of detecting amyloids in real time from which kinetic data can be determined (*see* **Note 14**).

3.4.2 DLS

1. Prepare samples for each condition with 60 μM SOD1 in 50 mM TRIS pH 7.4. $CaCl_2$ may be added as two fold molar equivalents over SOD1.
2. Filter the samples with a 0.45 μm filter and incubate at 37 °C overnight, in a thermoshaker setting agitation at 600 rpm.
3. Measure light scattering intensity after 24, 48, 68 and 90 h incubation. Perform around 17 runs for each measurement.
4. Evaluation of the scattering intensity should be performed combining the analysis of mean hydrodynamic diameter (*Z*-average size) and mean count rate parameter. Visualizing an increase in these parameters suggests aggregates formation (Fig. 3b) and the relative distribution of species at a given particle size cut off can be determined (Fig. 3c).

4 Notes

1. The SOD1 expression plasmid (a kind gift from M. Oliveberg, Sweden) was designed to include the yeast copper chaperone, yCCS, for coexpression with SOD1 in order to avoid incorrect and undermetallization of SOD1 production. This coexpression strategy yields SOD1 with high levels of metallation for both Cu^{2+} and Zn^{2+} (85–95%) [34].
2. Supplementation with $ZnSO_4$ and $CuSO_4$ solutions also helps with expression of metallated SOD1 [34].
3. To confirm overexpression, one should run SDS-PAGE of the supernatant and observe two coexpressed bands, corresponding to SOD1 (~16 kDa) and yCCS (~30 kDa).
4. The fractions containing SOD1 are easy to identify by their characteristic cyan/blueish colour arising from copper absorption.
5. After elution the column should be cleaned with 1 M NaOH and/or 70% ethanol to remove precipitated proteins and contaminants bound to the matrix. For long-term storage columns should be kept in 20% ethanol.
6. Usually a yield around 80 $mg.l^{-1}$ of culture and >95% of pure protein is obtained.
7. Make sure to use cold buffer, refrigerating the buffer at 4 °C for 2 h before dialysis. Metal release is due to a slow conformational change promoted by low pH.
8. Metal content of apo-SOD1 can be confirmed using the colorimetric Zincon assay [35].
9. To make sure that the cuvette is clean, first record a spectrum of the empty cell and then with buffer. The CD spectrum of the protein is obtained after subtracting the spectrum of the buffer.
10. For a list of buffers compatible with CD *see* [13].
11. Usually good quality spectra can be obtained with 0.1 $mg.ml^{-1}$ (far-UV) and 1 $mg.ml^{-1}$ (near UV or visible) protein concentration. CD signals obtained at voltages >600 V are to be ignored.
12. The morphology of formed aggregates may be analyzed using atomic force microscopy or electron transmission microscopy.
13. Low binding surface tubes should be used to avoid adsorption of aggregates to the surface of the eppendorf.
14. Fluorescence aggregation curves can indicate the apparent rate of the reaction (K_{app}) and the time needed to form the nuclei (t_{lag}). To calculate these parameters the curve should

be fitted to the equation $\Upsilon = (y_i + m_i x) + \frac{(v_f + m_i x)}{(1 + \exp^{-\left(x - \frac{X_0}{\tau}\right)})} t$,

where Υ corresponds to ThT fluorescence intensity, x is time, and X_0 corresponds to the time of half-height of fluorescence intensity. The lag phase (t_{lag}) is calculated by $t_{lag} = X_0 - 2\tau$ and the apparent rate $k_{app} = 1/\tau$.

Acknowledgments

This work was partially supported by the Fundação para a Ciência e Tecnologia (FCT/MCTES, Portugal) through fellowships to J.S.C. (SFRH/BD/101171/2014) and B.J.H. (SFRH/BPD/74475/2010), and grant PTDC/BBB-BQB/5366/2014 (to B.J.H.) and PTDC/NEU-NMC/2138/2014 (to C.M.G.). The Gomes laboratory is partly supported by grant UID/MULTI/04046/2013 from FCT/MCTES/PIDDAC (to BioISI). Bial Foundation is acknowledged through grant PT/FB/BL-2014-343 (to CMG). Joana S. Cristóvão and Bárbara J. Henriques contributed equally to this work.

References

1. Rowland LP, Shneider NA (2001) Amyotrophic lateral sclerosis. N Engl J Med 344(22): 1688–1700
2. Nordlund A, Leinartaite L, Saraboji K, Aisenbrey C, Grobner G, Zetterstrom P, Danielsson J, Logan DT, Oliveberg M (2009) Functional features cause misfolding of the ALS-provoking enzyme SOD1. Proc Natl Acad Sci U S A 106(24):9667–9672. https://doi.org/10.1073/pnas.0812046106
3. Cristovao JS, Santos R, Gomes CM (2016) Metals and neuronal metal binding proteins implicated in Alzheimer's disease. Oxidative Med Cell Longev 2016:9812178. https://doi.org/10.1155/2016/9812178
4. Leal SS, Botelho HM, Gomes CM (2012) Metal ions as modulators of protein conformation and misfolding in neurodegeneration. Coordin Chem Rev 256(19–20):2253–2270. https://doi.org/10.1016/j.ccr.2012.04.004
5. Leal SS, Cardoso I, Valentine JS, Gomes CM (2013) Calcium ions promote superoxide dismutase 1 (SOD1) aggregation into non-fibrillar amyloid: a link to toxic effects of calcium overload in amyotrophic lateral sclerosis (ALS)? J Biol Chem 288(35):25219–25228. https://doi.org/10.1074/jbc.M113.470740
6. Leal SS, Cristovao JS, Biesemeier A, Cardoso I, Gomes CM (2015) Aberrant zinc binding to immature conformers of metal-free copper-zinc superoxide dismutase triggers amorphous aggregation. Metallomics 7(2):333–346. https://doi.org/10.1039/c4mt00278d
7. Leal SS, Gomes CM (2015) Calcium dysregulation links ALS defective proteins and motor neuron selective vulnerability. Front Cell Neurosci 9:225. https://doi.org/10.3389/fncel.2015.00225
8. Estacio SG, Leal SS, Cristovao JS, Faisca PF, Gomes CM (2015) Calcium binding to gatekeeper residues flanking aggregation-prone segments underlies non-fibrillar amyloid traits in superoxide dismutase 1 (SOD1). Biochim Biophys Acta 1854(2):118–126. https://doi.org/10.1016/j.bbapap.2014.11.005
9. Greenfield NJ (2006) Using circular dichroism spectra to estimate protein secondary structure. Nat Protoc 1(6):2876–2890. https://doi.org/10.1038/nprot.2006.202

10. Henriques BJ, Lucas TG, Rodrigues JV, Frederiksen JH, Teixeira MS, Tiranti V, Bross P, Gomes CM (2014) Ethylmalonic encephalopathy ETHE1 R163W/R163Q mutations alter protein stability and redox properties of the iron Centre. PLoS One 9(9):e107157. https://doi.org/10.1371/journal.pone.0107157
11. Kelly SM, Price NC (2006) Circular dichroism to study protein interactions. Curr Protoc Protein Sci. Chapter 20:Unit 20 10. doi:https://doi.org/10.1002/0471140864.ps2010s46
12. Clarke DT (2012) Circular dichroism in protein folding studies. Curr Protoc Protein Sci. Chapter 28:Unit 28 23. doi:https://doi.org/10.1002/0471140864.ps2803s70
13. Kelly SM, Jess TJ, Price NC (2005) How to study proteins by circular dichroism. Biochim Biophys Acta 1751(2):119–139. https://doi.org/10.1016/j.bbapap.2005.06.005
14. Radko SP, Khmeleva SA, Suprun EV, Kozin SA, Bodoev NV, Makarov AA, Archakov AI, Shumyantseva VV (2015) Physico-chemical methods for studying amyloid-beta aggregation. Biochem Mosc Suppl S 9(3):258–274. https://doi.org/10.1134/S1990750815030075
15. Barth A (2007) Infrared spectroscopy of proteins. Biochim Biophys Acta 1767(9):1073–1101. https://doi.org/10.1016/j.bbabio.2007.06.004
16. Sarroukh R, Goormaghtigh E, Ruysschaert JM, Raussens V (2013) ATR-FTIR: a "rejuvenated" tool to investigate amyloid proteins. Biochim Biophys Acta 1828(10):2328–2338. https://doi.org/10.1016/j.bbamem.2013.04.012
17. Ladokhin AS (2006) Fluorescence spectroscopy in peptide and protein analysis. In: Encyclopedia of analytical chemistry. John Wiley & Sons, Ltd, Hoboken, NJ. https://doi.org/10.1002/9780470027318.a1611
18. Hawe A, Sutter M, Jiskoot W (2008) Extrinsic fluorescent dyes as tools for protein characterization. Pharm Res 25(7):1487–1499. https://doi.org/10.1007/s11095-007-9516-9
19. Leal SS, Gomes CM (2007) Studies of the molten globule state of ferredoxin: structural characterization and implications on protein folding and iron-sulfur center assembly. Proteins 68(3):606–616. https://doi.org/10.1002/prot.21448
20. Henriques BJ, Saraiva LM, Gomes CM (2005) Probing the mechanism of rubredoxin thermal unfolding in the absence of salt bridges by temperature jump experiments. Biochem Biophys Res Commun 333(3):839–844. https://doi.org/10.1016/j.bbrc.2005.06.004
21. Hagmeyer S, Cristovao JS, Mulvihill JJE, Boeckers TM, Gomes CM, Grabrucker AM (2017) Zinc binding to S100B affords regulation of trace metal homeostasis and excitotoxicity in the brain. Front Mol Neurosci 10:456. https://doi.org/10.3389/fnmol.2017.00456
22. Biancalana M, Koide S (2010) Molecular mechanism of Thioflavin-T binding to amyloid fibrils. Biochim Biophys Acta 1804(7):1405–1412. https://doi.org/10.1016/j.bbapap.2010.04.001
23. Cristovao JS, Leal SS, Cardoso I, Gomes CM (2013) Small molecules present in the cerebrospinal fluid metabolome influence superoxide dismutase 1 aggregation. Int J Mol Sci 14(9):19128–19145. https://doi.org/10.3390/ijms140919128
24. Botelho HM, Leal SS, Cardoso I, Yanamandra K, Morozova-Roche LA, Fritz G, Gomes CM (2012) S100A6 amyloid fibril formation is calcium-modulated and enhances superoxide dismutase-1 (SOD1) aggregation. J Biol Chem 287(50):42233–42242. https://doi.org/10.1074/jbc.M112.396416
25. Matos AM, Cristovao JS, Yashunsky DV, Nifantiev NE, Viana AS, Gomes CM, Rauter AP (2017) Synthesis and effects of flavonoid structure variation on amyloid-beta aggregation. Pure Appl Chem 89(9):1305–1320. https://doi.org/10.1515/pac-2017-0201
26. Klingstedt T, Aslund A, Simon RA, Johansson LBG, Mason JJ, Nystrom S, Hammarstrom P, Nilsson KPR (2011) Synthesis of a library of oligothiophenes and their utilization as fluorescent ligands for spectral assignment of protein aggregates. Org Biomol Chem 9(24):8356–8370. https://doi.org/10.1039/c1ob05637a
27. Younan ND, Viles JH (2015) A comparison of three fluorophores for the detection of amyloid fibers and prefibrillar oligomeric assemblies. ThT (Thioflavin T); ANS (1-Anilinonaphthalene-8-sulfonic acid); and bisANS (4,4′-Dianilino-1,1′-binaphthyl-5,5′-disulfonic acid). Biochemistry 54(28):4297–4306. https://doi.org/10.1021/acs.biochem.5b00309
28. Royer CA (2006) Probing protein folding and conformational transitions with fluorescence. Chem Rev 106(5):1769–1784. https://doi.org/10.1021/cr0404390
29. Hassan PA, Rana S, Verma G (2015) Making sense of Brownian motion: colloid characterization by dynamic light scattering. Langmuir 31(1):3–12. https://doi.org/10.1021/la501789z
30. Lorber B, Fischer F, Bailly M, Roy H, Kern D (2012) Protein analysis by dynamic light scattering: methods and techniques for students. Biochemistry and molecular biology education :

a bimonthly publication of the international union of. Biochem Mol Biol 40(6):372–382. https://doi.org/10.1002/bmb.20644

31. Khodabandehloo A, Chen DD (2017) Particle sizing methods for the detection of protein aggregates in biopharmaceuticals. Bioanalysis 9(3):313–326. https://doi.org/10.4155/bio-2016-0269
32. den Engelsman J, Garidel P, Smulders R, Koll H, Smith B, Bassarab S, Seidl A, Hainzl O, Jiskoot W (2011) Strategies for the assessment of protein aggregates in pharmaceutical biotech product development. Pharm Res 28(4):920–933. https://doi.org/10.1007/s11095-010-0297-1
33. Filipe V, Hawe A, Jiskoot W (2010) Critical evaluation of nanoparticle tracking analysis (NTA) by NanoSight for the measurement of nanoparticles and protein aggregates. Pharm Res 27(5):796–810. https://doi.org/10.1007/s11095-010-0073-2
34. Ahl IM, Lindberg MJ, Tibell LA (2004) Coexpression of yeast copper chaperone (yCCS) and CuZn-superoxide dismutases in Escherichia coli yields protein with high copper contents. Protein Expr Purif 37(2):311–319. https://doi.org/10.1016/j.pep.2004.06.006
35. Sabel CE, Neureuther JM, Siemann S (2010) A spectrophotometric method for the determination of zinc, copper, and cobalt ions in metalloproteins using Zincon. Anal Biochem 397(2):218–226. https://doi.org/10.1016/j.ab.2009.10.037

Chapter 2

Ultrasensitive RT-QuIC Seed Amplification Assays for Disease-Associated Tau, α-Synuclein, and Prion Aggregates

Eri Saijo, Bradley R. Groveman, Allison Kraus, Michael Metrick, Christina D. Orrù, Andrew G. Hughson, and Byron Caughey

Abstract

The abnormal assembly of tau, α-synuclein (αSyn), or prion protein into oligomers and multimers underpins the molecular pathogenesis of multiple neurodegenerative diseases. Such pathological aggregates can often grow by seeded polymerization mechanisms. We and others have taken advantage of these mechanisms to amplify seeding activities in vitro and devise ultrasensitive, specific and quantitative assays for these etiological biomarkers. Real-time quaking-induced conversion (RT-QuIC) assays are performed in multiwell plates with fluorescent readouts, facilitating efficient throughput. Prion RT-QuIC assays on cerebrospinal fluid (CSF) samples are being widely used for *antemortem* diagnosis of human prion diseases. Recently, we have also described a tau RT-QuIC prototype that has been optimized for Pick disease (with predominant 3R tau pathology) that detects 3R tau seeds in *postmortem* CSF, and brain tissue dilutions as extreme as a billion-fold. αSyn RT-QuIC prototypes have also been developed, providing ~92% diagnostic sensitivity and 100% specificity for Parkinson's disease and dementia with Lewy bodies using *antemortem* CSF. Here we provide detailed protocols for our 3R tau and αSyn RT-QuIC assays and refer the reader to published up-to-date protocols for prion RT-QuIC assays (Orru et al. Methods Mol Biol 1658:185–203, 2017; Schmitz et al. Nat Protoc 11:2233–2242, 2016).

Key words Tau, Synuclein, Prion, Parkinson, Alzheimer, Biomarkers, Cerebrospinal fluid, Seed amplification, Diagnosis, Quantitation

1 Introduction

Protein misfolding diseases such as Alzheimer's, Parkinson's, and related tauopathies, and synucleinopathies represent a major disease burden on society. It can often be difficult to accurately diagnose such neurodegenerative diseases while the patient is still alive, especially early in the clinical phase. The diagnosis and treatment of these diseases may be aided by the ability to detect and discriminate the protein aggregates that accumulate in the brain to cause neurological dysfunction. Such pathological biomarker detection

Cláudio M. Gomes (ed.), *Protein Misfolding Diseases: Methods and Protocols*, Methods in Molecular Biology, vol. 1873,
https://doi.org/10.1007/978-1-4939-8820-4_2,

in accessible specimens such as CSF or blood might also improve cohort selection in clinical trials and serial monitoring of the efficacy of therapeutics aimed at reducing the body burden of specific misfolded protein aggregates.

Considerable progress has been made in recent years to detect prions, i.e., the infectious, self-propagating misfolded prion protein (PrP) aggregates of prion diseases such as human Creutzfeldt–Jakob disease (CJD) [1, 2]. Among the most practical and widely implemented of the prion assays are the real-time quaking-induced conversion (RT-QuIC) assays (for multiple published examples, *see* [1–15]). RT-QuIC assays involve shaking a small amount of test specimen with a vast excess of monomeric recombinant PrP in wells of multiwell plates. As shown initially with the "amyloid seeding assay" [16], any prions in the specimen will seed the assembly of the recombinant PrP (the "substrate") into amyloid fibrils that enhance the fluorescence of a dye called thioflavin T (ThT) [4, 5]. Frequent "real-time" readings of ThT fluorescence allow kinetic distinctions to be made between prion-seeded and unseeded fibrillization reactions. Unseeded reactions are slower because they require spontaneous nucleation of the substrate into seeds before fibril elongation can occur. In practice, prion-seeded reactions have shorter lag-phases, that is, the time prior to detectable growth of amyloid fibrils. For example, 15 μL samples of CSF from sporadic CJD patients usually give enhanced ThT fluorescence in <10 h, while control samples remain negative for at least 55 h [6]. As a diagnostic test for sporadic CJD, second-generation RT-QuIC assays of CSF have been ~95% sensitive and virtually 100% specific [13, 14, 17]. By combining RT-QuIC analyses of CSF and/or nasal swab samples, diagnostic sensitivity and specificity of 100% for sporadic Creutzfeldt–Jakob disease has been obtained [9, 11]. Importantly, RT-QuIC assays can also provide quantitation of relative amounts of seeding activity [4, 18–22]. Detailed up-to-date protocols for prion RT-QuIC assays have been published recently, and we refer the reader to those accounts [3, 23].

The RT-QuIC approach has also been adapted recently to the detection of certain disease-associated forms of tau (tau^D) [18] and α-synuclein ($αSyn^D$) [19, 24, 25]. The first tau RT-QuIC assay selectively detects tau^D of Pick disease (PiD) [18]. Humans express six isoforms of tau that result from differential splicing, three with three microtubule binding domains (3R isoforms) and three with four microtubule binding domains (4R isoforms). In PiD, the abnormal tau deposits are composed predominantly of 3R tau isoforms, whereas in Alzheimer disease roughly equivalent amounts of 3R and 4R isoforms are deposited. Still other tauopathies accumulate predominantly 4R isoforms. Below we first describe our most highly developed 3R tau RT-QuIC assay prototype [18].

Green and colleagues reported a prototypic CSF-based αSyn RT-QuIC assay which has given 92–95% diagnostic sensitivity for Parkinson disease (PD) and dementia with Lewy bodies (DLB) and 100% specificity with respect to Alzheimer disease and other controls [24]. Soto and colleagues reported a related CSF-based assay, called αSyn PMCA, with 89% sensitivity for PD and 97% specificity [25]. These prototypic assays for αSyn^D in CSF take 5–13 days to perform. In contrast, our recently described αSyn RT-QuIC assay can be performed in 1–2 days with quantitation and diagnostic accuracy similar to that of the Green assay [19]. Thus, after our 3R tau RT-QuIC protocol, we also describe our αSyn RT-QuIC protocol in detail below.

2 Materials

2.1 3R Tau RT-QuIC (See Note 1)

1. K19CFh expression vector: K19 cysteine-free (K19CF) is a three-repeat (3R) construct with a point mutation at residue 322 cysteine to serine to prevent intermolecular disulfide bond formation [26]. The K19CF encoding sequence was inserted between two restriction endonuclease enzyme sites *NdeI* at the 5′ and *XhoI* at the 3′ end in a bacterial expression vector pET-28a (prepared by GenScript). This resulted in the addition of a polyhistidine tag at the N terminus, yielding a construct designated K19CFh [18].
2. Kan^+ LB agar plate (with 20 μm/mL of kanamycin).
3. 20× M buffer, pH 6.8 (per 1 L stock): 71 g sodium phosphate, dibasic anhydrous, 68 g potassium phosphate monobasic, 53.6 g ammonium chloride, 14.2 g sodium sulfate.
4. 100× 505 solution (per 1 L stock): 50 g glucose, 500 g glycerol.
5. 50× 5052 solution (per 1 L stock): 250 g glycerol, 25 g glucose (dextrose), 100 g alpha-lactose monohydrate.
6. 1 M $MgSO_4$.
7. Bacterial culture medium: 500 mL ZY medium, 25 mL 20× M buffer, 1 mL 500× $MgSO_4$ (1 M), 0.5 mL 1000× kanamycin (50 mg/mL stock).
8. Culture flasks (2.5 L) with half-baffle for bacteria growth (Sigma Z710822).
9. High *g*-force resistant 50 mL polypropylene tubes (Falcon).
10. 250 mL conical centrifuge tubes (e.g., Corning).
11. Sonicator (e.g., Ultrasonic).
12. HisTrap FF™ 5 mL nickel column (GE Healthcare 17-5255-01).

13. FPLC: e.g., ÄKTA Pure chromatography system (GE).
14. Buffer A: 5 mM imidazole, 10 mM Tris, pH 8.0, 500 mM NaCl.
15. Buffer B: 200 mM imidazole, 10 mM Tris, pH 8.0, 500 mM NaCl.
16. PD-10 desalting columns (GE Healthcare 17-0851-01).
17. 10× PBS pH 6.9 (for 1 L stock): 26.8 g sodium phosphate, dibasic heptahydrate, 13.8 g sodium phosphate, monobasic monohydrate, 75.9 g sodium chloride. When diluted to 1×, pH is 7.4, which is then adjusted to 7.0.
18. Microvolume spectrophotometer.
19. 96-well plate optical black bottom plate (Nunc 265301).
20. Sealing tape (Nunc 236366).
21. 100 kDa NANOSEP spin filter (Pall, OD100C35).
22. N_2/HEPES: 1× N_2 (Gibco 17502-048, 100× stock stored at −20 °C); 10 mM HEPES, pH 7.4.
23. KO/N_2/HEPES: 10% tau Knockout (KO) mouse brain homogenate is prepared in 137 mM NaCl, 2.7 mM KCl, and 25 mM Tris, pH 7.4 (at room temperature) using a bead beater. Dilute 5 μL of tau KO mouse brain homogenate into 90 μL N_2/HEPES. Vortex.
24. Plate reader: FLUOstar OPTIMA fluorescence reader (BMG Labtech).

2.2 αSyn RT-QuIC (See Note 1)

1. αSyn expression vectors: DNA sequence coding for human α-synuclein sequence (Accession No. NM_000345.3) amino acid residues 1–140 (wildtype) were amplified and ligated into the pET28 vector with an *N*-terminal His-tag (EMD Biosciences) and sequence was confirmed. The α-synuclein K23Q mutation [27] was engineered using Q5 site-directed mutagenesis (NEB) using the primers CCACACCCTGTT GGGTTTTCTCAG and CAGAAGCAGCAGGAAAGAC. The plasmids were transformed into BL21(DE3) *Escherichia coli* (EMD Biosciences).
2. LB-M broth media (sterilized by autoclaving).
3. 1000× kanamycin stock (Sigma): 50 mg/mL kanamycin.
4. 20× M buffer, pH 6.8 (per 1 L): 71 g sodium phosphate, dibasic anhydrous, 68 g potassium phosphate monobasic, 53.6 g ammonium chloride, 14.2 g sodium sulfate. QS volume to 1 L and filter-sterilize.
5. 50× 5052 solution (per 1 L): 250 g glycerol, 25 g glucose, 100 g alpha-lactose monohydrate, QS volume to 1 L and filter-sterilize.

6. 1 M $MgSO_4$.
7. Culture flasks, 2.5 L, with half-baffle for bacteria growth (Sigma Z710822).
8. High *g*-force resistant 50 mL polypropylene tubes (e.g., Falcon).
9. 250 mL conical centrifuge tubes (Corning).
10. Osmotic shock buffer: 40% sucrose, 30 mM Tris, pH 7.2.
11. Saturated $MgCl_2$.
12. pH-indicator strips (EMD Biosciences)
13. FPLC (e.g., ÄKTA Pure chromatography system GE).
14. Nickel column (e.g., His Trap™ FF, GE).
15. Anion exchange column (e.g., HiTrap™ Q-HP, GE).
16. Dialysis membrane: 3.5 kDa MWCO (e.g., SnakeSkin, Fisher).
17. 100 kDa MWCO NANOSEP filter (Pall OD100C35).
18. Silica beads (0.8 mm, OPS Diagnostics).
19. Plate reader: FLUOstar OPTIMA fluorescence reader (BMG Labtech).

3 Methods

3.1 3R Tau RT-QuIC

3.1.1 Recombinant Tau K19CFh Purification

1. K19CFh is expressed in BL21(DE3) strain of *Escherichia coli* following the previously described Overnight Express autoinduction method (Current Protocols in Protein Science 5.23.1–5.23.18, April 2009). Once the BL21(DE3) cells are shown to express the target gene with a good yield of the recombinant protein, the cells are stored in glycerol and kept at −80 °C until use.
2. Plate cells from frozen bacterial glycerol stock. Scrape from top of glycerol-*E. coli* stock, plate onto a kan^+ LB agar plate, and incubate plate at 37 °C overnight.
3. Inoculation of starter culture from plate colonies:
   (a) Prepare 500 mL bacterial culture medium.
   (b) Transfer 3 mL culture medium into fresh 15 mL culture tube.
   (c) Add 30 μL 100× 505. Inoculate this starter culture with a single *E. coli* colony from overnight kanamycin plate (*see* **step 2**).
   (d) Grow at 37 °C for 6–8 h with a vigorous shaking (225 rpm) with loose cap to allow air exchange.
4. Inoculation of overnight autoinduction culture:

(a) Add 10 mL 50× 5052 to remaining bacterial culture medium mixture (from **step 3**) in 2.5 L culture flask.

(b) Prewarm culture flask with culture medium for 30 min at 37 °C. Add 1 mL of starter culture (from **step 3**) to prewarmed culture medium.

(c) Grow overnight in shaker at 37 °C with vigorous shaking. This combination of flask, volume and vigorous shaking will result in frothing of the culture.

5. Preparation of cell extracts for FPLC chromatography:

(a) Split 500 mL culture media into 2 × 250 mL conical centrifuge tubes.

(b) Spin tubes at 3273 × g for 35 min at 4 °C.

(c) Decant and discard supernatants (*see* **Note 2**).

(d) Resuspend each pellet in ~25 mL prechilled Buffer A by pipetting up and down with serological pipette.

(e) Transfer resuspensions into two 50 mL polypropylene tubes and keep tubes on ice for remaining steps.

(f) Adjust each to a final volume of 35 mL with prechilled Buffer A. This volume works well with our probe sonication system; however, the ideal final volume might be different for other equipment.

(g) Probe sonicate the contents of each tube in ethanol–ice bath for a total of 3 min (three cycles of 45 s sonication +15 s pause) at a power setting of 25%; *see* **Note 3**.

(h) Centrifuge 50 mL tubes at 10,000 × *g* for 1 h at 4 °C.

(i) Retain and filter supernatant using 0.45 μm syringe filter + 30 mL syringe. You now have two tubes of filtrate (~35 mL each). Move on to **step 6** with one tube and keep the other at 4 °C to be loaded on the same column after the first filtrate is fractionated.

6. Prepare 5 mL nickel column on FPLC:

(a) Wash FPLC system and preload with Buffer A and Buffer B as appropriate.

(b) Attach column in line on the FPLC.

(c) Preequilibrate column with ten column volumes (50 mL) of Buffer A at a flow rate of 1 mL/min.

7. Prepare fraction collector to collect 2 mL fractions.

8. Loading sample onto column:

(a) Flow one of the filtrates (from **step 5**) over the column using a loading pump or equivalent and a flow rate of 1 mL/min.

(b) Wash column with Buffer A until the A280 returns to baseline (~10 column volumes).

(c) Wash column at 13% Buffer B (total imidazole concentration is 30 mM) with seven column volumes (35 mL) to elute contaminants peak (*see* Supplementary Fig. 2 in [18] and **Note 4**).

(d) Wash column with 21% Buffer B (total imidazole concentration is 46 mM) for 5 column volumes (25 mL) to elute contaminants peak (*see* Supplementary Fig. 2 in [18] and **Note 4**).

(e) Elute K19CFh peak with a linear gradient from 21% Buffer B to 100% Buffer B over 8 column volumes (40 mL).

(f) Collect 2 mL fractions across the K19CFh elution peak; *see* Supplementary Fig. 2 in [18] for a representative elution. Immediately add 2 μL of 2 M dithiothreitol (DTT) to each fraction give a final concentration of 2 mM.

9. Acetone precipitation of K19CFh fractions:

(a) With reference to the elution profile above, collect fractions analogous to fractions 38–45 (totaling 16 mL) shown in Supplementary Fig. 2 of [18]. Pool and gently mix these fractions in prerinsed and prechilled 50 mL polypropylene tube. The assigned fraction numbers are the result of an extended fraction collection starting at the beginning of the 13% Buffer A wash. The target for K19CFh collection is the front two-thirds of the elution peak, which has the highest specific activity in 3R tau RT-QuIC assays.

(b) 4 mL of this pool is then transferred to each of four prerinsed and pre-chilled 50 mL polypropylene tubes, which are kept on ice.

(c) 16 mL of prechilled 100% acetone is added to each of the four tubes bringing the volume to 20 mL/tube (higher volumes may lead to tube cracking).

(d) Place tubes in −80 °C freezer for 10 min.

(e) Place in 4 °C refrigerator overnight (*see* **Note 5**).

(f) Spin fractions in 50 mL tubes at 10,000 ×*g*, 20 min, 4 °C (*see* **Note 6**).

(g) Discard supernatant. (Usually the first pellet is well attached to be decanted.)

(h) Add to each pellet in 10 mL acetone containing 2 mM DTT.

(i) Spin the 50 mL polypropylene tubes at 10,000 ×*g*, 20 min, 4 °C.

(j) Discard supernatant. This pellet is usually detached from the tube wall; therefore the supernatant should be removed with a pipette.

(k) Add 2 mL of 8 M guanidine hydrochloride in PBS to each pellet and allow to dissolve without agitation. This is a convenient break point for up to a few hours.

10. Desalting using PD-10 columns.

(a) Follow the gravity protocol provided in the GE PD-10 manual.

(b) Use a separate PD-10 column for each of four resolubilized pellets generated in **step 9**.

(c) Cut sealed bottom end of each column and tip to decant remaining above the column.

(d) Fill each column with PBS (elution buffer) to the brim and wash with >25 mL (five column volumes) of elution buffer.

(e) To each column apply 2 mL of the resolubilized pellet generated in **step 9** into the center of a column.

(f) Allow column to drip into a flow-through collector.

(g) Add 500 μL PBS to column and let column drip into flow-through collector. This application is equilibrating the column; no protein is lost here.

(h) Elute the K19CFh by successively adding 8 × 500 μL PBS and collecting the eluant as separate fractions in 1.5 mL tubes. For each of these fractions, add 500 μL PBS and wait for the dripping into the fraction tube to stop before adding the next 500 μL volume.

(i) Measure the A280 fractions 1–8 using a spectrophotometer to ensure that the guanidine HCl is washed away. The baseline should be flat in fraction 1, and again in fraction 7, with a high, guanidinium peak in fraction 8.

(j) The purified protein should be present in fractions 2–5.

(k) Assuming expected elution, pool fractions 2–5 and adjust immediately to <1 mg/mL with 1× PBS pH 7.0.
K19CFh molecular weight: 12714.4 g/mole.
Extinction coefficient: 1490 M^{-1} cm^{-1}.
Absorbance of 0.1% (i.e., 1 g/L) solution: 0.117.

(l) Aliquot substrate pool in 0.5–1 mL volumes and freeze immediately at −80 °C.

11. Repeat **steps 6–10** using the second filtrate from **step 5**.

3.1.2 3R Tau RT-QuIC Assay Procedure

1. Start the plate reader and computer.

(a) Open the "Omega Mars" data analysis program first before opening the "Omega" control program. (However, this sequence may not be required for newer software.)

(b) Enter the following settings into your script: target temperature, 37 °C; shaking speed, 500 rpm; gain, 1100; cycle

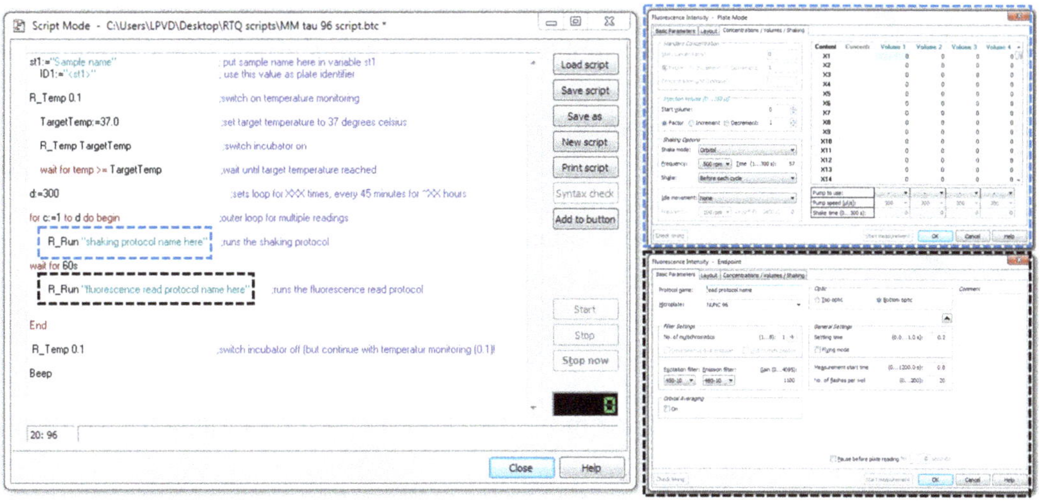

Fig. 1 Screen shot from Omega Mars program showing example of a tau RT-QuIC script with associated shaking and reading protocols for the fluorescence plate reader

time, 1800 s; read every ~45 min; total measurement time as required (Fig. 1). The gain setting should be determined for each plate reader to avoid detector saturation with strongly positive reactions.

(c) Set the temperature to 37 °C, and let the machine heat up.

2. Thaw frozen test samples (brain homogenate, CSF, etc.) and N_2 stock (100×) on ice.
3. Prepare serial dilutions of test samples (seeds): Transfer 18 μL of the KO/N_2/HEPES into each tube. Add the volume to the bottom of a tube (not on the wall of the tube) using a pipettor. Vortex the KO/N_2/HEPES mix periodically to keep the KO/N_2/HEPES well mixed. Add 2 μL of test sample and vortex, avoiding splashing onto tube walls. Repeat to generate serial tenfold dilutions.
4. Preparation of K19CFh tau substrate:

(a) Thaw substrate at room temperature. Thawing at higher or lower temperature is not recommended, as it might promote aggregation. Since the experiment is prepared at room temperature, it is strongly recommended to keep substrates at the same temperature once thawed.

(b) Once thawed, immediately filter the substrate with a 100 KDa spin filter at 3300 × *g* for 5 min at room temperature (*see* **Note 7**). Use within 2 h of filtration to minimize nascent spontaneous seed formation at this high concentration.

(c) Transfer filtered substrate into a new tube and pool together if needed.

(d) Measure 280 nm absorbance (A280) with a NanoDrop spectrophotometer and calculate concentration of substrate:

5. Preparation of tau RT-QuIC reactions:

(a) Prepare 1 mM ThT by adding 20 μL of 10 mM ThT stock into 180 μL of Milli-Q H_2O and vortexing. These volumes may be adjusted as needed.

(b) Add all components of the reaction buffer (*see* Table 1) except substrate and water into a prerinsed clean tube and vortex. Prerinsing removes particulates.

(c) Add appropriate amounts of water and substrate to the master mix (Table 1) and gently mix by pipetting (*see* **Note 8**).

(d) Depending on the volume of the test samples to be assayed (e.g., 2 μL of brain homogenate dilution or 5 μL of CSF; [18]), transfer into each well a volume of the master mix needed to make the final reaction volume 100 μL.

(e) Add 2–5 μL (or other as appropriate) test sample to each well.

(f) Seal the plate with sealing tape.

(g) The plate is placed in the fluorescence plate reader and incubated at 37 °C with intermittent shaking, consisting of 60 s of orbital shaking at 500 rpm and no shaking for 60 s, with a 1-min pause to measure the fluorescence.

(h) The kinetics of fibril formation are monitored by the bottom reading of the fluorescence intensity every 45 min using 450–10 nm excitation and 480–10 nm emission. Instrument gains are set to avoid frequent saturation of the detector. We aim for baseline fluorescence values of 600–1000 PFU

6. Data analysis.

Refer to Subheading 3.3 for general considerations in RT-QuIC data analysis and our prior study for specific examples of analysis of 3R tau RT-QuIC results [18].

3.2 αSyn RT-QuIC

3.2.1 Purification of Recombinant K23Q αSyn

1. Bacterial cultures:

(a) Prepare a 5 mL LB-M broth culture containing 50 μg kanamycin (Sigma) in a 15 mL breathable culture tube to be used as a starter culture.

(b) In the morning, inoculate the 5 mL LB-M broth starter culture from a frozen bacterial glycerol stock previously transformed with the K23Q αSyn containing pET28 vector (*see* Subheading 2.2, item 1 list). Alternatively, this starter culture may be inoculated with colonies from an agarose plate culture.

Table 1
3R tau RT-QuIC reaction buffer master mix

		Per reaction		Per plate (105[a] reactions)		Final concentration in 100 μL reaction
	MQ H_2O	*x*[b]	μL	105*x*[b]	μL	
50 mM	HEPES pH 7.4	20	μL	2100	μL	10 mM
2 M	NaCl	10	μL	1050	μL	200 mM
10 mg/mL	Heparin	4.5	μL	473	μL	80 μM
1 mM	ThT	1	μL	105	μL	10 μM
	K19CFh	*y*[b]	μL	105*y*[b]	μL	20 μM
	Total	98	μL	10,290	μL	

[a]Including extra volume to cover pipetting losses
[b]*y* is based on [K19CFh] determined in **step 4**d. *x* is then adjusted based on y to give the indicated final volume

(c) Grow starter culture in a shaking incubator at 37 °C with vigorous shaking.

(d) Prepare and warm a 1 L autoinduction media culture (1 L LB-M broth media with 50 μg kanamycin, 50 mL 20× M buffer, 20 mL 50× 5052 solution and 2 mL 1 M $MgSO_4$) to 37 °C in a 2.5 L culture flask.

(e) Following 4–5 h of incubation the starter culture should be cloudy with significant bacterial growth. Inoculate prepared 1 L autoinduction media culture with the entire 5 mL of starter culture.

(f) Grow the cells in a shaking incubator at 37 °C with vigorous shaking overnight. This combination of flask, volume, and vigorous shaking should result in frothing of the culture.

(g) The next morning, harvest cells by splitting the culture volume into fourths using 250 mL conical centrifuge tubes and centrifuge at 3200 × *g*, 4 °C, 10 min.

(h) Decant and discard supernatant.

2. Osmotic treatment of cells: To each of the four cell pellets add 25 mL of osmotic shock buffer and gently resuspend using a 25 mL serological pipette. Incubate resuspended pellets at room temperature for 10 min.
3. Centrifuge treatment cell suspension: 9000 × *g*, 20 °C, 20 min.
4. Discard supernatants and gently resuspend each pellet in 10 mL ice-cold water using a 25 mL serological pipette, keeping tubes on ice to release contents of periplasmic space.
5. Pool into two high g-force resistant 50 mL tubes with 20 mL each and then add 20 μL of saturated $MgCl_2$ to each tube.
6. Mix and incubate with gentle rocking for 3 min on ice.
7. Centrifuge tubes to pellet suspension: 9000 × *g*, 4 °C, 30 min.
8. Collect supernatants into a single 100 mL beaker.
9. Adjust the supernatant to pH 3.5:

(a) Add a stir bar to the beaker and set stirring rate such that the solution is mixing strongly but not enveloping air.

(b) Add 800 μL of 1 M HCl and then check the pH using a pH-indicator strip. The pH will likely be acidic (e.g., 5.5), but further adjustment will be necessary to get to 3.5. As the pH drops a large amount of white precipitate will be generated.

(c) Continue by adding volumes of 50 μL (or 25 μL) of 1 M HCl and then checking the pH using a pH-indicator strip until pH 3.5 is achieved. Try to adjust the pH rapidly but with continuous monitoring to avoid going past the pH of 3.5.

(d) Continue to incubate with stirring at room temperature for 10 min.

10. Centrifuge to pellet precipitate: 9000 × *g*, 4 °C, 30 min and then transfer the supernatant to a clean 100 mL beaker.
11. Adjusting the supernatant to pH 7.5:
    (a) Again, add a stir bar and adjust the rate of stirring such that the solution is mixing strongly but not enveloping air.
    (b) Add a volume of 1 M NaOH that is 50 μL less than the total volume of 1 M HCl added in **step 9** above and then check the pH using a pH-indicator strip.
    (c) Continue by adding volumes of 25 μL of 1 M NaOH and then checking the pH using a pH-indicator strip until pH 7.5 is achieved. As the pH drops a small amount of white precipitate may be generated. Continue to incubate with stirring at room temperature for 10 min.
12. Filter protein extract through a filter of pore size ≤0.45 μm.
13. Extract can now be stored at 4 °C for an hour or so if needed.
14. Nickel column: The extract is loaded onto a 5 mL column on an FPLC and washed with 20 mM Tris, pH 7.5 at room temperature. The column is further washed with 50 mM imidazole, 20 mM Tris, pH 7.5 which generates a peak that is not collected. A linear gradient up to 500 mM imidazole in 20 mM Tris, pH 7.5, is performed and a peak is collected between 150 and 375 mM imidazole. (*See* Additional file 1 in [19] for elution profile and gel analysis).
15. Anion exchange column: The peak fractions from **step 14** are pooled and loaded onto a 5 mL column on FPLC and washed with 20 mM Tris, pH 7.5. The column is further washed with 100 mM NaCl, 20 mM Tris, pH 7.5. A linear gradient up to 500 mM NaCl in 20 mM Tris pH 7.5 is performed and a peak recovered between 300 and 350 mM NaCl. The peak fractions are pooled and filtered through a 0.22 μm filter. (*See* Additional file 1 in [19] for elution profile and gel analysis.)
16. Dialyze pooled fractions against 4 L of water overnight at 4 °C using 3.5 kDa MWCO dialysis membrane and then again for 4 h against another 4L of fresh water.
17. Determine the protein concentration with a UV–VIS spectrophotometer using a theoretical extinction coefficient at 280 nm of 0.36 $(mg/mL)^{-1}$ cm^{-1}.
18. Flash freeze and lyophilize protein in aliquots appropriate for use for single experiments. In our case, this is typically aliquots that when resuspended in 500 μL, the concentration is ~1.0 mg/mL. Store lyophilized aliquots at −80 °C.

3.2.2 αSyn RT-QuIC Protocol

1. Plate preparation: RT-QuIC reactions are performed in black 96-well plates (*see* Subheading 2.1, **item 19**). Preload plates with six silica beads per well (*see* **Note 9**).
2. For brain homogenate samples: To each well with beads add 98 μL of a reaction mix to give final concentrations of 40 mM phosphate buffer (pH 8.0), 170 mM NaCl, 0.1 mg/mL of the designated rαSyn (filtered through a 100 kDa MWCO filter immediately prior to use), and 10 μM ThT. Add 2 μL of brain homogenate appropriately diluted in PBS as seed. The plates are then sealed with a plate sealer film (*see* Subheading 2.1, **item 20**) and incubated at 42 °C in the plate reader with cycles of 1 min shaking (400 rpm double orbital) and 1 min rest throughout the overall reaction time. ThT fluorescence measurements (450 ± 10 nm excitation and 480 ± 10 nm emission; bottom read) are taken every 45 min.
3. For CSF samples: Wells preloaded with six silica beads are then loaded with 85 μL of a reaction mix adjusted to give final reaction concentrations of 40 mM phosphate buffer (pH 8.0), 170 mM NaCl, 0.1 mg/mL rαSyn, 10 μM ThT, and 0.0015% sodium dodecyl sulfate (SDS) and then 15 μL CSF, or dilutions thereof, in normal pooled CSF. The reaction plates are subjected to shake–rest cycles as described for brain homogenate samples in **step 2**.
4. Data analysis: For general strategy, *see* below. For specific examples, *see* [19].

3.3 RT-QuIC Data Analysis: General Considerations

A matter of kinetics. The primary influence of preexisting tau, αSyn or prion seeds in a sample should be to accelerate the formation of amyloid from the initially monomeric substrate protein relative to any spontaneous nucleation and elongation of fibrils in the absence of preexisting seeds. This results in a shortening of the time to a chosen fluorescence threshold reading to an extent that, in the absence of other confounding factors, depends upon the initial seed concentration (Fig. 2). Most substrate proteins such as prion protein, tau, or αSyn, or fragments thereof, will eventually form fibrils spontaneously under various RT-QuIC conditions in the absence of preformed seeds if given enough time. Thus, we usually distinguish between appropriately matched samples with differing seed concentrations and characteristics by comparing reaction kinetics, lag phases or "times to threshold" [4, 16] (Fig. 2). In some prion RT-QuIC assays, distinctions between prion seed strains have also be made based on maximal fluorescence readings [13, 28] or SDS-PAGE analyses of protease-treated RT-QuIC products [8, 29].

Positive vs. *negative.* Multiple approaches can be envisioned to compare RT-QuIC data from different samples in order to make binary positive/negative decisions about a given sample. Depending

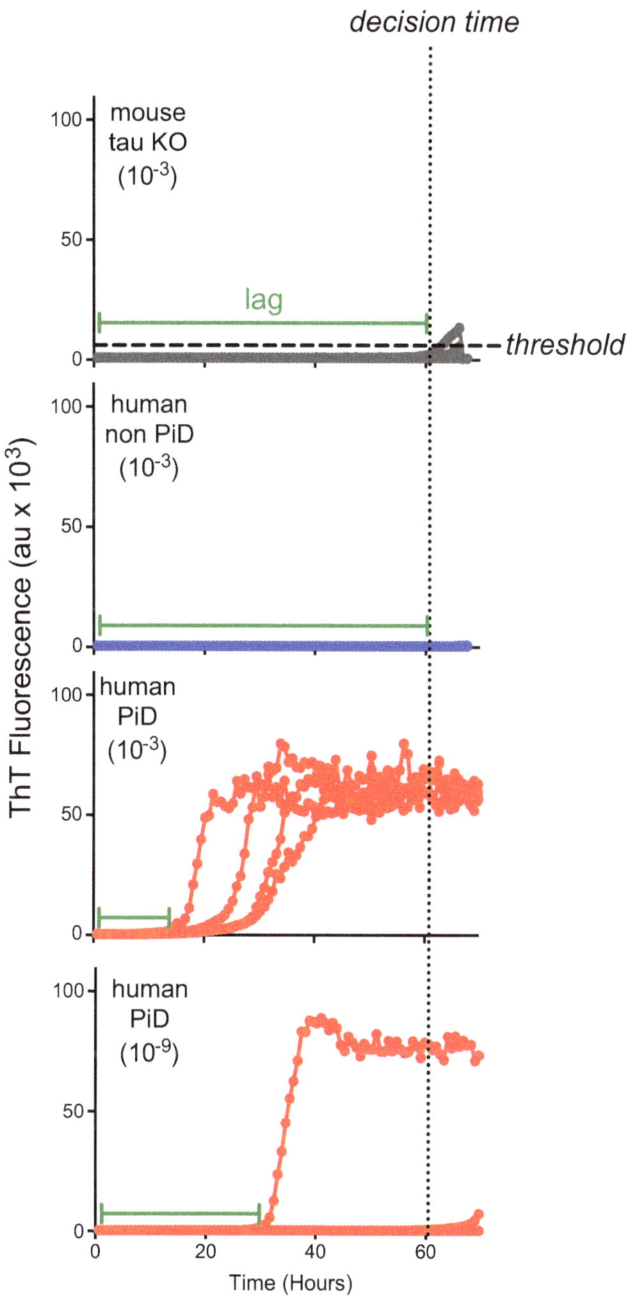

Fig. 2 Representative 3R tau RT-QuIC data. 3R tau RT-QuIC results are shown in the presence of mouse tau knockout (KO) and human non-PiD control brain homogenates at 10^{-3} dilutions and human PiD brain homogenates at 10^{-3} and 10^{-9} dilutions. Traces for 4 replicate wells are shown in each panel. The dotted line indicates the decision time before which individual wells with positive ThT fluorescence readings are designated as "positives." The dashed line indicates the threshold that must be surpassed prior to the cutoff for a "positive" designation. The lag time of each reaction is indicated by the green line

on the goals of the testing, a balance must be struck between sensitivity (the proportion of true positive samples that give positive tests) versus specificity (the proportion of true negative samples that give negative tests). Our typical approach is summarized as follows:

1. We first obtain data from a substantial number of known positive and negative control samples derived in the same way from the same tissues so that the matrices of the samples are as closely matched as possible. The plate readers measure ThT fluorescence and the data can be plotted and analyzed using GraphPad Prism 6 or other graphing software (Fig. 2).
2. Baseline analysis: We select a suitable window of time prior to any observable increases in fluorescence in the positive control samples to measure the average ±SD baseline fluorescence values for each reaction. Then we establish a fluorescence threshold for positivity that represents the average baseline fluorescence plus a sufficient number of SDs to prevent any artefactual drifts in the baseline during the lag phase from being called positive (Fig. 2).
3. Decision point (or cut-off time): A time point prior to when the vast majority of negative control reactions surpass the threshold fluorescence is usually selected as the decision point of the experiment, i.e., the time at which positive/negative determinations for the reactions is made (Fig. 2). The closer the decision point is to the onset of false (spontaneous) positive responses from the negative controls, the higher the sensitivity and the lower the specificity of the assay. The decision point is selected based on the investigators' relative preference for sensitivity vs specificity, and their empirical results obtained for each type of sample under their given assay conditions and instruments.
4. Overall positive/negative determination for sample: The ratio of positive/total replicate (usually quadruplicate) reactions for the sample is determined. In most applications we consider a sample positive overall when half, or more, of replicate wells cross the fluorescence threshold by the decision time. When a single replicate, or less than half, cross the threshold, the analysis is repeated. If, with repeat testing, all replicate wells are negative, we call the sample negative. If another well is positive, then we view the sample as questionable/borderline. In diagnostic applications in particular, we, and others, have applied this further "half-positive replicates" criterion because individual negative control wells can (rarely, presumably stochastically) become positive spontaneously at almost any time, so we do not want to call a sample positive based on any single

reaction. The logic for repeating the sample is that if the assay conditions only rarely produce spontaneous reactions, then it would be improbable that such a rare event would occur twice in a row for a particular sample.

Quantitation. Relative amounts of seeding activity can be assessed using end-point dilution [4, 18, 19] or kinetic [20–22] analyses as has been described elsewhere for prion, 3R tau, or αSyn RT-QuIC reactions.

4 Notes

1. Whenever feasible, buffers and reagents are filtered through a prerinsed 0.22 μm filter into prerinsed vessels prior to storage in order to sterilize and minimize lint contamination. Filters, syringes and tubes are also prerinsed.
2. Although it is common to store such cell pellets at this stage at −20 °C until use, we have avoided this for fear that freezing and thawing might reduce yield and/or promote tau aggregation or seed formation.
3. This step may need adjusting to available equipment: it is important to complete cell lysis without oversonication, which can deteriorate the sample.
4. Although the Supplementary Fig. 2 indicates 15% and 23% Buffer B, now we recommend 13% Buffer B for step c and 21% Buffer B for step d.
5. Ensure that tubes are not crooked such that the protein precipitate is concentrated at the bottom of the tube.
6. Check for cracks in polypropylene tubes. If cracked from acetone exposure, transfer to new tube.
7. A small portion (<10%) of the substrate might be lost during the filtration step while eliminating any preformed aggregates.
8. Never vortex after the substrate has been added to the master mix. Vortexing or vigorous mixing might promote aggregate (artificial seed) formation.
9. 1 mm "glass" beads (BioSpec products) have also been used successfully for this application. As we experimented with increasing the number of beads per well, we found that there was a general increase in ThT fluorescence enhancement in seeded reactions containing up to ~12 beads, at which point the fluorescence began to drop due to light scattering in the well. We have chosen to use the six silica beads as a compromise between reasonable ThT signal strength and the utility of setting up plates.

Acknowledgments

This work was supported by the Intramural Research Program of the NIAID.

References

1. Zanusso G, Monaco S, Pocchiari M, Caughey B (2016) Advanced tests for early and accurate diagnosis of Creutzfeldt-Jakob disease. Nat Rev Neurol 12(6):325–333. https://doi.org/10.1038/nrneurol.2016.65
2. Caughey B, Orru CD, Groveman BR, Hughson AG, Manca M, Raymond LD, Raymond GJ, Race B, Saijo E, Kraus A (2017) Amplified detection of prions and other amyloids by RT-QuIC in diagnostics and the evaluation of therapeutics and disinfectants. Prog Mol Biol Transl Sci 150:375–388. https://doi.org/10.1016/bs.pmbts.2017.06.003
3. Orru CD, Groveman BR, Hughson AG, Manca M, Raymond LD, Raymond GJ, Campbell KJ, Anson KJ, Kraus A, Caughey B (2017) RT-QuIC assays for prion disease detection and diagnostics. Methods Mol Biol 1658:185–203. https://doi.org/10.1007/978-1-4939-7244-9_14
4. Wilham JM, Orrú CD, Bessen RA, Atarashi R, Sano K, Race B, Meade-White KD, Taubner LM, Timmes A, Caughey B (2010) Rapid end-point quantitation of prion seeding activity with sensitivity comparable to bioassays. PLoS Path 6(12):e1001217. https://doi.org/10.1371/journal.ppat.1001217
5. Atarashi R, Satoh K, Sano K, Fuse T, Yamaguchi N, Ishibashi D, Matsubara T, Nakagaki T, Yamanaka H, Shirabe S, Yamada M, Mizusawa H, Kitamoto T, Klug G, McGlade A, Collins SJ, Nishida N (2011) Ultrasensitive human prion detection in cerebrospinal fluid by real-time quaking-induced conversion. Nat Med 17(2):175–178. https://doi.org/10.1038/nm.2294.nm.2294 pii
6. Orru CD, Groveman BR, Hughson AG, Zanusso G, Coulthart MB, Caughey B (2015) Rapid and sensitive RT-QuIC detection of human Creutzfeldt-Jakob disease using cerebrospinal fluid. MBio 6(1). https://doi.org/10.1128/mBio.02451-14
7. Orru CD, Yuan J, Appleby BS, Li B, Li Y, Winner D, Wang Z, Zhan YA, Rodgers M, Rarick J, Wyza RE, Joshi T, Wang GX, Cohen ML, Zhang S, Groveman BR, Petersen RB, Ironside JW, Quinones-Mateu ME, Safar JG, Kong Q, Caughey B, Zou WQ (2017) Prion seeding activity and infectivity in skin samples from patients with sporadic Creutzfeldt-Jakob disease. Sci Transl Med 9(417). https://doi.org/10.1126/scitranslmed.aam7785
8. Orru CD, Groveman BR, Raymond LD, Hughson AG, Nonno R, Zou W, Ghetti B, Gambetti P, Caughey B (2015) Bank vole prion protein as an apparently universal substrate for RT-QuIC-based detection and discrimination of prion strains. PLoS Path 11(6):e1004983. https://doi.org/10.1371/journal.ppat.1004983
9. Orru CD, Bongianni M, Tonoli G, Ferrari S, Hughson AG, Groveman BR, Fiorini M, Pocchiari M, Monaco S, Caughey B, Zanusso G (2014) A test for Creutzfeldt-Jakob disease using nasal brushings. N Engl J Med 371(6):519–529
10. McGuire LI, Peden AH, Orru CD, Wilham JM, Appleford NE, Mallinson G, Andrews M, Head MW, Caughey B, Will RG, Knight RSG, Green AJE (2012) RT-QuIC analysis of cerebrospinal fluid in sporadic Creutzfeldt-Jakob disease. Ann Neurol 72(2):278–285
11. Bongianni M, Orrù CD, Groveman BR, Sacchetto L, Fiorini M, Tonoli G, Triva G, Capaldi S, Testi S, Ferrari S, Cagnin A, Ladogana A, Poleggi A, Colaizzo E, Tiple D, Vaianella L, Castriciano S, Marchioni D, Hughson AG, Imperiale D, Cattaruzza T, Fabrizi G, Pocchiari M, Monaco S, Caughey B, Zanusso G (2017) Diagnosis of human prion disease using real-time quaking-induced conversion testing of olfactory mucosa and cerebrospinal fluid samples. JAMA Neurol 74(2):1–8
12. Cramm M, Schmitz M, Karch A, Mitrova E, Kuhn F, Schroeder B, Raeber A, Varges D, Kim YS, Satoh K, Collins S, Zerr I (2015) Stability and reproducibility underscore utility of RT-QuIC for diagnosis of Creutzfeldt-Jakob disease. Mol Neurobiol 53(3):1896–1904. https://doi.org/10.1007/s12035-015-9133-2
13. Foutz A, Appleby BS, Hamlin C, Liu X, Yang S, Cohen Y, Chen W, Blevins J, Fausett C, Wang H, Gambetti P, Zhang S, Hughson A, Tatsuoka C, Schonberger LB, Cohen ML, Caughey B, Safar JG (2017) Diagnostic and prognostic value of human prion detection in

cerebrospinal fluid. Ann Neurol 81(1):79–92. https://doi.org/10.1002/ana.24833

14. Franceschini A, Baiardi S, Hughson AG, McKenzie N, Moda F, Rossi M, Capellari S, Green A, Giaccone G, Caughey B, Parchi P (2017) High diagnostic value of second generation CSF RT-QuIC across the wide spectrum of CJD prions. Sci Rep 7(1):10655. https://doi.org/10.1038/s41598-017-10922-w
15. Haley NJ, Siepker C, Hoon-Hanks LL, Mitchell G, Walter WD, Manca M, Monello RJ, Powers JG, Wild MA, Hoover EA, Caughey B, Richt JA (2016) Seeded amplification of chronic wasting disease prions in nasal brushings and recto-anal mucosa associated lymphoid tissues from elk by real time quaking-induced conversion. J Clin Microbiol 54(4):1117–1126. https://doi.org/10.1128/JCM.02700-15
16. Colby DW, Zhang Q, Wang S, Groth D, Legname G, Riesner D, Prusiner SB (2007) Prion detection by an amyloid seeding assay. Proc Natl Acad Sci U S A 104(52):20914–20919
17. Groveman BR, Orru CD, Hughson AG, Bongianni M, Fiorini M, Imperiale D, Ladogana A, Pocchiari M, Zanusso G, Caughey B (2017) Extended and direct evaluation of RT-QuIC assays for Creutzfeldt-Jakob disease diagnosis. Ann Clin Transl Neurol 4(2):139–144. https://doi.org/10.1002/acn3.378
18. Saijo E, Ghetti B, Zanusso G, Oblak A, Furman JL, Diamond MI, Kraus A, Caughey B (2017) Ultrasensitive and selective detection of 3-repeat tau seeding activity in pick disease brain and cerebrospinal fluid. Acta Neuropathol 133(5):751–765. https://doi.org/10.1007/s00401-017-1692-z
19. Groveman BR, Orru CD, Hughson AG, Raymond LD, Zanusso G, Ghetti B, Campbell KJ, Safar J, Galasko D, Caughey B (2018) Rapid and ultra-sensitive quantitation of disease-associated alpha-synuclein seeds in brain and cerebrospinal fluid by alphaSyn RT-QuIC. Acta Neuropathol Commun 6(1):7. https://doi.org/10.1186/s40478-018-0508-2
20. Henderson DM, Davenport KA, Haley NJ, Denkers ND, Mathiason CK, Hoover EA (2015) Quantitative assessment of prion infectivity in tissues and body fluids by real-time quaking-induced conversion. J Gen Virol 96(Pt 1):210–219. https://doi.org/10.1099/vir.0.069906-0
21. Shi S, Mitteregger-Kretzschmar G, Giese A, Kretzschmar HA (2013) Establishing quantitative real-time quaking-induced conversion (qRT-QuIC) for highly sensitive detection and quantification of PrPSc in prion-infected tissues. Acta Neuropathol Commun 1(1):44. https://doi.org/10.1186/2051-5960-1-44
22. Gray JG, Graham C, Dudas S, Paxman E, Vuong B, Czub S (2016) Defining and assessing analytical performance criteria for transmissible spongiform encephalopathy-detecting amyloid seeding assays. J Mol Diagn 18(3):454–467. https://doi.org/10.1016/j.jmoldx.2016.01.005
23. Schmitz M, Cramm M, Llorens F, Muller-Cramm D, Collins S, Atarashi R, Satoh K, Orru CD, Groveman BR, Zafar S, Schulz-Schaeffer WJ, Caughey B, Zerr I (2016) The real-time quaking-induced conversion assay for detection of human prion disease and study of other protein misfolding diseases. Nat Protoc 11(11):2233–2242. https://doi.org/10.1038/nprot.2016.120
24. Fairfoul G, McGuire LI, Pal S, Ironside JW, Neumann J, Christie S, Joachim C, Esiri M, Evetts SG, Rolinski M, Baig F, Ruffmann C, Wade-Martins R, Hu MT, Parkkinen L, Green AJ (2016) Alpha-synuclein RT-QuIC in the CSF of patients with alpha-synucleinopathies. Ann Clin Transl Neurol 3(10):812–818. https://doi.org/10.1002/acn3.338
25. Shahnawaz M, Tokuda T, Waragai M, Mendez N, Ishii R, Trenkwalder C, Mollenhauer B, Soto C (2017) Development of a biochemical diagnosis of parkinson disease by detection of alpha-synuclein misfolded aggregates in cerebrospinal fluid. JAMA Neurol 74(2):163–172. https://doi.org/10.1001/jamaneurol.2016.4547
26. Dinkel PD, Siddiqua A, Huynh H, Shah M, Margittai M (2011) Variations in filament conformation dictate seeding barrier between three- and four-repeat tau. Biochemistry 50(20):4330–4336. https://doi.org/10.1021/bi2004685
27. Koo HJ, Lee HJ, Im H (2008) Sequence determinants regulating fibrillation of human alpha-synuclein. Biochem Biophys Res Commun 368(3):772–778. https://doi.org/10.1016/j.bbrc.2008.01.140
28. Cramm M, Schmitz M, Karch A, Zafar S, Varges D, Mitrova E, Schroeder B, Raeber A, Kuhn F, Zerr I (2015) Characteristic CSF prion seeding efficiency in humans with prion diseases. Mol Neurobiol 51(1):396–405. https://doi.org/10.1007/s12035-014-8709-6
29. Masujin K, Orru CD, Miyazawa K, Groveman BR, Raymond LD, Hughson AG, Caughey B (2016) Detection of atypical H-type bovine spongiform encephalopathy and discrimination of bovine prion strains by real-time quaking-induced conversion. J Clin Microbiol 54(3):676–686. https://doi.org/10.1128/JCM.02731-15

Chapter 3

Vesicle-Based Assays to Study Membrane Interactions of Amyloid Peptides

Ravit Malishev, Sofiya Kolusheva, and Raz Jelinek

Abstract

The growing interest in membrane interactions of amyloidogenic peptides and proteins emanates from the realization that lipid bilayers and membranes play central roles in the toxicity and pathological pathways of amyloid diseases. This chapter presents experimental schemes designed to study membrane interactions and membrane-induced fibrillation of amyloid peptides.

Key words Amyloid peptides, Membranes, Amyloid-membrane interactions, Fibrillation, Membrane dynamics, Lipid bilayer fluidity

1 Introduction

Membranes and membrane interactions play important roles in the biological functions and pathologies of amyloidogenic peptides and proteins. While varied hypotheses link toxicity to membrane interactions of amyloid peptides, their oligomeric species, or mature fibrils, research is still required to decipher such interactions. In particular, biomimetic vesicle systems allow investigation of different structural and dynamical parameters pertaining to peptide–membrane interactions. Herein we present experimental protocols for using lipid vesicles, including multilamellar dispersion (MLD), small unilamellar vesicles (SUVs; <100 nm), large unilamellar vesicles (LUVs; 100–1000 nm), and giant unilamellar vesicles (GUVs; >1000 nm) [1]. The preparation of the different vesicle species depend upon the types of lipids used and experimental methods.

The protocols below describe applications of several biophysical techniques in conjunction with the vesicle systems. Specifically, ***Förster resonance energy transfer (FRET)*** record the energy transferred between fluorescence donor and acceptor embedded

Cláudio M. Gomes (ed.), *Protein Misfolding Diseases: Methods and Protocols*, Methods in Molecular Biology, vol. 1873, https://doi.org/10.1007/978-1-4939-8820-4_3,

within the vesicle bilayer. FRET modulations are correlated to changes in lipid organization and dynamics within the bilayer [2, 3]. ***Fluorescence anisotropy*** of bilayer-embedded fluorophore provides information upon lipid mobility in bilayers and the effects of membrane-active compounds [4, 5]. ***Fluorescence quenching*** experiments probe the effects of membrane-active compounds upon the surface properties of lipid bilayers [6, 7]. ***Vesicle leakage*** analyses examine the induction of release of vesicle-encapsulated fluorescent dye upon the activity of bilayer-disruptive molecules [8, 9]. ***Calorimetric measurements,*** specifically ***differential scanning calorimetry (DSC)*** and ***isothermal titration calorimetry (ITC),*** are applied to probe the ordering and molecular packing of lipid bilayers prior and following interactions with membrane-active molecules [10, 11]. ***Electron spin resonance (ESR)*** is useful for evaluating modification of bilayer fluidity upon binding and insertion of membrane-active species [1]. ***Microscopy*** methods enable *visualizing* the impact of membrane processes, generally through the use of giant vesicles [12, 13]. Overall, this chapter summarizes biophysical methods and biomimetic vesicle systems applied for studying membrane interactions of amyloid peptides.

2 Materials

2.1 Amyloid Protein\Peptide

Dissolve amyloid-β (Aβ) 40 or 42 (1 mg/mL), prion protein fragment 106-126 [termed PrP(106-126)] (1 mg/mL) and islet amyloid polypeptide (IAPP) (0.5 mg/mL) in 1,1,1,3,3, 3-hexafluoro-2-propanol (HFIP) ≥99% and store at −20 °C (*see* **Note 1**).

2.2 Lipids

1,2-dioleoyl-sn-glycero-3-phosphocholine (DOPC), 1,2-dioleoyl-sn-glycero-3-phospho-(1′-rac-glycerol) (DOPG), 1,2-dioleoyl-sn-glycero-3-phospho-L-serine (DOPS), cholesterol (ovine wool, >98%), sphingomyelin (brain, porcine). Store at −20 °C. Biophysical experiments can be carried out with other synthetic\natural lipids, unless otherwise specified. It is recommended to order lipids from Avanti Polar Lipids, Inc.

2.3 Buffer Solutions

- 10 mM phosphate buffer, pH 7.4; 0.2 M sodium phosphate, dibasic dehydrate, 0.2 M sodium phosphate, monobasic monohydrate (*see* **Note 2**).
  Biophysical methods: Dissolving liposomes and amyloid proteins.
- Sodium dithionite buffer, pH 11; 0.6 M sodium dithionite, 50 mM Tris base (*see* **Note 3**).
  Biophysical methods: Fluorescence quenching.

- Carboxyfluorescein (CF) buffer, pH 7.4; 50 mM CF, 50 mM Hepes, 110 mM NaCl, 0.02% NaN_3 (*see* **Note 4**).
  Biophysical methods: Leakage liposomes.

2.4 Fluorescent and Dye Probes

- Dissolve 1,2-dimyristoyl-sn-glycero-3-phosphoethanolamine-N-(7-nitro-2–1,3 benzoxadiazol-4-yl (N-NBD-PE) or 1,2-dimyristoyl-sn-glycero-3-phosphoethanolamine-N-(lissamine Rhodamine B sulfonyl) (N-Rh-PE) in chloroform\ethanol 1:1 mixture at a 1 mg/mL concentration. Store at −20 °C.
  Biophysical methods: Förster resonance energy transfer (FRET).
- Dissolve 1,6-diphenylhexatriene (DPH) or 1-(4-trimethylamm oniumphenyl)-6-phenyl-1,3,5-hexatriene (TMA-DPH) in tetrahydrofuran (THF) at a concentration of 1 mg/mL. Store at 4 °C.
  Biophysical methods: Fluorescence anisotropy.
- Dissolve 5-doxyl-stearic acid (5-DS) or 12-doxyl-stearic acid (12-DS) in ethanol at 1 mg/mL concentration. Store at 4 °C.
  Biophysical methods: Electron spin resonance (ESR).

3 Methods

3.1 Peptide Preparation

1. Dissolve the peptide in HFIP solvent and store at −20 °C.
2. For each experiment, HFIP-dissolved peptide should be dried in vacuo to constant weight. Before using, dried peptide samples are dissolved in 10 mM phosphate buffer, pH 7.4 (*see* **Note 2**) to the required final concentration (Table 1).

3.2 Vesicles

3.2.1 Stock Solutions

1. Prepare lipid stock solutions in chloroform–ethanol (1:1 v/v) (Table 2) and store at −20 °C (*see* **Note 5**).

Table 1
Conditions used for amyloid peptide experiments[a]

Disease	Peptide	Final concentration	Lipid composition[b]	Ref.
Alzheimer's disease	Aβ40, Aβ42	10–20 μM	PC; PC/Chol; PC/Chol/SM; PC/PG	[13, 15]
Spongiform encephalopathies	PrP(106–126)	50 μM	PC/PG; PC/Chol/SM	[16, 17]
Type II diabetes	IAPP	15 μM	PC/PG; PC/PS	[16, 18]

[a]Recommended temperature range 25–37 °C
[b]*PC* phosphocholine, *Chol* cholesterol, *SM* sphingomyelin, *PG* phospho-(1′-rac-glycerol), *PS* phosphatidylserine

Table 2
Lipids employed in vesicle formation (*see* Note 6)

Phospholipid	Charge	Molecular weight[a]	Stock solutions 50 mM (mg/mL)[b]
PC	Zwitterionic	786.113	39
PG	Anionic	797.026	40
PS	Anionic	810.025	40.5
Cholesterol	Uncharged	386.654	19.3
Sphingomyelin	Zwitterionic	760.223	38

[a]The presented M.W (g\mol) of PC, PG, and PS refer to the 1,2-dioleoyl-sn-glycero-3 structure
[b](*see* **Note 7**)

3.2.2 Prepreparation

1. Combine the lipids in the appropriate ratio in a glass test tube to a final total lipid concentration of 1 mM (total volume 2 mL).
2. Remove the organic solvents by placing the tube in a rotary evaporator or under N_2 purge until constant weight.
3. Add 2 mL buffer to the dried lipid mixture (*see* **Note 6**).

3.2.3 Multilamellar Dispersion (MLD)

1. Follow prepreparation steps (Subheading 3.2.2).
2. Add glass beads and thoroughly shake the sample using a vortex.
3. The resulting MLD solution should exhibit a "milky" white color and should be stored at 4 °C.

3.2.4 Small Unilamellar Vesicles (SUVs), Size Range <100 nm

1. Follow prepreparation steps (Subheading 3.2.2).
2. Place the tip of a probe-sonicator (power: 130 w, frequency: 20 kHz) in the glass tube and sonicate at 20% amplitude for 10 min (1 min on\1 min off) above the T_m of the main lipid species in the vesicle (*see* **Note 8**).
3. Incubate the vesicle suspensions for 1 h at room temperature prior to usage.
4. Store the prepared SUVs at 4 °C for up to 1 week.

3.2.5 Large Unilamellar Vesicles (LUVs) 100–1000 nm

1. Follow prepreparation steps (Subheading 3.2.2).
2. Vortex lipid samples for 2 h.
3. Extrude the lipid suspensions through a polycarbonate filter with 400 nm pore size and incubate for 30 min at room temperature (*see* **Note 9**).
4. After extrusion, wash the LUVs three times with buffer by centrifugation at 20,000 × *g* and resuspend, yielding a stock solution of 0.5 mM total lipids.

5. Store the LUVs at 4 °C, these samples can be used within 3–4 days.

3.2.6 Giant Unilamellar Vesicles (GUVs) >1000 nm

1. Combine lipids in the desired ratios (final lipid concentration 1 mM and total volume 2 mL) and place in a 50-mL round-bottom flask containing 980 μL chloroform and 200 μL methanol.
2. Carefully add 5 mL of distilled water or buffer. Apply the solutions onto the flask walls (*see* **Note 10**).
3. Remove the organic solvent in a rotary evaporator under reduced pressure (final pressure 10 mbar) at room temperature and 40 rpm.
4. After evaporating for 4–5 min, an opalescent fluid will be obtained with a volume of approximately 5 mL.
5. The resulting vesicle solution will exhibit a turbid appearance and can be used on the day of preparation.

3.3 Fluorescence Measurements

3.3.1 Förster Resonance Energy Transfer (FRET)

1. Follow the described protocol to prepare SUVs, LUVs or GUVs, Subheadings 3.2.4, 3.2.5 or 3.2.6 respectively.
2. Prior to drying, add to the lipid mixtures NBD-PE (materials Subheading 2.4) and N-Rh-PE (materials Subheading 2.4) in a 500:1:1 mole ratio (phospholipid:N-NBD-PE:N-Rh-PE) (Fig. 1).
3. Mix the following solutions in 96-well-plate. Final volumes should be 100 μL in each well (carry out at least three replications from each sample):
   - Control: 30 μL vesicles and 70 μL 10 mM phosphate buffer, pH 7.4.
   - Complete vesicle dissolution: 30 μL vesicles, 5 μL 20% Triton-100 (20% solution in water) and 65 μL 10 mM phosphate buffer, pH 7.4.
   - Effect of amyloid peptide: 30 μL vesicles, X μL amyloid peptide and (70-X) μL 10 mM phosphate buffer, pH 7.4.
4. Measure the fluorescence emission spectra during the reaction time at $\lambda_{ex} = 469$ nm, and λ_{em} in the range of 490–650 nm using a 96-well plate reader.
5. Calculate the FRET efficiency values by the following equation:

$$\text{Efficiency} = \frac{R_i - R_{100\%}}{R_0 - R_{100\%}} \times 100\%$$

   - R is a ratio of fluorescence emissions NBD-PE (536 nm)/ Rhodamine B-PE (586 nm).

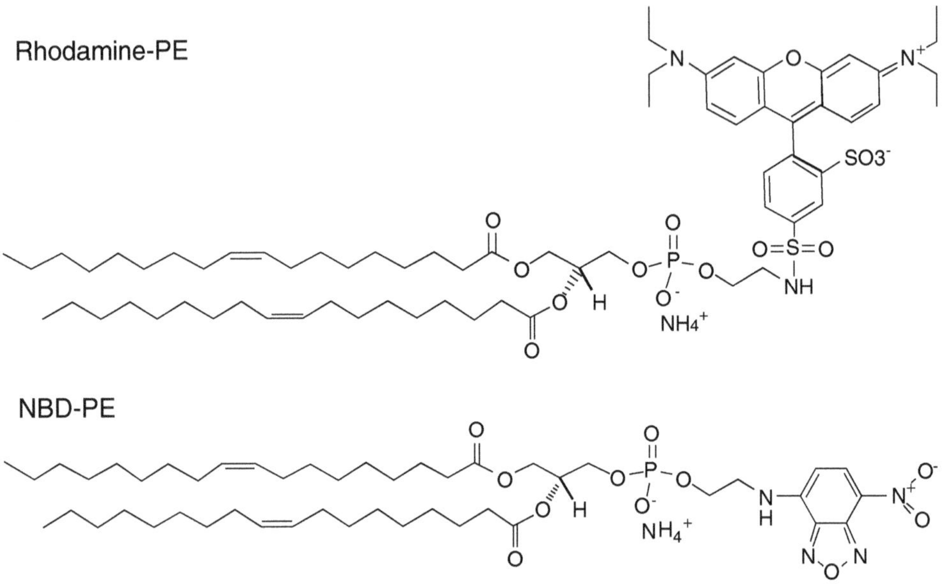

Fig. 1 Schematic structures of the fluorescent probes Rhodamine-PE and NBD-PE. The scheme shows phosphatidylethanolamine (PE) lipids modified at the headgroup with the fluorescent dyes Rhodamine B or NBD. The fluorophores are used in FRET experiments

- R_i is the ratio in the amyloid/vesicles mixtures.
- $R_{100\%}$ is ratio following the addition of 20% Triton X-100 to the vesicles (Triton X-100 is a detergent causing complete dissolution of the vesicles).
- R_0 corresponds to the ratio recorded for vesicles without any additives (control).

3.3.2 Fluorescence Anisotropy

1. Follow the described protocol to prepare SUVs, LUVs or GUVs, Subheadings 3.2.4, 3.2.5 or 3.2.6 respectively.
2. Add the fluorescence probe TMA-DPH (Subheading 2.4) or DPH (Subheading 2.4) into the prepared vesicles stock (1 mM) at final concentration of 1.25 μM and incubate for 30 min at room temperature (Fig. 2 **top**) (*see* **Note 11**).
3. Mix the following solutions in a quartz cuvette (path length 10 mm) and fill the remaining volume up to total of 1 mL with 10 mM phosphate buffer, pH 7.4:
   - Control: 30 μL vesicles.
   - Amyloid peptide: 30 μL vesicles and X μL amyloid protein\ peptide.
   - Measure the fluorescence anisotropy of TMA-DPH or DPH probes at λ_{ex} = 360 nm and λ_{em} = 430 nm using a spectrofluorimeter with polarized options.

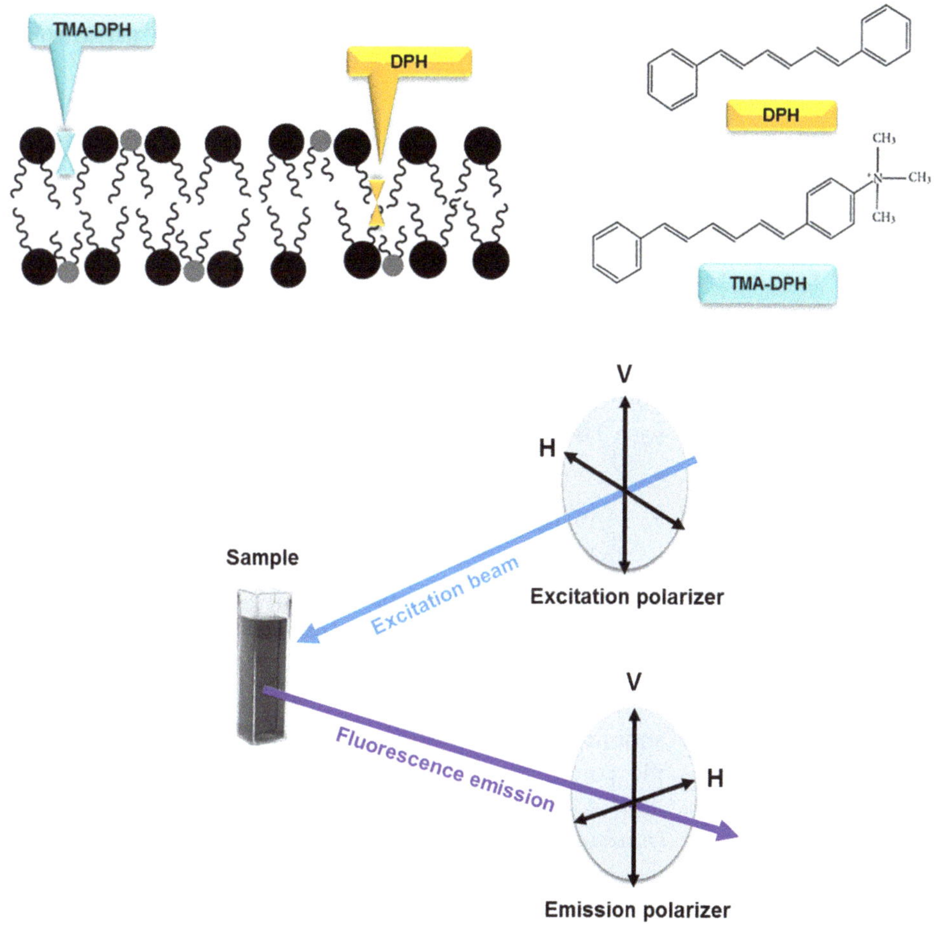

Fig. 2 Fluorescence anisotropy experiments utilizing membrane-embedded probes. (Top) Schematic structure and localization of the fluorescent probes DPH and TMA-DPH. The scheme shows that owing to its charged group, TMA-DPH is anchored at the lipid–water interface, while DPH is inserted into the hydrophobic core of the bilayer. (Bottom) Diagram of fluorescence polarization. Vertical (V) and horizontal (H) orientation of each polarizer are presented

4. The anisotropy values are usually calculated by the spectrofluorimeter software using the equation:

$$r = \left(I_{\mathrm{VV}} - GI_{\mathrm{VV}}\right) / \left(I_{\mathrm{VV}} + 2GI_{\mathrm{VH}}\right)$$

$$G = \frac{I_{\mathrm{HV}}}{I_{\mathrm{HH}}}$$

in which I_{VV} is with excitation and emission polarizers mounted vertically; I_{HH} corresponds to the excitation and emission polarizers mounted horizontally; I_{HV} is the excitation polarizer horizontal and the emission polarizer vertical; I_{VH} requires the excitation polarizer vertical and emission polarizer horizontal (Fig. 2 **bottom**).

5. Repeat 8–10 times and calculate the average values and standard deviations.

3.3.3 Fluorescence Quenching

1. Follow the described protocol to prepare SUVs, LUVs or GUVs, Subheadings 3.2.4, 3.2.5 or 3.2.6 respectively.
2. Prior to drying, add to the lipids NBD-PE (Subheading 2.4) at a 500:1 mole ratio (phospholipid:N-NBD-PE).
3. Mix the following solutions in a quartz cuvette (path length 10 mm) and fill the remaining volume up to total of 1 mL with 10 mM phosphate buffer, pH 7.4:
   - Control: 30 μL vesicles containing the fluorescent probe.
   - Amyloid peptide: 30 μL vesicles and X μL amyloid peptide.
4. Measure the decrease in fluorescence emission during the time reaction at room temperature using λ_{ex} = 469 nm and λ_{em} = 530 nm on a spectrofluorimeter.
5. Initiate the quenching reaction by adding 1 μL of 0.6 M sodium dithionite (stock solution concentration) to the sample containing the vesicles (prior and after addition of the amyloid peptide). The final concentration of the sodium dithionite in the quartz cuvette should be 0.06 mM (*see* **Note 3**).
6. The first point of the measurement (before adding the sodium dithionite) is the reference (100%). Normalize the values of the quenched samples according to this value.

3.3.4 Vesicle Leakage Assay

1. Follow the described protocol to prepare LUVs, Subheading 3.2.5 and use carboxyfluorescein (CF) solution (pH 7.4) for leakage measurements (Fig. 3) (*see* **Note 4**).
2. Mix the following solutions in quartz cuvette (path length 10 mm) and fill the remaining volume up to total of 1 mL with 10 mM phosphate buffer, pH 7.4:
   - Control: 30 μL vesicles containing the fluorescent probe.
   - Complete leakage due to vesicle dissolution: 30 μL vesicles, 20 μL 20% Triton-100.
   - Amyloid peptide: 30 μL vesicles and X μL amyloid peptide.
3. Measure the fluorescence emission of CF at 517 nm using an excitation wavelength of 490 nm on a spectrofluorimeter.
4. Calculate the % leakage values by the following equation:

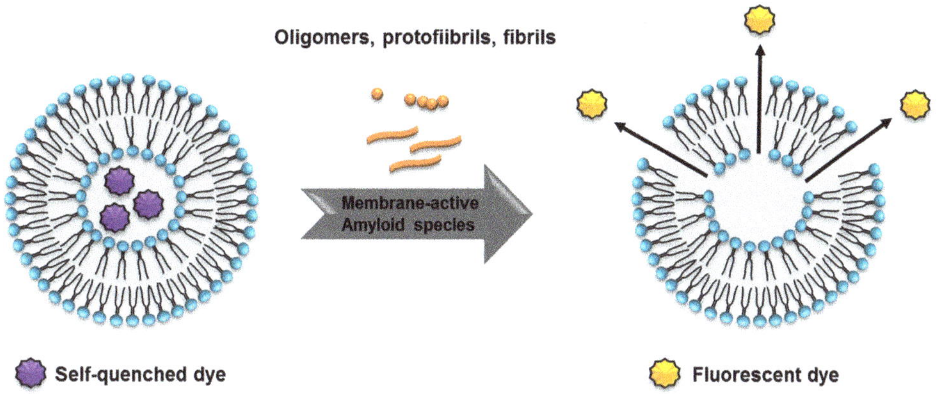

Fig. 3 Vesicle leakage experiment. Schematic description of dye-release experiments indicating pore formation induced by amyloid peptides. Fluorophores encapsulated inside of lipid vesicles self-quench due to high local concentration. Addition of membrane-active compounds induces bilayer disintegration and/or formation of discrete pores, consequently leading to dye release and increase of fluorescence intensity

$$\%\text{Leakage} = \frac{I_{\text{sample}} - I_0}{I_{100} - I_0} \times 100\%$$

- I_0 is the fluorescence intensity of vesicles alone.
- I_{100} is the fluorescence intensity after addition of 20% Triton X-100 (which induces vesicle disintegration).
- I_{sample} is the fluorescence intensity after addition of the amyloid peptide and 1-h incubation at room temperature.

3.4 Calorimetric Measurements

3.4.1 Differential Scanning Calorimetry (DSC)

1. Follow the described protocol to prepare MLD, Subheading 3.2.3 (*see* **Note 12**).
2. Mix the following solutions with the MLD:
   - Control: MLD.
   - Amyloid peptide: MLD with X μL amyloid peptide.
3. DSC experiments are performed at a temperature range of 10–50 °C using instrument such as; VP-DSC microcalorimeter (Malvern) or Nano-DSC (TA).
4. Run heating scans at a rate of 60 °C/h.
5. Measure at least three heat scans per sample.

3.4.2 Isothermal Titration Calorimetry (ITC)

1. Follow the described protocol to prepare SUVs (final concentration of 5 mM, total volume 2 mL) Subheading 3.2.4.
2. Prepare the following solutions:
   - Cell (Nano ITC low volume cell): Amyloid peptide dissolved in 10 mM phosphate buffer, pH 7.4 (choose the total volume according to the manufacturer's specifications).

- Syringe: 5 mM SUVs dissolved in 10 mM phosphate buffer, pH 7.4 (choose the total volume according to the manufacturer's specifications).

3. ITC experiments are performed on instruments such as Nano-ITC (TA) low volume or VP-ITC (Malvern). After reaching equilibrium, inject 2.5 μL of 5 mM SUVs every 3 min for a total of 75 min.
4. Use analysis software to identify the model of the interaction between the liposomes and the amyloid peptide.

3.5 Electron Spin Resonance (ESR)

1. Follow the described protocol to prepare SUVs, LUVs or GUVs, Subheadings 3.2.4, 3.2.5 or 3.2.6 respectively.
2. Add the spin-probe fatty acid (5-DS or 12-DS) to the prepared vesicle samples in a molar ratio of 500:1 (phospholipid: spin-probe).
3. Test the following solutions:
   - Control: vesicles labeled with the spin-probe.
   - Amyloid peptide: mix the labeled vesicles with X μL amyloid peptide in Eppendorf and follow by adding the required amount to the capillary.
4. Place the sample in a 20-mm length, 1 mm id quartz capillary.
5. Record the ESR spectra using an X-band ESR spectrometer at 25 °C.
6. Select amplitudes of 12.5 kHz and 100 kHz, modulation, and a microwave power level at sub-critical values (0.5 G and 20 mW, respectively) to reach the best signal-to-noise ratios.

3.6 Microscopy

1. Prepare GUVs according to the procedure described in Subheading 3.2.6 (*see* **Note 10**).
2. Place GUVs suspension on a glass bottom confocal dish and subsequently add the amyloid peptide at the required concentration.
3. Microscopy images of the GUVs should to be taken during the incubation time, before and after the addition of tested peptide (Fig. 4).
4. Bright field or fluorescent microscopy images of labeled giant vesicles are recorded; excitation/emission wavelengths are determined according to the fluorescent dye (*see* **Note 13**).

4 Notes

1. Usually the amyloid peptides are dissolved in an organic solvent such as HFIP and stored at −20 °C until use to prevent aggregation.

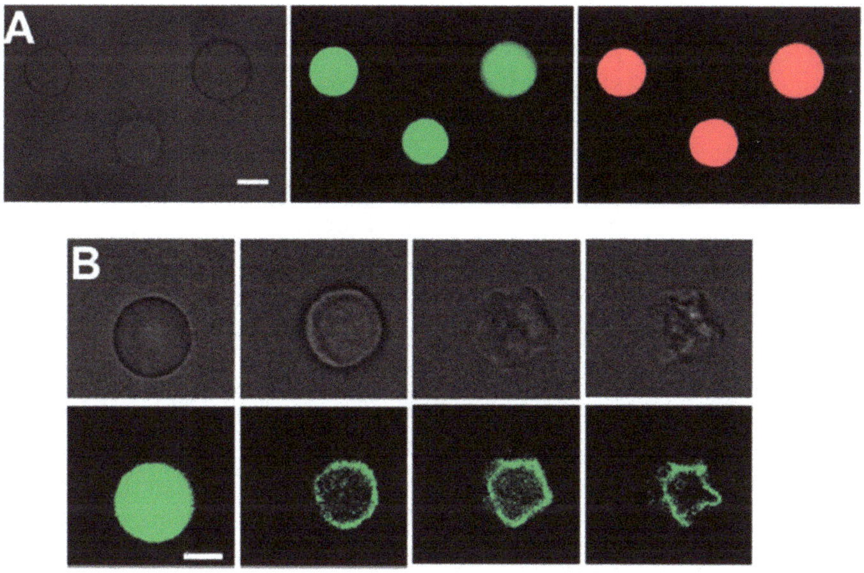

Fig. 4 Fluorescence imaging of giant vesicles labeled with the amphiphilic carbon dots [12]. (**a**) Bright field microscopy (top left), and confocal fluorescence microscopy images recorded upon excitation using a 440 nm emission filter EM 477/45 (green); excitation using a 488 nm emission filter EM 525/50 (magenta). Scale bar corresponds to 10 μm. (**b**) Bright field (top) and fluorescent images (excitation at 440 nm) of giant vesicles labelled with carbon dots, following addition of Aβ40. From left: before addition (control); 1 min after addition; 10 min after addition and 20 min after addition. Scale bar corresponds to 5 μm

2. 10 mM phosphate buffer, pH 7.4. (Solution A) 0.2 M sodium phosphate, dibasic dihydrate; 3.56 mg/100 mL deionized water (M.W. 178.05). (Solution B) 0.2 M sodium phosphate, monobasic monohydrate; 2.76 mg/100 mL deionized water (M.W. 138.01). Mix 40.5 mL from solution (A) and 9.5 mL from solution (B). Adjust the final pH to 7.4 using a pH meter and adjust to a final volume of 1 L with deionized water. Store the buffer in 4 °C.
3. Solution for fluorescence quenching methods. Prior to the experiment, prepare a set of Eppendorfs as follows (taking into account the number of the required samples): Weight 10.5 mg sodium dithionite (M.W 174.12) and dissolve it in 50 mM Tris base (M.W. 121.0) buffer, pH 11 to final volume of 100 μL with 0.6 M concentration. This quenching solution is valid for 40 min only; make sure to store it in a dark place.
4. To prepare the CF solution mix the following components: 4 g CF (M.W. 376.32), 6 g Hepes (M.W. 238.30), 3.2 g NaCl (M.W. 58.44), and 0.02% NaN_3 (M.W. 65.01). Adjust the final pH to 7.4 using a pH meter and adjust to final volume of 500 mL with deionized water. Make sure to store in a dark place.

5. Properties of lipid formulations vary, depending on the composition (cationic, anionic, neutral lipid species); however, the same preparation method can be used for all lipid vesicles regardless of composition.
6. If possible, use the same solution employed to dissolve the amyloid peptide (10 mM phosphate buffer, pH 7.4) in the prepreparation step of the vesicles.
7. Typically, lipid stock solutions are prepared at 10–25 mg lipid/mL (chloroform–ethanol, 1:1 volume ratio), although higher concentrations may be used if the lipid solubility and mixing are acceptable.
8. In case of 1:1 lipid mixture, choose the sonication temperature according to the lipid with the higher Tm.
9. The extruder procedure is provided by the manufacturer's specifications. For example, *see* Avanti's extruder flow: https://goo.gl/PyzTXa.
10. Relevant buffer for GUVs: Hepes buffer (10 mM, pH adjusted to 7.4 with NaOH), sodium phosphate buffer (10 mM, with 11 mM MgCl2, pH 7.4), and sodium borate buffer (50 mM, pH 9.2) [14].
11. Bilayer localization of the DPH dye in each vesicle system enables probing the effects of amyloid peptides in different membrane regions. Specifically, the TMA substituent in TMA-DPH acts as a surface anchor, localizing the probe close to the membrane interface, while the highly hydrophobic DPH resides primarily deeper within the bilayer interior, sensing the dynamic properties of the acyl chain environments [4].
12. When performing DSC experiment, use synthetic lipids comprising dimyristoyl or dipalmitoyl acyl chains. When using natural lipids, make sure that the maximum concentrations are 2% of the total amount (e.g., 98% of 1,2-dimyristoyl-sn-glycero-3-phosphocholine (DMPC) with 2% cardiolipin).
13. Visualization of the GUVs by fluorescence microscopy can be carried out by addition of NBD-PE or N-Rh-PE dye (500:1) to the lipid mixture prior to the addition of 5 mL buffer.

References

1. Jelinek R, Kolusheva S (2005) Membrane interactions of host-defense peptides studied in model systems. Curr Protein Pept Sci 6(1):103–114
2. Malishev R, Nandi S, Kolusheva S, Levi-Kalisman Y, Klärner F-G, Schrader T, Bitan G, Jelinek R (2015) Toxicity inhibitors protect lipid membranes from disruption by Aβ42. ACS Chem Neurosci 6(11):1860–1869
3. Nandi S, Malishev R, Bhunia SK, Kolusheva S, Jopp J, Jelinek R (2016) Lipid-bilayer dynamics probed by a carbon dot-phospholipid conjugate. Biophys J 110(9):2016–2025
4. do Canto AM, Robalo JR, Santos PD, Carvalho AJP, Ramalho JP, Loura LM (2016) Diphenylhexatriene membrane probes DPH and TMA-DPH: a comparative molecular

dynamics simulation study. Biochim Biophys Acta 1858(11):2647–2661
5. Lentz BR (1993) Use of fluorescent probes to monitor molecular order and motions within liposome bilayers. Chem Phys Lipids 64(1–3): 99–116
6. McIntyre JC, Sleight RG (1991) Fluorescence assay for phospholipid membrane asymmetry. Biochemistry 30(51):11819–11827
7. Sokolovski M, Sheynis T, Kolusheva S, Jelinek R (2008) Membrane interactions and lipid binding of casein oligomers and early aggregates. Biochim Biophys Acta 1778(10):2341–2349
8. Yasuda T, Naito Y, Tsumita T, Tadakuma T (1981) A simple method to measure anti-glycolipid antibody by using complement-mediated immune lysis of fluorescent dye-trapped liposomes. J Immunol Methods 44(2):153–158
9. Engel MF, Khemtémourian L, Kleijer CC, Meeldijk HJ, Jacobs J, Verkleij AJ, de Kruijff B, Killian JA, Höppener JW (2008) Membrane damage by human islet amyloid polypeptide through fibril growth at the membrane. Proc Natl Acad Sci 105(16):6033–6038
10. Johnson CM (2013) Differential scanning calorimetry as a tool for protein folding and stability. Arch Biochem Biophys 531(1):100–109
11. Callies O, Daranas AH (2016) Application of isothermal titration calorimetry as a tool to study natural product interactions. Nat Prod Rep 33(7):881–904
12. Nandi S, Malishev R, Kootery KP, Mirsky Y, Kolusheva S, Jelinek R (2014) Membrane analysis with amphiphilic carbon dots. Chem Commun 50(71):10299–10302
13. Milanesi L, Sheynis T, Xue W-F, Orlova EV, Hellewell AL, Jelinek R, Hewitt EW, Radford SE, Saibil HR (2012) Direct three-dimensional visualization of membrane disruption by amyloid fibrils. Proc Natl Acad Sci 109(50): 20455–20460
14. Moscho A, Orwar O, Chiu DT, Modi BP, Zare RN (1996) Rapid preparation of giant unilamellar vesicles. Proc Natl Acad Sci 93(21): 11443–11447
15. Gorbenko GP, Kinnunen PK (2006) The role of lipid–protein interactions in amyloid-type protein fibril formation. Chem Phys Lipids 141(1):72–82
16. Gal N, Morag A, Kolusheva S, Winter R, Landau M, Jelinek R (2013) Lipid bilayers significantly modulate cross-fibrillation of two distinct amyloidogenic peptides. J Am Chem Soc 135(36):13582–13589
17. Kazlauskaite J, Sanghera N, Sylvester I, Vénien-Bryan C, Pinheiro TJ (2003) Structural changes of the prion protein in lipid membranes leading to aggregation and fibrillization. Biochemistry 42(11):3295–3304
18. Jayasinghe SA, Langen R (2007) Membrane interaction of islet amyloid polypeptide. Biochim Biophys Acta 1768(8):2002–2009

Chapter 4

Differential Scanning Fluorimetry and Hydrogen Deuterium Exchange Mass Spectrometry to Monitor the Conformational Dynamics of NBD1 in Cystic Fibrosis

Naoto Soya, Ariel Roldan, and Gergely L. Lukacs

Abstract

Cystic fibrosis (CF) is one of the most common, lethal autosomal recessive diseases in Caucasians with a life expectancy of 37–47 years. The CF transmembrane conductance regulator (CFTR) is a plasma membrane ion channel, confined to apical membrane of epithelia, and ensures transepithelial water and solute movement across secretory epithelia in several organs. Numerous CF mutations, including the most prevalent deletion of F508 (ΔF508) in the nucleotide binding domain 1 (NBD1) leads to CFTR global misfolding and premature intracellular degradation at the endoplasmic reticulum (ER). To better understand the misfolding mechanism caused by CF-causing point mutations in the NBD1, which is poorly understood, differential scanning fluorimetry (DSF) and hydrogen deuterium exchange coupled with mass spectrometry (HDX-MS) are the choice of techniques. These established methods can measure the conformational dynamics of the NBD1 globally and at peptide resolution level by monitoring backbone amide HDX, respectively, and will be instrumental to evaluate the mechanism of action of CF mutations and folding correctors that rescue CFTR folding defects via stabilizing the mutant NBD1.

Key words Cystic fibrosis, Conformational diseases, Missense mutation, NBD1, Unfolding kinetic, Thermal stability, Melting temperature

1 Introduction

1.1 Cystic Fibrosis and CFTR Folding

Cystic fibrosis (CF) is caused by the functional expression defect of the cystic fibrosis transmembrane conductance regulator (CFTR), a phosphorylation regulated anion selective channel that belongs to the ATP binding cassette transport family [1]. Impaired functional expression of CFTR at the apical plasma membrane of secretory epithelia causes CF, the most prevalent, lethal genetic disease in the Caucasian population [2]. While the channel functional expression defect can be attributed to mutations that impair CFTR transcription, splicing, translation, folding, function, or stability, combinations of these defects have also been documented [3, 4]. Channel misfolding represents the most frequent cause of

Cláudio M. Gomes (ed.), *Protein Misfolding Diseases: Methods and Protocols*, Methods in Molecular Biology, vol. 1873, https://doi.org/10.1007/978-1-4939-8820-4_4,

CF, which renders the newly synthesized polypeptide susceptible to endoplasmic reticulum (ER) associated degradation [5]. Immunoblot analysis of 54 mutations revealed that steady-state expression of the mature, complex-glycosylated form of 44 CFTR variants were significantly reduced in Fisher rat thyroid (FRT) epithelia [6]. Impaired conformational maturation of newly synthesized channels at the ER similar to that described for the most common point mutation, deletion of F508 (ΔF508) in the nucleotide binding domain 1 (NBD1) accounts for the loss-of-expression phenotype.

CFTR contains two symmetrical halves, and each consists of a membrane spanning domain (MSD1 or MSD2) that forms the ion permeation pathway and a cytosolic NBD (NBD1 or NBD2). The dimerization of NBDs is coupled to CFTR channel opening, formed by the MSD1-2, via association of the NBDs to the cytosolic loops of MSDs [7]. The two symmetrical halves are connected with a largely unstructured regulatory (or R) region. The cAMP-dependent phosphorylation of the R region is a prerequisite for NBD1-2 dimerization and channel opening [7]. From the 44 processing mutants, 14 were confined to the NBD1 in FRT epithelia [6]. The underlying mechanism of CFTR misfolding, initiated by these point mutations in NBD1, however, remains enigmatic. Although the X-ray crystal structures of the human wild-type (WT) and ΔF508 NBD1, containing seven (F409L, F429S, F433L, G550E, R553Q, R555K, and H667R) and two (F494N and Q637R) second site solubilizing or suppressor mutations, respectively, were very similar [8], more recent analysis showed substantial difference between the mutant and WT NBD1 local and global conformational stability [9, 10].

The ΔF508 is located in the alpha subdomain of the NBD1 and interferes with both the thermodynamic and kinetic stability of NBD1, as well as uncouples the NBD1-CL4 interface that contributes to CFTR cooperative domain unfolding [11, 12]. This paradigm may be relevant for elucidating the underlying cause of other CF mutations associated with the NBD1. While the isolated NBD1 with ΔF508 has been intensively studied using conventional structural biology techniques, such as X-ray crystallography and NMR, the dynamic perturbation of the NBD1 by other CF mutations remain elusive. This can be partly attributed to the low purification yield and the thermal instability of the mutant NBD1s, leading to increased unfolding and aggregation propensity. Here, we describe hydrogen deuterium exchange with mass spectrometry (HDX-MS) technique to establish the conformational dynamics of isolated NBD1 in the absence and presence of a CF mutation. As an initial screening assay to uncover the impact of CF mutation on the conformation stability of NBD1, differential scanning fluorimetry (DSF) is also described.

2 Materials

2.1 NBD1 Preparation

1. pET26b-derived vectors with N-terminal His_6–SUMO fusion tag containing the human WT-CFTR with the crystallization domain boundaries (389–678 amino acid residues) (*see* **Note 1**).
2. BL21-CodonPlus (DE3)-RIL *E. coli* (Agilent).
3. Isopropyl β-D-1-thiogalactopyranoside (IPTG).
4. Lysis buffer: 100 mM Tris–HCl, pH 7.4, 150 mM NaCl, 10 mM $MgCl_2$, 5 mM ATP, 5 mM imidazole, 2 mM β-mercaptoethanol, 10% ethylene glycol, 10% glycerol.
5. Protease inhibitor cocktail P8849 (Sigma-Aldrich).
6. cOmplete, EDTA-free protease inhibitor cocktail (Roche).
7. IMAC Sepharose 6 Fast Flow (GE Healthcare).
8. Storage buffer: 50 mM $NaPO_4$, pH 7.5, 150 mM NaCl, 2 mM $MgCl_2$, 1 mM ATP, 1 mM TCEP, 10% glycerol.
9. PD10 desalting column (BioRad).
10. Superdex 200 size-exclusion chromatography column (GE Healthcare).
11. Amicon Ultra 10 kDa cutoff (Millipore).

2.2 Differential Scanning Fluorimetry (DSF)

1. 96-well plate with round or conical-bottom wells for preparing the stock solutions.
2. White, nonskirted, low-profile 96-well Real-Time (RT)-PCR plate (used to run on the machine).
3. Self-adhesive optical seal for PCR (BioRad).
4. Purified NBD1 protein at ~4 mg/ml concentration (*see* Subheading 3.1).
5. SYPRO orange dye, 5000× concentration in DMSO (Thermo Fisher Scientific).
6. Screening buffer 3×: 300 mM HEPES, pH 7.5, 450 mM NaCl, 15 mM $MgCl_2$.
7. 250 mM ATP, pH 7.4.
8. Stratagene Mx3005P RT-PCR (Agilent).
9. Refrigerated centrifuge for 96-well plates.

2.3 HDX-MS

1. Deuterium oxide (D_2O).
2. Deuterium chloride (DCl).
3. Sodium deuteroxide (NaOD).
4. Urea and urea-d_4.
5. Water with 0.1% formic acid (FA) (LC/MS grade).

6. Acetonitrile (ACN) with 0.1% FA (LC/MS grade).
7. Quenching buffer: 300 mM glycine–HCl, pH 2.4, 8 M urea.
8. H_2O-based buffer: 10 mM HEPES, pH 7.5, 150 mM NaCl, 1 mM ATP, 2 mM $MgCl_2$, 1 mM TCEP, 1% glycerol.
9. D_2O-based buffer: same components as H_2O-based buffer dissolved in D_2O, pD 7.5 (*see* **Note 2**).
10. Pepsin column (*see* **Note 3**).
11. Pepsin column washing cocktails: (a) 10 mM glycine–HCl, pH 2.5, 2 M guanidine hydrochloride and (b) 5% ACN, 5% isopropanol (IPA), 20% acetic acid (AcOH) [13].
12. Trap and analytical column washing solution: 80% IPA in ACN.
13. A 1260 Infinity II isocratic pump (Agilent) and a 1290 Infinity II binary pump (Agilent).
14. A 0.2 μm precolumn filter (Thermo Fisher Scientifc), a Hypersil GOLD C_{18} column, 1 mm × 50 mm, 1.9 μm (Thermo Fisher Scientifc) and an EXP C_{18} trap column (Optimize Technologies).
15. A LTQ Orbitrap XL mass spectrometer (Thermo Fisher Scientific) and Proteome Discoverer software (Thermo Fisher Scientific).
16. A model 8125 low dispersion sample injector with a 10 μl sample loop (Rheodyne).
17. A 1290 Infinity valve drive and six-port valve head (Agilent).
18. HDExaminer software (Sierra Analytics).
19. PeakFit (Systat Software).

3 Methods

3.1 Isolated NBD1 Expression and Purification (See Note 4)

1. Revert two of the three solubilizing mutations (F429S and Q637R) in the pET26b-derived vectors to WT sequence by PCR mutagenesis.
2. Introduce the ΔF508 mutation using the same technique above.
3. Express the WT- and ΔF508-NBD1 containing the F494N second site solubilizing mutations (NBD1-1S) in BL21-CodonPlus (DE3)-RIL cells.
4. Scale up an overnight starter culture to 6–8 l at 30 °C. At 0.2–0.4 optical density, cool cells to 12 °C (2 h) and then induce with 0.1 mM IPTG.
5. After 16 h of IPTG-induction at 12 °C, centrifuge cells (10 min, 6690 × *g*) at 4 °C. Resuspend cells in lysis buffer,

supplemented with protease inhibitor cocktail, and sonicate for 12 × 10 s with 10 s intervals on ice.

6. Recover the soluble fraction after centrifugation (30 min, 27200 × *g*) at 4 °C and apply the fraction onto an equilibrated 1 ml IMAC Sepharose 6 Fast Flow. Wash the IMAC column with eight column volumes (CV) of lysis buffer, including 30 mM imidazole. Elute with lysis buffer containing 500 mM imidazole.
7. Pool aliquots showing the highest protein concentration and perform buffer exchange with storage buffer using a PD10 desalting column.
8. Cleave His_6-Sumo tag overnight at 4 °C using 50 μg/ml of Ulp protease. His_6-Sumo Tag and Ulp protease can be pulled down by incubating samples with 100 μl of nickel beads.
9. Purify cleaved protein using Superdex 200 size-exclusion chromatography column preequilibrated in storage buffer.
10. Concentrate purified NBD1 to 3–5 mg/ml using Amicon Ultra 10 kDa cutoff.

3.2 NBD1 Melting Temperature Determination with DSF

The advantage of monitoring NBD1 thermal unfolding by DSF over spectroscopic techniques (UV absorbance and circular dichroism) is that it requires considerable smaller amount of protein. DSF for NBD1 is performed in the presence of SYPRO orange [14], a fluorescent dye that has significantly increased quantum efficiency upon binding to the exposed hydrophobic patches of unfolded polypeptides as compared to it low binding affinity to the native polypeptides, largely exposed of hydrophilic residues. A polypeptide stability is related to its unfolding Gibbs free energy (ΔG_u), which usually decreases with increasing temperature. At the melting temperature (T_m), the ΔG_u value is zero and the solution contains equal amounts of folded and unfolded protein. Thus, monitoring augmented SYPRO orange fluorescence as a function of increasing temperature can reveal the impact of CF causing mutation on the isolated native-like NBD1 thermal stability using a RT-PCR machine (*see* **Note 5**). Introduction of the ΔF508 mutation into NBD1-1S causes a ~8 °C reduction in the T_m as compared to that of the WT-NBD1-1S as illustrated by the DSF plots (Fig. 1). In contrast, 10% glycerol, a chemical chaperone, stabilized the ΔF508-NBD1-1S by ~4 °C (Fig. 1). Similar measurements could identify the impact of CF mutations, second site suppressor mutations and specific ligands (or correctors) on the NBD1 domain stability as shown before [15–17].

3.2.1 Determining the Melting Temperature of NBD1 by DSF

1. Before starting, make sure that the RT-PCR machine is ready to go (*see* **Note 6**) and that centrifuge is at 4 °C. Keep the 96-well plate on ice during preparation.

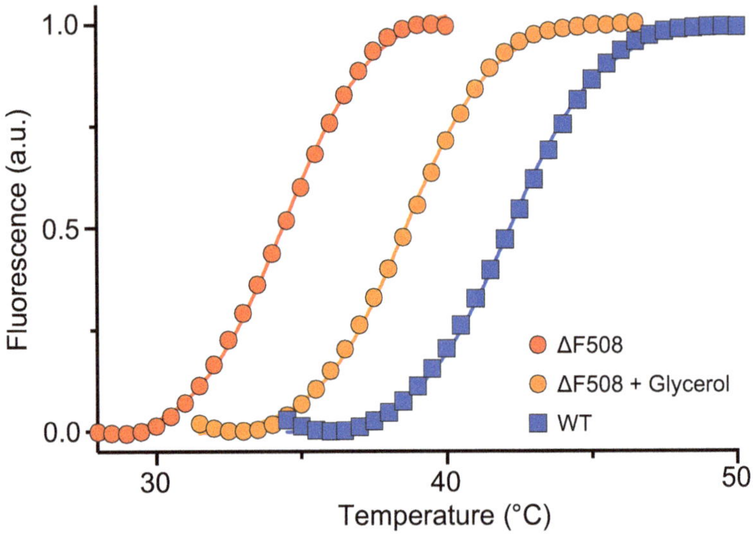

Fig. 1 Thermal stability of the isolated human WT-NBD1 and ΔF508-NBD1-1S. Unfolding of the domains was monitored by differential scanning fluorimetry (DSF) in the presence of SYPRO orange as described. The fluorescence intensity was expressed as the percentage of the maximum as a function of temperature. As indicated, ΔF508-NBD1-1S thermal melt was also determined in the presence of 10% glycerol

2. Prepare test compound(s) (e.g., glycerol) at 2× concentration in 1× Screening buffer (100 mM HEPES, pH 7.5, 150 mM NaCl, 5 mM $MgCl_2$).
3. Pipette 5 μl of each condition to be tested into the 96-well plate at least in triplicate.
4. Prepare the protein mix, containing NBD1-1S at ~0.4 mg/ml, 8× SYPRO orange, 2 mM ATP in 1× screening buffer. Add NBD1 last, mix carefully but thoroughly.
5. Distribute 5 μl of the protein mix into the RT-PCR 96-well plate.
6. Cover plate with optical seal, and centrifuge it for 2 min at 930 × *g* at 4 °C.
7. Insert plate into PCR machine and start the run.

3.2.2 Determining the Thermal Melting Temperature (T_m)

1. To retrieve the fluorescence data from the RT-PCR machine, go to "Analysis" (top right) and choose the "Results" tab.
2. In the "Fluorescence" scroll down and choose "R (multicomponent view)".
3. From "File" menu, choose "Export chart data," then "Export data to excel" and finally "Format 2—Horizontally grouped by plot."

4. This will generate a read-only Excel file, including two tables. The top table will include the fluorescence data from wells A1 to D12, and the bottom table the data for wells E1 to H12. In each table, the cycle number is indicated in column A, which corresponds to the plate temperature starting from 25 °C in 0.5 °C increment. The fluorescence intensity data of individual wells are shown in the following columns (e.g., fluorescence of A1 well will be in column B, A2 well in column C, and so on).
5. Change the cycle number in column A for the actual temperature (e.g., replace 1 by 25 °C, 2 by 25.5 °C and 3 by 26 °C).
6. Copy the raw data into GraphPad Prism and analyze them using "non-linear regression" from the "xy analysis" group, and choose "Boltzmann sigmoidal" equation for fitting according to

$$Y = \text{Bottom} + (\text{Top} - \text{Bottom}) / 1 + \exp(\text{V50} - X / \text{slope}),$$

where Top and Bottom are the extrapolated maximum and minimum SYPRO orange fluorescence, respectively, X denotes the measured temperature, and V50 is the T_m.

3.3 HDX-MS

The principle of this technique is based on the impeded HDX kinetics of backbone amide hydrogens forming H-bonds or structurally buried in natively folded proteins, relative to flexible or unfolded segments that are more solvent accessible [18, 19]. HDX can be initiated by diluting protein samples from stock solutions into high concentration D_2O containing medium (Fig. 2a) and monitoring deuterium incorporation at backbone amides as a function of incubation time by mass determinations of proteolytic peptides. On-line proteolysis is achieved by immobilized pepsin column, under conditions (pH 2.5, 0 °C) that minimize deuterium back exchange. Considering that the back-exchange kinetics of hydrogen at side-chains is very rapid even under quenching conditions, HDX-MS only monitors the backbone amide hydrogen deuteration, which is considered as the primary determinant of the mass shift detected by HDX-MS. The difference in the centroid of isotopic envelopes between deuterated and nondeuterated samples indicates the deuterium incorporation at the peptide level.

The advantage of HDX-MS for the NBD1 unfolding studies is the high sensitivity of the method that permits monitoring the domain conformation dynamics at relatively low NBD1 concentration (5–10 μM) that largely prevents domain aggregation during thermal unfolding. Unfolded and destabilized regions can be identified by comparing HDX kinetics between WT- and ΔF508-NBD1-1S and projected onto the X-ray crystal structures (e.g., PDB: 2BBT from Protein Data Bank). Backbone amides confined

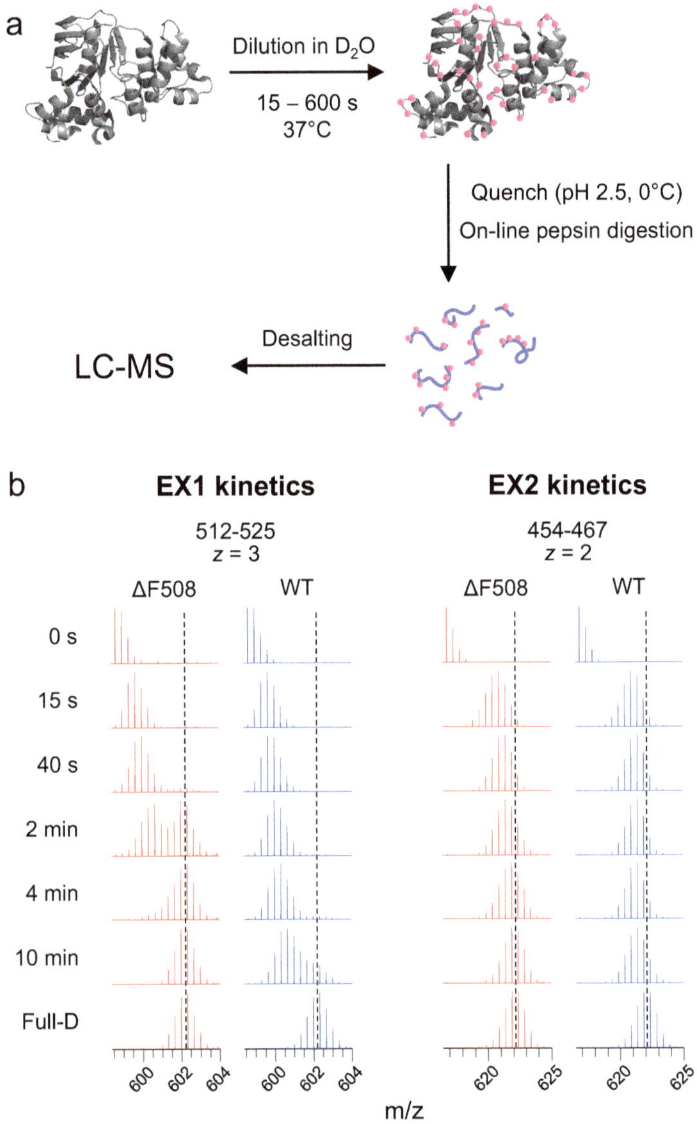

Fig. 2 (**a**) The schematic work flow HDX-MS experiment to determine the NBD1 conformational dynamics. (**b**) Mass spectra of peptides showing the bimodal peak distribution of the EX1 signature and unimodal EX2 signature peak distribution. The blue dotted lines indicate the centroids of isotopic envelopes from unfolded NBD1

to secondary structural elements (i.e., α-helices and β-strands) during NBD1 unfolding are exposed to solvent display accelerated HDX. The hallmark of peptide/protein unfolding is the EX1 type kinetics (Fig. 2b). The characteristic bimodal peak of peptide mass distributions can be observed for the EX1 kinetics, as a consequence of the presence of both folded and unfolded peptide populations during the course of thermal unfolding. In contrast, the unimodal isotopic distributions of EX2 kinetics indicate a gradual increase of the deuteration of individual peptide population as a consequence of thermal conformational fluctuation of the polypeptide with regional differences (Fig. 2b). Using comparison of deuterium incorporation kinetics, the impact of CF-causing mutations on NBD1 dynamics can be studied in comparison to that of its WT counterpart (*see* Fig. 2b).

3.3.1 Pepsin Digestion Pattern of NBD1 for HDX Analysis

Detailed conformational dynamic analysis requires full coverage of the entire primary sequence of the NBD1 by its proteolytically digested peptides. To define the proteolytic coverage and facilitate the identification of the deuterated peptides, retention time and the mass-to-charge ratio (m/z) of peptic peptides generated from WT- and ΔF508-NBD1-1S on a pepsin column are identified by tandem MS (MS/MS) analysis.

1. Prepare 1 μM of NBD1 solution in quenching buffer. Aspirate 20 μl of the solution and inject onto the sample loop using a syringe (Fig. 3a, *see* **Note 7**). At this moment, the injector valve is set at "LOAD" position.
2. The flow rate of the isocratic pump is set to 30 μl/min. Change the injector valve to "INJECT" position. The NBD1 solution passes through an on-line pepsin column, resulting in peptide generation. Generated peptides are captured at the trap column.
3. At 1.5 min, a flow rate of the isocratic pump is changed to 200 μl/min to wash out any salts and urea.
4. At 3 min, load peptides onto a C_{18} analytical column by changing the position of the switching valve (Fig. 3b). Peptides are separated using a linear gradient of 5–40% ACN, containing 0.1% FA for 60 min at a 65 μl/min flow rate (*see* **Note 8**).

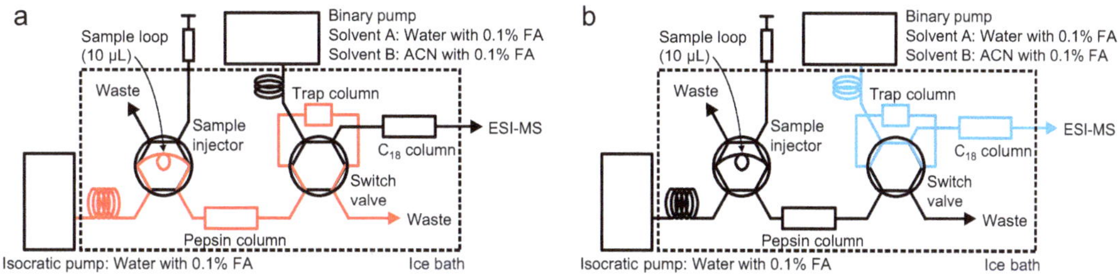

Fig. 3 The configuration of the liquid chromatography setup for HDX-MS experiments. (**a**) Injected samples are loaded on the pepsin column and resulting peptic peptides are captured by a trap column. (**b**) After washing out salts and urea, the captured peptides are eluted from the trap column and fractionated on a C18 analytical column. Separated peptides are directly injected to an ESI ion source

5. The analytical column is directly connected to an electrospray ionization (ESI) source of a mass spectrometer. Identify peptides by running MS/MS analysis in data-dependent acquisition and searching peptides with Proteome Discoverer. Save the results in Excel format (*see* **Note 9**).

3.3.2 Backbone Amide HDX

1. Place ~50–100 μM of NBD1 stock solution on ice. Aliquot 27 μl of D_2O-based buffer onto a 1.5 ml micro tube and pre-incubate at 37 °C for 10 min. Meanwhile, prepare 5 micro tubes containing 16–19 μl of quenching solution (depending on the concentration of an NBD1 stock solution), and place them on ice.
2. HDX is initiated by mixing 3 μl of NBD1 stock solution with the prewarmed D_2O-based buffer. Incubate for 15, 40, 120, 240, and 300 s at 37 °C, and then the HDX reaction is terminated by diluting 1–4 μl of the reaction mixture (pending on the NBD1 concentration) into quenching solutions 0 °C. The final concentration of NBD1 should be ~1 μM. The quenched protein solution is flash-frozen in either liquid nitrogen or methanol containing dry ice, and is stored at −80 °C until use (*see* **Note 10**).
3. For nondeuterated controls, the initial dilution is made in H_2O-based buffer. Quench, flash-freeze, and store as above.
4. Prepare fully deuterated samples by diluting NBD1 in D_2O containing 8 M urea-d_4 at room temperature for 3 h. Quench, flash-freeze, and store as above (*see* **Note 11**).
5. Prior to ultrahigh-performance liquid chromatography (UHPLC)-MS analysis, place a sample injection syringe on ice for 5 min. Deuterated samples are thawed and immediately loaded onto the injection valve.
6. Perform UHPLC-MS analysis. The LC settings are same as described in Subheading 3.2, except a 5–40% linear gradient of ACN, containing 0.1% formic acid for 10 min. Mass spectra of peptides are acquired in full-scan mode with *m/z* 200–2000.
7. After each LC-MS analysis, wash a pepsin column with pepsin column wash cocktails, as well as trap and analytical columns with 80% IPA in ACN in order to eliminate peptide carryover (*see* **Note 12**).

3.3.3 Data Analysis

1. Prepare the NBD1 amino acid sequence in the FASTA file format. Import the NBD1 sequence in HDExaminer software.
2. Import a peptide source created in Subheading 3.2.1. Retention time should be adjusted at this stage (*see* **Note 8**).

3. Import mass spectrometry data files of undeuterated controls, followed by mass spectrometry data files of deuterated samples in each time point.
4. Export HDX results (the number of deuterium uptake and the percentage of deuteration in each time point) in Excel format (*see* **Note 13**).
5. Optionally, export deuterium uptake plots (in PDF or EMF format), a deuteration heat map (EMF format for the primary structure, Pymol or Chimera format for the 3D structure, *see* Fig. 4a, b), and peptide spectra (Excel format) generated by HDExaminer.
6. Locate regions showing the difference in deuterium uptake by comparing different states (Fig. 4c).
7. Identify unfolded regions by inspecting EX1 kinetics in peptide mass spectra (Fig. 2b).
8. To determine unfolding kinetics, perform deconvolution of bimodal isotopic distributions to separate folded and unfolded populations using PeakFit software (Fig. 5a). The Gaussian distribution of folded and unfolded populations (A_f and A_u, respectively) is used to determine the fraction of unfolded population (F_u) according to.

$$F_u = A_u / (A_f + A_u)$$

The natural logarithm of the F_u is plotted against D_2O exposure time (Fig. 5b). Rate constant of unfolding is determined by fitting the data using an equation.

$$\mathrm{Ln}(F_u) = -k_u t$$

where k_u is the rate constant of unfolding, and t is the D_2O exposure time.

4 Notes

1. The pET26b-derived vectors were kindly provided by H. Lewis (Structural GenomiX Inc. and CFFT Inc.) and C. Lima (Columbia University).
2. Adjust pD based on pD = pH_{read} + 0.40 [20] using DCl or NaOD.
3. The online pepsin column provides fast and reproducible protein digestion, as well as less autolysis. Although immobilized pepsin resin can be prepared and packed in columns as described [21], they are also commercially available, e.g., StyrosZyme (Orachrom Inc.), Poroszyme immobilized pepsin

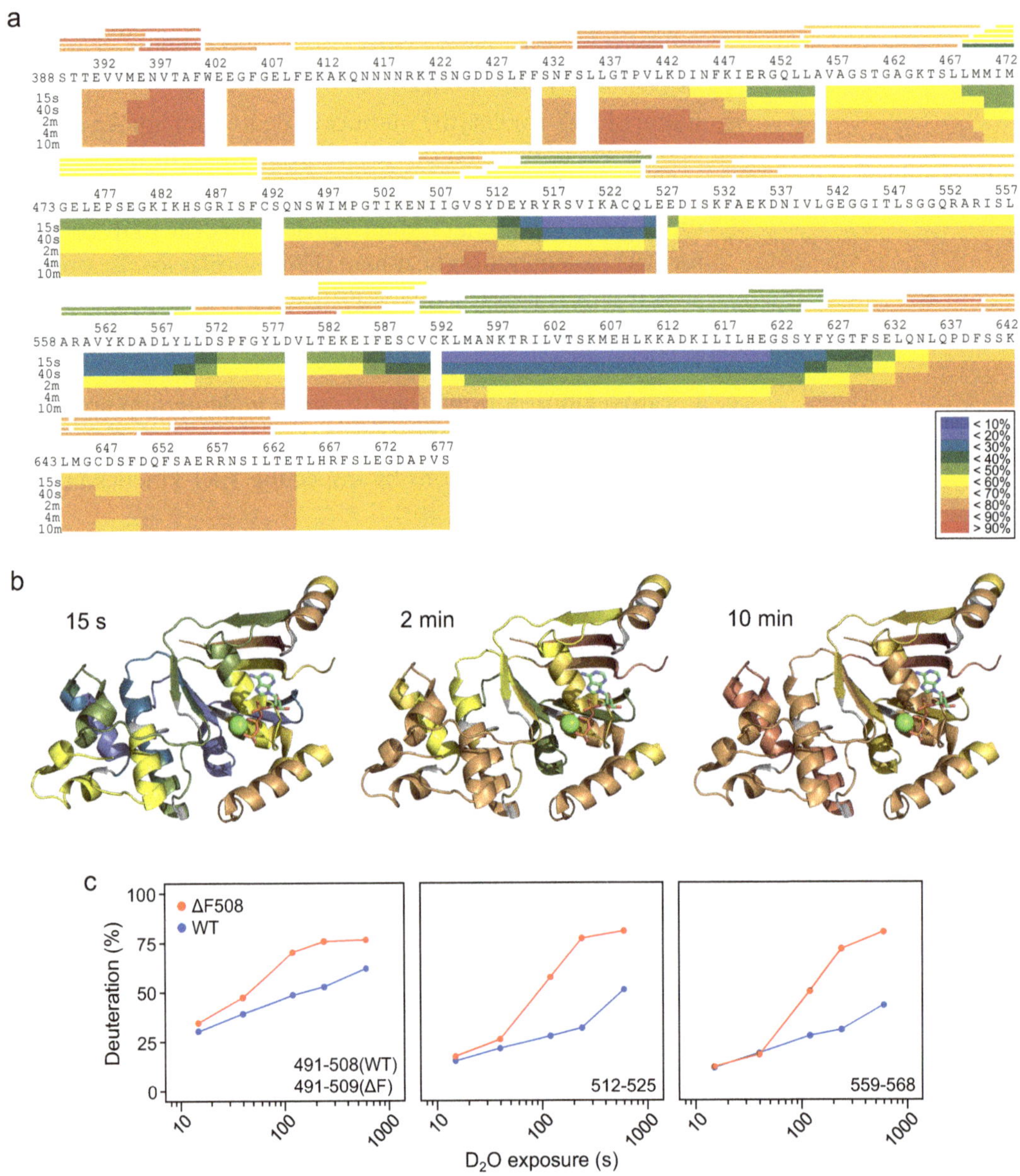

Fig. 4 Visualization of the HDX of the isolated NBD1s. (**a**) Heat map of human NBD1-1S in the presence of 1% glycerol, representing the deuterium uptake over D_2O exposure time. Blue (0%) to red (100%) indicates the extent of deuterium uptake expressed as % of the theoretical maximum. The horizontal bars above the sequence represent peptic peptides generated by on-line pepsin digestion. (**b**) Heat map projection of ΔF508-NBD1-1S HDX after 15 s, 2 min, and 10 min incubation in D_2O on the X-ray crystal structure of the mutant NBD1 (PDB: 2BBT). The 3D structure of NBD1 was created by the the PyMOL Molecular Graphics System, Version 1.8 Schrödinger, LLC (http://pymol.org). (**c**) Deuteration kinetics of representative peptides from WT- and ΔF508-NBD1-1S. Percentage deuteration of peptides relative to the theoretical maximum are plotted for the ΔF508 (red) and WT-NBD1-1S (blue) as a function of D_2O exposure time at 37°C

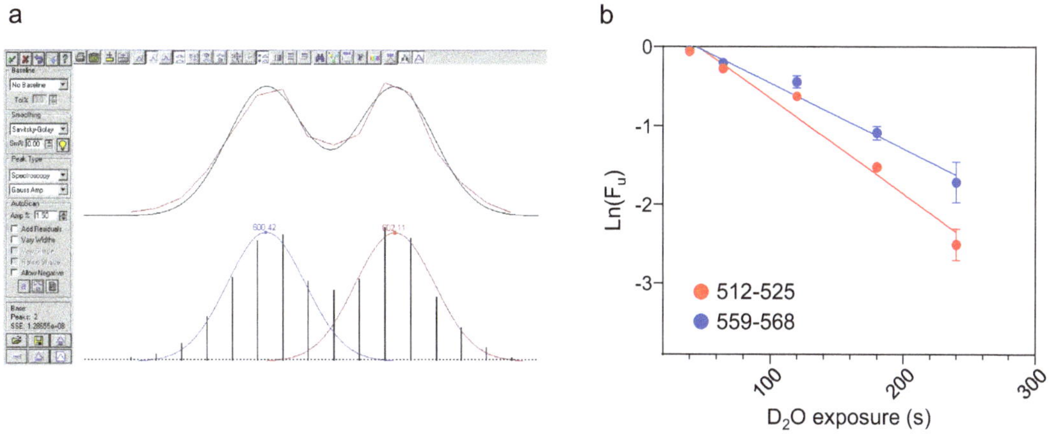

Fig. 5 Unfolding rates (k_u) of ΔF508-NBD1-1S peptides. (**a**) Deconvolution of bimodal peak distribution was performed for each incubation time point using the PeakFit software and plotted in panel **b**. (**b**) The k_u of peptides 512–525 and 559–568 derived from ΔF508-NBD1-1S is determined by plotting the natural logarithm of unfolding fraction (Ln(F_u)) against the D_2O exposure time at 37 °C

cartridge (Thermo Fisher) or Enzymate BEH pepsin column (Waters).

4. The expression and purification of recombinant NBD1 of human CFTR were previously described [12, 22].
5. We assume that the mutant NBD1s can be isolated in their native-like conformation by inducing protein expression at permissive temperature. However, it is possible that mutations that cause severe structural defects in NBD1 cannot be rescued at reduced temperature and/or by chemical chaperons, and enhanced aggregation tendency will prevent the purification of native-like domain. This hurdle may be overcome by engineering sufficient number of second site stabilizing mutations in the NBD1 alone or in combination with chemical chaperones inclusion in the culture.
6. Storing the samples longer than 20–30 min before reading may negatively affect the experiment. The Stratagene Mx3005P RT-PCR can only start from 25 °C. If lower temperature is required to prevent the unfolding of severely destabilized domain, one has to rely on an RT-PCR with cooling capacity e.g., QuantStudio™ 7 Flex System (Thermo Fisher Scientific). This RT-PCR can start the temperature ramp at 10 °C. In preparation for running RT-PCR machine the UV lamp has to warm up and plate holder cooled down for ~20 min.
7. The columns, solvent delivery lines, injector and other accessries should be placed in an ice bath to minimize back-exchange of deuterated analytes. Use stainless steel capillary tubing for solvent delivery lines in order to keep mobile phase chilled.

8. For LC-MS/MS analysis, slower gradient provides more peptide identification. Retention times of peptides should be adjusted to faster gradient for HDX analysis using the function "Use retention time adjustment" of HDExaminer.
9. At least three columns for peptide sequence, chage states and retention times must be included in Excel spreadsheets. A column for Xcorr values determined by Proteome Discoverer is optional.
10. Flash-frozen samples can be stored at −80 °C up to 3 weeks.
11. Fully deuterated sample is used to verify the deuteration level of fully unfolded states. By comparing the deuteration level, ΔF508-NBD1-1S was found to be completely unfolded at 37 °C at 10 min.
12. Peptide carryover contributes to "false EX1" in mass specta and causes false-netagive in deuterium level [13, 23]. Washing a pepsin column with washing cocktails (100 μl of each cocktail) as well as flashing trap and analytical columns using 80% IPA in ACN effectively eliminate peptide carry over.
13. Although HDExaminer generates deuterium uptake plots, the kintecs of deuterium uptake can be visualized by other software as well (e.g., Excel, GraphPad Prism, or Sigmaplot).

Acknowledgment

Experiments described were performed in GL lab and supported by CIHR, CFFT Inc., Cystic Fibrosis Canada, NIH-NIDDK, and CFI. NS acknowledges partial financial support by Groupe de Recherche Axé sur la Structure des Protéines (GRASP). GL is the recipient of a Canada Research Chair.

References

1. Riordan JR (2008) CFTR function and prospects for therapy. Annu Rev Biochem 77:701–726. https://doi.org/10.1146/annurev.biochem.75.103004.142532
2. Rowe SM, Miller S, Sorscher EJ (2005) Cystic fibrosis. N Engl J Med 352(19):1992–2001. https://doi.org/10.1056/NEJMra043184 352/19/1992 [pii]
3. Welsh MJ, Smith AE (1993) Molecular mechanisms of CFTR chloride channel dysfunction in cystic fibrosis. Cell 73(7):1251–1254 doi:0092-8674(93)90353-R [pii]
4. Veit G, Avramescu RG, Chiang AN, Houck SA, Cai Z, Peters KW, Hong JS, Pollard HB, Guggino WB, Balch WE, Skach WR, Cutting GR, Frizzell RA, Sheppard DN, Cyr DM, Sorscher EJ, Brodsky JL, Lukacs GL (2016) From CFTR biology toward combinatorial pharmacotherapy: expanded classification of cystic fibrosis mutations. Mol Biol Cell 27(3):424–433. https://doi.org/10.1091/mbc.E14-04-0935
5. Younger JM, Chen L, Ren HY, Rosser MF, Turnbull EL, Fan CY, Patterson C, Cyr DM (2006) Sequential quality-control checkpoints triage misfolded cystic fibrosis transmembrane conductance regulator. Cell 126(3):571–582. https://doi.org/10.1016/j.cell.2006.06.041
6. Van Goor F, Yu H, Burton B, Hoffman BJ (2014) Effect of ivacaftor on CFTR forms with

missense mutations associated with defects in protein processing or function. J Cyst Fibros 13(1):29–36. https://doi.org/10.1016/j.jcf.2013.06.008

7. Hwang TC, Kirk KL (2013) The CFTR ion channel: gating, regulation, and anion permeation. Cold Spring Harb Perspect Med 3(1):a009498. https://doi.org/10.1101/cshperspect.a009498
8. Lewis HA, Wang C, Zhao X, Hamuro Y, Conners K, Kearins MC, Lu F, Sauder JM, Molnar KS, Coales SJ, Maloney PC, Guggino WB, Wetmore DR, Weber PC, Hunt JF (2010) Structure and dynamics of NBD1 from CFTR characterized using crystallography and hydrogen/deuterium exchange mass spectrometry. J Mol Biol 396(2):406–430. https://doi.org/10.1016/j.jmb.2009.11.051
9. Protasevich I, Yang Z, Wang C, Atwell S, Zhao X, Emtage S, Wetmore D, Hunt JF, Brouillette CG (2010) Thermal unfolding studies show the disease causing F508del mutation in CFTR thermodynamically destabilizes nucleotide-binding domain 1. Protein Sci 19(10):1917–1931. https://doi.org/10.1002/pro.479
10. Wang C, Protasevich I, Yang Z, Seehausen D, Skalak T, Zhao X, Atwell S, Spencer Emtage J, Wetmore DR, Brouillette CG, Hunt JF (2010) Integrated biophysical studies implicate partial unfolding of NBD1 of CFTR in the molecular pathogenesis of F508del cystic fibrosis. Protein Sci 19(10):1932–1947. https://doi.org/10.1002/pro.480
11. Mendoza JL, Schmidt A, Li Q, Nuvaga E, Barrett T, Bridges RJ, Feranchak AP, Brautigam CA, Thomas PJ (2012) Requirements for efficient correction of DeltaF508 CFTR revealed by analyses of evolved sequences. Cell 148(1–2):164–174. https://doi.org/10.1016/j.cell.2011.11.023
12. Rabeh WM, Bossard F, Xu H, Okiyoneda T, Bagdany M, Mulvihill CM, Du K, di Bernardo S, Liu Y, Konermann L, Roldan A, Lukacs GL (2012) Correction of both NBD1 energetics and domain interface is required to restore DeltaF508 CFTR folding and function. Cell 148(1–2):150–163. https://doi.org/10.1016/j.cell.2011.11.024
13. Majumdar R, Manikwar P, Hickey JM, Arora J, Middaugh CR, Volkin DB, Weis DD (2012) Minimizing carry-over in an online pepsin digestion system used for the H/D exchange mass spectrometric analysis of an IgG1 monoclonal antibody. J Am Soc Mass Spectrom 23(12):2140–2148. https://doi.org/10.1007/s13361-012-0485-9
14. Uniewicz KA, Ori A, Xu R, Ahmed Y, Wilkinson MC, Fernig DG, Yates EA (2010) Differential scanning fluorimetry measurement of protein stability changes upon binding to glycosaminoglycans: a screening test for binding specificity. Anal Chem 82(9):3796–3802. https://doi.org/10.1021/ac100188x
15. Okiyoneda T, Veit G, Dekkers JF, Bagdany M, Soya N, Xu H, Roldan A, Verkman AS, Kurth M, Simon A, Hegedus T, Beekman JM, Lukacs GL (2013) Mechanism-based corrector combination restores DeltaF508-CFTR folding and function. Nat Chem Biol 9(7):444–454. https://doi.org/10.1038/nchembio.1253
16. He L, Aleksandrov AA, An J, Cui L, Yang Z, Brouillette CG, Riordan JR (2015) Restoration of NBD1 thermal stability is necessary and sufficient to correct F508 CFTR folding and assembly. J Mol Biol 427(1):106–120. https://doi.org/10.1016/j.jmb.2014.07.026
17. Hall JD, Wang H, Byrnes LJ, Shanker S, Wang K, Efremov IV, Chong PA, Forman-Kay JD, Aulabaugh AE (2016) Binding screen for cystic fibrosis transmembrane conductance regulator correctors finds new chemical matter and yields insights into cystic fibrosis therapeutic strategy. Protein Sci 25(2):360–373. https://doi.org/10.1002/pro.2821
18. Wales TE, Engen JR (2006) Hydrogen exchange mass spectrometry for the analysis of protein dynamics. Mass Spectrom Rev 25(1):158–170. https://doi.org/10.1002/mas.20064
19. Percy AJ, Rey M, Burns KM, Schriemer DC (2012) Probing protein interactions with hydrogen/deuterium exchange and mass spectrometry-a review. Anal Chim Acta 721:7–21. https://doi.org/10.1016/j.aca.2012.01.037
20. Glasoe PK, Long FA (1960) Use of glass electrodes to measure acidities in deuterium oxide. J Phys Chem 64(1):188–190. https://doi.org/10.1021/j100830a521
21. Wang L, Pan H, Smith DL (2002) Hydrogen exchange-mass spectrometry: optimization of digestion conditions. Mol Cell Proteomics 1(2):132–138
22. Lewis HA, Zhao X, Wang C, Sauder JM, Rooney I, Noland BW, Lorimer D, Kearins MC, Conners K, Condon B, Maloney PC, Guggino WB, Hunt JF, Emtage S (2005) Impact of the deltaF508 mutation in first nucleotide-binding domain of human cystic fibrosis transmembrane conductance regulator on domain folding and structure. J Biol Chem 280(2):1346–1353. https://doi.org/10.1074/jbc.M410968200
23. Fang J, Rand KD, Beuning PJ, Engen JR (2011) False EX1 signatures caused by sample carryover during HX MS analyses. Int J Mass Spectrom 302(1–3):19–25. https://doi.org/10.1016/j.ijms.2010.06.039

Chapter 5

A Multipronged Method for Unveiling Subtle Structural–Functional Defects of Mutant Chaperone Molecules Causing Human Chaperonopathies

Donatella Bulone, Pier Luigi San Biagio, Tatiana Quiñones-Ruiz, Manuel Rosario-Alomar, Igor K. Lednev, Frank T. Robb, Everly Conway de Macario, and Alberto J. L. Macario

Abstract

Chaperonopathies are diseases in which abnormal chaperones play an etiopathogenic role. A chaperone is mutated or otherwise abnormal (e.g., modified by an aberrant posttranslational modification) in structure/function. To understand the pathogenic mechanisms of chaperonopathies, it is necessary to elucidate the impact of the pathogenic mutation or posttranslational modification on the chaperone molecule's properties and functions. This impact is usually subtle because if it were more than subtle the overall effect on the cell and organism would be catastrophic, lethal. This is because most chaperones are essential for life and, if damaged in structure/function too strongly, there would be death of the cell/organism, and no phenotype, i.e., there would be no patients with chaperonopathies. Consequently, diagnostic procedures and analysis of defects of the abnormal chaperones require a multipronged method for assessing the chaperone molecule from various angles. Here, we present such a method that includes assessing the intrinsic properties and the chaperoning functions of chaperone molecules.

Key words Chaperonopathies, Genetic chaperonopathies, Mutant chaperones, Subtle changes, Multipronged method, Molecular properties, Chaperoning functions, Calorimetry, Heat protection, Fibril dispersion

1 Introduction

Chaperonopathies are diseases caused by abnormal chaperones [1, 2]. Chaperonopathy-causing mutations affecting human Chaperonin of Group II have been described [3–5]. There is a pressing need to elucidate the effects of mutations on chaperone molecules in chaperonopathies to better understand protein misfolding diseases and develop specific treatment agents. Likewise, there is a need for experimental models to speed up research in this field.

Cláudio M. Gomes (ed.), *Protein Misfolding Diseases: Methods and Protocols*, Methods in Molecular Biology, vol. 1873,
https://doi.org/10.1007/978-1-4939-8820-4_5,

A pathogenic mutation can affect the intrinsic properties (e.g., resistance to stressors such as heat or brusque pH changes, flexibility, and ability to form functional tertiary and quaternary structures) of the chaperone molecule. Various methods can be utilized to measure the biophysical–biochemical properties of the chaperone molecule, comparing the wild type with that which bears a mutation causing disease in humans. We will describe differential scanning calorimetry (DSC) and isothermal titration calorimetry (ITC).

A pathogenic mutation can also affect the chaperoning functions of a chaperone molecule. Several methods can be applied to measure the chaperoning ability of mutant chaperone molecules and comparing them with the wild-type counterparts. We will describe heat protection assays using malate dehydrogenase (MDH) and shrimp alkaline phosphatase (SAP); and Fibril Dispersion Assays and Atomic Force Microscopy.

Experimental models to examine the impact of mutations on chaperone molecules have been described. There are a few eukaryotic experimental models to study human chaperonopathies but prokaryotic models offer some advantages, for example the *Pyrococcus furiosus* (Pf) model [5]. We have standardized this model and used it to study a pathogenic mutation in the human CCT5 subunit that causes a severe chaperonopathy [3, 4]. The molecules used in the method described here are the Pf chaperonin Cpn60 (designated Pf-CD1) and two mutants: Pf-CD1 Ile138His (Pf-H), and Pf-CD1 Ile138Arg (Pf-R).

1.1 Assessing Molecular Properties

Differential scanning calorimetry (DSC). Power compensation DSC monitors the heat capacity of a sample as a function of temperature by operating in differential mode. Two identical cells loaded with equal volumes of the sample solution and solvent alone, are heated or cooled quasi-adiabatically at constant rate and the temperature of each cell is continuously monitored. Since the heat capacity of the sample solution is larger than that of the solvent alone, a given amount of extra power is continuously applied to the sample cell to keep its temperature equal to that of the reference cell. The differential power applied normalized for the scan rate and the solute molar mass is a direct measure of the solute molar heat capacity. The occurrence of temperature-induced endothermic or exothermic transition manifests itself in the appearance of a heat absorption (or release) peak whose integral over temperature corresponds to the enthalpy change of the process. Theoretical and experimental bases of application of DSC technique to the study of biological systems date back to the 1990s [6–8]. Starting from the first instrument models, technical improvements over time have resulted in high sensitivity instruments [9]. Nowadays, the technique is largely employed in various sectors for investigating the thermodynamic properties of different biomolecules such as, biopolymers, proteins, peptides, and lipids.

Isothermal titration calorimetry (ITC). ITC is the best method to measure the free energy, enthalpy, heat capacity, and stoichiometry of a binding reaction between ligand and biomolecule [10, 11]. The experiment is performed at constant temperature and consists in the repetitive injection of small aliquots of a solution containing the "titrant" into a cell containing a solution of reacting biomolecule, or "titrand." The quantity measured is the heat released (exothermic reaction) or absorbed (endothermic reaction) upon to the formation of titrant–titrand complex. The dependence of heat of reaction on the titrant–titrand molar ratio provides information on the energy of intermolecular interaction.

Since its first applications in the 1970s, ITC has steadily gained attention for determining the thermodynamic parameters of a variety of biochemical binding processes. More recently, the technique has been applied to investigate the self-association process of complex proteins [11], and micellization [12].

1.2 Assessing Chaperoning Functions. Protection Against Heat Denaturation

A CCT subunit ortholog (Cpn) from a hyperthermophilic archaeon (Pf-CD1 and its mutants Pf-H and Pf-R) has been used in mimicking human CCT chaperonopathies because the archaeal model is reduced eightfold in complexity compared to vertebrate CCT [13]. This is because the archaeal model Cpn has a single subunit that forms homo-oligomers. All CCT complexes appear to be conserved in their basic mechanism, in which repetitive opening and closing of the individual rings through cooperative conformational changes is necessary for their folding activity.

In preparation for the use of this hyperstable archaeal chaperonin at temperatures more suitable for enzymatic and biophysical measurements, we constructed a cold-adapted mutant form of the CCT subunit from the hyperthermophile *Pyrococcus furiosus* (Pf) that forms rings with eight subunits and hexadecamers of two rings like in humans [14]. In another recent study, we have used a cold adapted mutant with a deletion of the C-terminus (Pf-CD1) which lowers the optimal activity of the archaeal CCT to 55 °C while maintaining a 16-mer complex structure in order to faithfully replicate the structure of human CCT [13]. To create an experimental platform for studying the effects of specific mutations that cause chaperonopathies in humans, we have introduced mutations into the archaeal (Pf) gene in the conserved positions that exist in the individuals affected by the inborn defects leading to a distal neuropathy. After creating the mutations and expressed the proteins in *E. coli*, we carried out experiments to detect the functional deficits introduced into the archaeal CCT complex. We assay ATPase activity and determine the efficiency of the protein folding functions of the archaeal Pf-CD1, and mutants Pf-H and Pf-R, by examining their stabilizing and refolding effects on the enzymes malate dehydrogenase (MDH) and shrimp alkaline phosphatase

(SAP). Here, we describe the methodology of these tests in detail to allow accurate replication of this study directed at CCT mutants that are known to cause human chaperonopathies.

1.3 Fibril Dispersion Ability

Amyloid fibrils are formed from misfolded proteins and possess a characteristic general morphology, which can be accurately revealed by means of Atomic Force Microscopy (AFM). Typical fibrils are up to several micrometers long with a diameter of the order of 10 nm. High resolution atomic force microscopy (AFM) allows for elucidating structural features of individual fibrils including super-helical twists and filament intertwining.

Amyloid fibrils are observed in a variety of neurodegenerative diseases including Alzheimer's disease [15], Parkinson's disease [16], and others. Many proteins such as lysozyme, insulin, and α-lactalbumin form amyloid fibrils in vitro under appropriate conditions including low pH and elevated temperature [17, 18]. The fact that proteins with very diverse structures can rearrange their conformation to form fibrillar aggregates led researchers to believe that these proteins could share the same mechanism of fibril formation. Partially unfolded proteins are considered to be a prerequisite to the formation of β-sheet-rich amyloids. Typically, several micrometer-long unbranched fibrils consist of 2–6 protofilaments of 2–5 nm in diameter. The protofilaments associate laterally and twist or coil together. Cross-β-sheet structure is the most characteristic polypeptide conformation in the protofilament core. In addition to their role in neurodegenerative diseases, amyloid fibrils are also utilized as scaffolds in biology, 3D imaging and drug delivery [19]. Amyloids fibrils show significant promise for the development of novel bionanomaterials since they are metastable, affordable, robust, and biocompatible.

AFM is a type of scanning probe microscopy, which offers a subnanometer spatial resolution that is more than 1,000 times better than the resolution of optical microscopes limited by light diffraction [19]. AFM was originally developed as a surface characterization technique for material science and quickly became a powerful tool for morphological studies in many areas of chemistry, biology, materials sciences, and nanotechnology in particular. The AFM principal component is a scanning flexible cantilever, at the end of which a sharp tip is mounted. This tip follows the sample surface profile due to attractive or repulsive forces, which occur in contact or noncontact mode. Normally, to measure the cantilever deflection, a laser beam is reflected off the back side of the cantilever toward a photodetector (Fig. 1a). The AFM can be operated in two imaging modes, static (contact mode) and dynamic (noncontact or tapping mode). The cantilever obeys Hooke's law, $F = k{\cdot}x$, where F is the force on the cantilever, k its spring constant, and x its deflection. This relationship describes the tip-sample

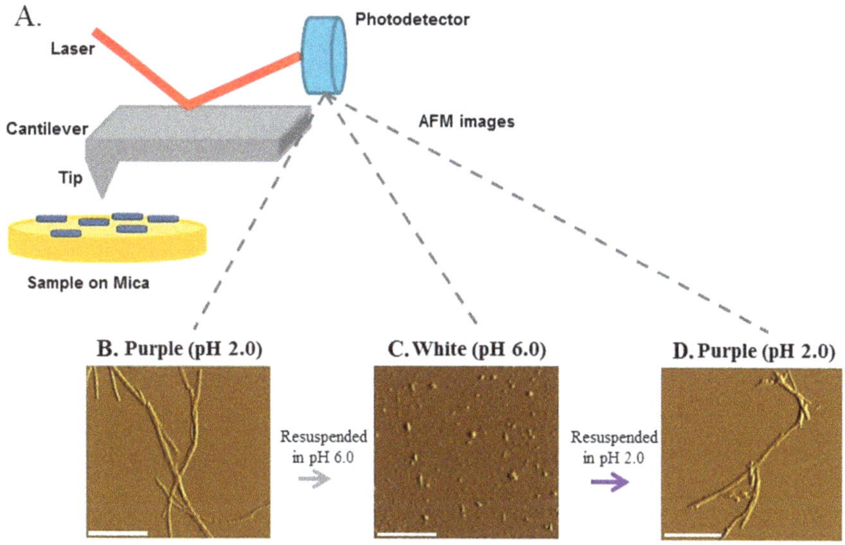

Fig. 1 AFM schematic diagram and AFM images reported for the purple fibrils structural rearrangements [21]. Panel (**A**) shows scale-up component parts of AFM instrument including tip, cantilever, laser beam, and photodetector. AFM images of purple fibrils (**B**), disintegrated purple fibrils at pH 6 (**C**), and reassembled purple fibrils at pH 2.0 (**D**). Scale bars are 500 nm. Information and a figure pertinent to Pf-CD1, Pf-H, and Pf-R can be found in ref. 13

interaction force. The cantilever deflection data are used to construct a three-dimensional representation of the surface profile that is called an AFM image [19]. AFM has emerged within the last 10 years as a powerful tool for studying the morphology of amyloid fibrils and the fibrillation mechanism through the characterization of aggregation intermediates at all stages of the process. A unique suitability of AFM for characterizing protein aggregation has been recently utilized for studying the inhibition of amyloid fibrillation with hydrogen sulfide [20], discovery of purple fibrils, a new type of protein chromophore [21] (Fig. 1B–D) and spontaneously refolding of apo-α-lactalbumin from one polymorph to another as a result of mild changes in the solution temperature and salinity [22]. The procedure described here was used with the *Pyrococcus furiosus* experimental model to test the fibril dispersion ability of the Cpns Pf-CD1, Pf-CD1 Ile138His (Pf-H), and Pf-CD1 Ile138Arg (Pf-R) [13].

2 Materials

2.1 Materials for DSC and ITC

1. Recombinant proteins Pf-CD1, Pf-CD1 Ile138His (Pf-H), and Pf-CD1 Ile138Arg (Pf-R) prepared as described in [13].
2. Buffer 1: 20 mM phosphate buffer, pH 7.5, 100 mM NaCl, 10% glycerol.

3. Buffer 2: 20 mM phosphate buffer, pH 7.5, 100 mM NaCl.
4. Amicon centrifugal filters, molecular weight cutoff 30 kD (Amicon Ultra-4, Millipore).
5. 0.22 μm sterile filters (Millipore).
6. Shimadzu UV-Vis spectrophotometer.
7. One Semi-Micro quarz cuvette, 10 mm path length (114-OS, Hellma).
8. 1N potassium hydroxide solution.
9. Super-Q water.
10. Nano differential scanning calorimeter (TA Instruments).
11. Cell cleaning adapter with silicone tubing.
12. 2.5-mL micropipette with tips and silicone tubing.
13. Silicone tubing for cleaning.
14. Vacuum degassing system.
15. Lab tissue.
16. Small caps to cover the access tube ends of DSC capillary cells.
17. Standard volume isothermal titration calorimeter (TA Instruments).
18. 2.5-mL filling syringe with 16-gauge, 8-in. long needle.
19. 100-μL burette syringe for titrant injection.
20. Stainless steel needle for the reference cell.
21. Ethanol 99% grade.

2.2 ATPase Activity Measurement

Ultrapure water (reverse osmosis-purified water is deionized by MilliQ instrument to a sensitivity of 18 MΩ-cm at 25 °C) is used for reagent stocks and preparing reaction mixtures. Solutions are stored at −20 °C in aliquots, unless otherwise noted.

1. ATP, made as a 100 mM stock in 50 mM Tris–HCl, (pH 8.5) (Sigma-Aldrich, St. Louis, MO).
2. Dithiothreitol (DTT), made as a 0.1 M stock in dH_2O (Sigma-Aldrich).
3. NAD, β NADH.oxaloacetate made as 10 mM stock solutions in dH_2O. Sigma-Aldrich.
4. Porcine heart malate dehydrogenase (MDH) (Amresco, Solon, OH). Prepared freshly on day of use.
5. Shrimp alkaline phosphatase (SAP). BioAssay Systems, Hayward, CA. Prepared freshly on the day of use.
6. Phosphatase Assay Kit (ENZO Life Sciences, New York, NY).
7. Perchloric acid (Merck, Kenilworth, NJ).
8. Mineral oil (Molecular Probes, Eugene, OR).

9. MyCycler thermal cycler (Bio-Rad, Hercules, CA).
10. Microplate reader (TECAN Infinite M200, Männedorf, Switzerland).
11. Microtiter plates (Whatman Uniplate Cat#7701-1350, Sigma-Aldrich).
12. Multichannel Pipettor (Brinkmann Transferpette-12, Loveland, CO).
13. BIOMOL Green reagent (Enzo Life Sciences, New York, NY).

2.3 For Fibril Formation

1. Fibrils sample.
2. Buffer used to form fibrils.
3. Centrifuge tubes—one tube for sample and a matching tube for counterbalance.
4. A petri dish.
5. A microscope slide.
6. Nail polish to temporarily attach the mica to a microscope slide.
7. A mica substrate for AFM sample preparation.
8. Scotch tape.
9. Clean forceps.
10. Double-sided tape.
11. AFM probe—we recommend Aspire CT300R.
12. AFM instrument—we utilize a SmartSPM™-1000 atomic force microscope from AIST-NT, Inc. (Novato, CA, USA).

3 Methods

3.1 DSC and ITC

3.1.1 DSC and ITC Measurements

The correct interpretation of both DSC and ITC results relies on the assumptions that proteins in solution are pure and homogeneous and their concentration is known precisely. Thus, a purification protocol is recommended to validate both assumptions. We describe here the procedure applied to Pf-CD1, Pf-CD1 Ile138His, and Pf-CD1 Ile138Arg. All three proteins were available in buffer #1 stock solutions at a concentration of 4 mg/mL.

1. Prerinse a centrifugal filter 30 kD (*see* **Note 1**) by adding 4 mL of buffer #2 (*see* **Note 2**) and spinning out at 7500 × *g*, 5 °C, for 15 min in a centrifuge with fixed rotor.
2. Add 3 mL of the buffer #2 to 1 mL of sample and put in the prerinsed centrifugal filter.
3. Spin out the sample at 7500 × *g*, 5 °C, for 25 min.

4. When the volume is reduced to 1 mL, add 3 mL of buffer and repeat the procedure four times (*see* **Note 3**).
5. Filter the final samples into sterilized Eppendorf tubes, using 0.22 mm sterile filters.
6. Dilute an aliquot of the final protein sample two or three times and record its absorbance spectrum from 600 to 200 nm.
7. Record the absorbance spectrum of pure buffer from 600 to 200 nm and subtract it from that of the protein.
8. Determine the protein's concentration from the absorbance value at 280 nm, using the extinction coefficient, epsilon = 13,410 (M^{-1} cm^{-1}) calculated by ExPASy ProTParam tool [23].
9. Store purified protein solutions at 4 °C and use in DSC and ITC experiments within 3 days at most.

3.1.2 DSC Measurements

A delicate balance of weak and opposite forces regulates the stability of protein native conformation. Typical values of free-energy change associated to the loss of native structure are in the range 5–15 kcal/mol, which is comparable with the energy of a few hydrogen bonds. DSC is the most suitable technique to investigate the thermal stability of biomolecules, since it is the only method allowing a direct and high resolution determination of the enthalpy associated with the unfolding process. The experiment consists in following the change of protein heat capacity while the temperature increases with a linear rate.

Here, we describe the procedure applied to study the thermal unfolding of Pf-CD1.

1. Dilute the protein sample with filtered buffer #2 to the chosen concentration (*see* **Note 4**).
2. Degas both the buffer #2 and sample by pulling a vacuum of 400–500 mm Hg for a period of 10–15 min (*see* **Note 5**). A degassing station is usually available as a calorimeter accessory.
3. With a micropipette accurately put 300 μL of buffer in both reference and sample cell (*see* **Note 6**).
4. Remove excess liquid from the surroundings of the access tubes, using lab tissue.
5. On each cell, cover one access tube with a small cap and leave the other open to allow pressurization (*see* **Note 7**).
6. Wait for the baseline to stabilize before beginning the experiment.
7. Pressurize to 3 atm and wait for baseline stability.
8. Set the instrument for four heating–cooling cycles from 25 to 120 °C, at 60 °C/h scan rate (*see* **Note 8**).

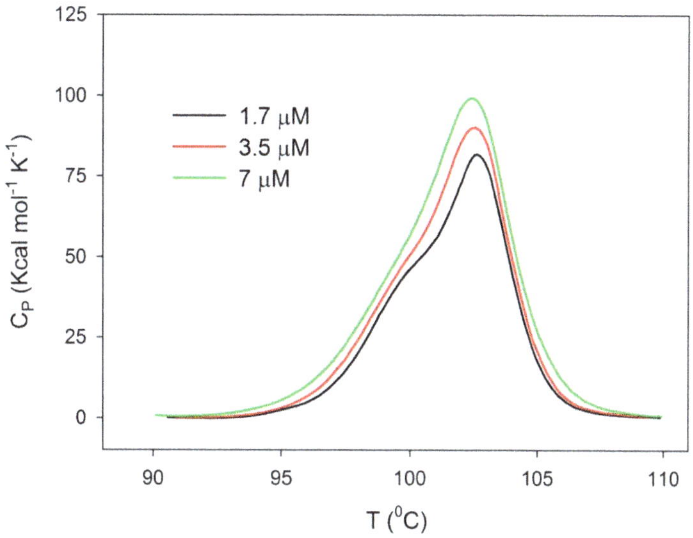

Fig. 2 DSC thermograms of Pf-CD1 in PB buffer (20 mM NaH_2PO_4, 100 mM NaCl, pH 7.5) at 1.7 μM (black), 3.5 μM (red), and 7 μM (green) protein concentration. Calorimetric traces are given after subtraction of the instrumental base line. The scan rate was 60 °C/h. Adapted from ref. 24 (Spigolon et al., Biochem. Biophys. Rep.), according to Creative Commons Attribution (CC BY-NC-ND 4.0) policy

9. Perform multiple buffer–buffer scans (*see* **Note 9**).
10. At the end of the heating–cooling cycles depressurize the cells, remove the buffer from the sample cell and load 300 μL of protein solution with a micropipette.
11. Wait again for baseline stabilization.
12. Pressurize to 3 atm and wait for baseline stability.
13. Perform a heating–cooling cycle over the same temperature interval at the same scan rate used for buffer–buffer scan. Results for Pf-CD1 are shown in Fig. 2.
14. At the end of protein–buffer scan, depressurize the instrument and remove both the protein solution and buffer from the cells.
15. Clean the reference cell by repeated additions of super-Q water.
16. Clean the sample cell with repeated additions of super-Q water followed by K-OH solutions (*see* **Note 10**).
17. Connect both cells with silicon tubes to a pump and flush about 1 L of super-Q water.

3.1.3 DSC Data Analysis

Temperature-induced denaturation of proteins is characterized by a large increase of heat capacity since the energy is continuously adsorbed to break the interactions that stabilize the native structure. With increasing temperature, the process proceeds more or less cooperatively, depending on the protein structure, until the protein is completely unfolded. The analysis of the endothermic peak provides the thermodynamic parameters that regulate the thermal stability. Here, we describe the analysis procedure carried out by using Nano-Analyze Data Analysis TA® and MicroCal Origin Pro 7.0 Software.

1. Subtract the baseline obtained from the average of consistent multiple buffer–buffer scans from the protein–buffer thermogram to obtain the excess heat capacity, C_P^{EX}.
2. Determine the temperature value of C_P^{EX} maximum, T_M.
3. Calculate the total enthalpy change, ΔH_{cal}, associated with the thermal process by integrating the excess heat capacity, C_P^{EX}, over temperature in the significant temperature interval:

$$\left[\Delta H_{\mathrm{cal}}\right] = \int_{T_1}^{T_2} \left[C_{\mathrm{P}}^{\mathrm{EX}}\right] dT$$

   where T_1 and T_2 represent the absolute temperature values before and after the thermal process, respectively (*see* **Note 11**).
4. Examine the thermogram for evaluating the occurrence of skewness, or evident multiple-peak behavior (Fig. 2). If so, use multipeak fitting models to describe the thermal transition, with the caveat that a multipeak transition may be due to a variety of different thermal processes (e.g., independent or sequential melting of different subunits of a single large multimeric molecule; unfolding coupled to oligomers dissociation in case of multimeric molecules; unfolding coupled to conformational transition or ligand dissociation). Measurements at varying protein concentrations will be essential to discriminate between these various possibilities (*see* **Note 12**).

3.1.4 ITC Measurements

Besides its common application to studies of protein–ligand binding, ITC can profitably be used for investigating the thermodynamics that dictates the oligomeric equilibrium of large, multicomponent proteins. The idea is to measure the energy of monomer–monomer interaction at increasing protein concentrations, starting from very low micromolar values. This procedure reveals the critical concentration for self-assembly, the energy involved, and the occurrence of intermediates. The experiment consists of measuring the heat of dilution of small volumes of protein solution sequentially added to the sample chamber loaded with the solvent alone. Here we describe the protocol applied to recombinant chaperonins Pf-CD1, Pf-CD1 Ile138His, and Pf-CD1 Ile138Arg.

1. Set the instrument temperature at 25 °C.
2. Degas both the buffer and protein solutions by pulling a vacuum of 400–500 mm Hg for a period of 10–15 min. A degassing station is usually available as a calorimeter accessory.
3. By using the filling syringe, flush several times the sample cell with degassed buffer.
4. Remove all buffer and load the sample cell with 1200 μL of buffer.
5. Flush several times the reference cell with degassed buffer.
6. Remove all buffer and load the reference cell with 1200 μL of buffer.
7. Insert a stainless steel needle inside the reference cell. This would mimic the configuration of the sample cell during the measurements.
8. Load the injection syringe with protein solution (*see* **Note 13**) taking care for obtaining a uniform filling without bubbles. The syringe should be loaded to 10 μL in excess.
9. Install the injection syringe into a burette. This will cause the leak of some drops of solution. Remove the fluid expelled from the syringe tip with lab-tissue.
10. Install the burette in the instrument and wait for thermal equilibration.
11. Turn on the stirrer at 300 rpm (*see* **Note 14**).
12. Wait for thermal equilibration.
13. Set the experimental parameters: initial delay, volume of each injection, number of injections, and time between injections (*see* **Note 15**). Illustrative data on Pf-CD1 are shown in Fig. 3a, and were obtained by using an initial delay of 600 s, a time between injections of 5 min, an injection volume of 5 μL for each of 19 successive injections.
14. At the end of the experiment, dismount the burette assembly and the injection syringe. Clean the syringe by manually rinsing super-Q water and 50% ethanol aqueous solution.
15. Withdraw the sample-cell content using the filling syringe. Manually rinse the cell with super-Q water. If necessary, use a detergent solution. Then rinse with the cleaning device using 1–2 L water of super-Q water.
16. Load again the reference cell with 1200 mL of degassed buffer. Load the injection syringe with degassed buffer and repeat the titration experiment by using the same experimental setting as before (*see* **Note 16**).

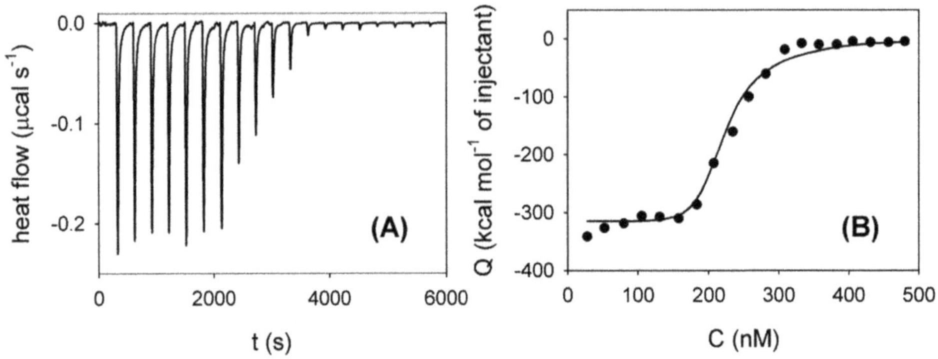

Fig. 3 Titration of 5 μM Pf-CD1 into buffer (20 mM in NaH_2PO_4, 100 mM NaCl, pH 7.5) at 25 °C in aliquots of 5 μL. Panel A shows the raw spikes of heat flow generated by the successive injections (with 300 s time interval). Each spike is integrated over time and, after subtraction of the heat of buffer–buffer dilution, normalized to the number of protein molecules added at each injection. Panel B shows the resulting heat values (symbol) and the fit (continuous line) to the hexadecamer–monomer equilibrium model described in the text. Adapted from ref. 24 (Spigolon et al., Biochem. Biophys. Rep.), according to Creative Commons Attribution (CC BY-NC-ND 4.0) policy

3.1.5 ITC Data Analysis

Raw output from ITC measurements is the power added or subtracted to the sample cell to keep its temperature equal to that of the reference cell during the injection sequence. At the end of the experiment, a series of spikes of heat flow (power) versus time are obtained, as shown in Fig. 3A. The first step of data analysis consists in calculating the total heat associated to each injection. As each injection corresponds to a different total amount of protein inside the cell, the dependence of the total heat on protein concentration can be analyzed to gain information on thermodynamic parameters of protein-protein interactions. Here, we describe the procedure followed for analyzing ITC data on Pf-CD1. We carried out data analysis and modeling by using Nano-Analyze Data Analysis TA® and MicroCal Origin Pro 7.0 Software.

1. Integrate the power peaks measured in the control (buffer–buffer) experiment (*see* **Note 17**), and calculate the average heat of buffer–buffer injection.
2. Integrate the power peaks measured for protein titration.
3. Subtract the buffer–buffer heat from each heat value of protein titration.
4. Calculate the amount of protein inside the cell after each injection (*see* **Note 18**).
5. Plot the heat of dilution as a function of protein concentration inside the sample cell.

6. Choose an appropriate model to gain information on the thermodynamic parameters (*see* **Note 19**). Results for Pf-CD1 are shown in Fig. 3B, and more information is available in [24].

3.2 ATPase Activity Measurement

3.2.1 Protocol to Measure ATPase Activity vs Temperature

1. Prepare a batch of reaction mixture containing 50 mM Tris–HCl (pH 8.5), 100 mM KCl, 5 mM $MgCl_2$, 2 mM DTT, and purified chaperonin to a final concentration of 2 nM.
2. Preincubate at a series of temperatures from 30, 40, 50, 60, 70, 80, and 90 °C for 3 min before adding ATP (50 mM) to initiate ATP hydrolysis. All reactions and controls are performed under light mineral oil (*see* **Note 20**).
3. After the addition of ATP, mixing and brief centrifugation, the tubes are transferred to a thermocycler (e.g., Bio-Rad MyCycler), for 15 min incubation at the scheduled temperature followed by perchloric acid addition to a final concentration of 2% v/v to quench the reaction (*see* **Note 21**).
4. A total volume of 100 μL of BIOMOL Green reagent (Enzo Life Sciences) is added to the reaction followed by 25 min incubation at 23 °C before measuring the absorbance at 630 nm, using a spectrophotometer equipped with a microcell holder (volume 0.1 mL, light path 1 cm). The assays are performed in triplicate to allow standard deviations to be calculated.
5. For more details on measuring phosphatase activity *see* ref. 25.

3.2.2 Protein Refolding Assay Using Malate Dehydrogenase (MDH)

CCT chaperonins (Cpn) can carry out catalytic refolding of denatured proteins in vitro. This assay allows one to compare refolding activity in different variants of Cpns, for example Pf-CD1, Pf-H, and Pf-R.

1. Preparation of denatured protein substrate. Dilute the porcine heart malate dehydrogenase (MDH) 1 mg/mL stock into denaturation buffer (25 mM HEPES (pH 7.2), 300 mM NaCl, 1 mM $MgCl_2$, 4 M guanidine–HCl, and 5 mM Dithiothreitol) in a final volume of 3 mL and final concentration of 0.1 mg/mL and incubate at 37 °C for 1 h.
2. Renaturation with Cpn and ATP. Dilute denatured MDH 100-fold into refolding buffer (25 mM HEPES (pH 7.2), 300 mM NaCl, 1 mM $MgCl_2$, 1 mM ATP, and 0.5 M ammonium sulfate) in the presence Cpn. The final concentration of denatured MDH during refolding should be 375 nM and the ratio of Cpn and to denatured MDH should be 1:10. The reactions are incubated at 42 °C for 1 h. Negative controls without ATP and Cpn should be included.
3. Measurement of MDH following refolding. Measure MDH activity after refolding by diluting the refolding mixture ten-

fold in MDH buffer (50 mM Tris–HCl (pH 8.0), 0.2 mM NADH, and 0.5 mM oxaloacetate) at 25 °C. The oxidation of β-NADH to β NAD is recorded by following decrease in OD_{340} in real time in a spectrophotometer. The final concentration of MDH should be 35 nM in the MDH assay and reactions are started by the addition of cofactor NAD.

4. The rate of oxidation of NADH is related to MDH activity which can be calculated as relative units (DeltaOD_{340}/min/mg of MDH added to the reaction). The extent of refolding is calculated as a percentage of the activity measured with the same mass of the nondenatured MDH. Typically refolding peaks at between 20 and 50% of the original activity.

3.2.3 Protein Heat Protection by Cpn Using MDH or SAP

Another measure of the ability of a chaperonin to refold proteins is to assay the ability of the Cpn to salvage a client protein that is in the process of denaturation. This assay allows one to determine reactivation of labile enzymes in destabilizing conditions, for example incubation at 50 °C. For optimal protection the incubation conditions of the denaturation test must accommodate Cpn stability and ongoing ATPase activity to allow refolding to take place.

1. Incubate MDH samples (1 mg/mL) in the reaction mixture (25 mM Hepes-KOH, pH 8.0, 300 mM KCl, 1 mM $MgCl_2$ and 0.5 M ammonium sulfate) plus 4 mM ATP at each temperature to be tested in the presence or absence of Cpn (0.2 mg/mL). Withdraw samples and maintain them at 4 °C after different incubation-time intervals (for examples 0, 3, 6, 9, 12, and 15 min) at 37 °C and 0, 1, 2, 3, 4, and 5 min at 42 °C (*see* **Note 22**).
2. The MDH activity of withdrawn samples should be measured within 1 h after sampling. Assays are at 25 °C in a 1 mL assay containing 90 mM Hepes-KOH, pH 8.0, 0.22 mM NADH, and 0.55 mM oxaloacetate. The time dependent oxidation of NADH should be measured by recording decreasing OD_{340} nm in a Beckman DU spectrophotometer.
3. Alternatively, SAP may be used as the enzyme to be protected, using the same sampling protocol as for MDH. Following heat treatment at 50 °C, the residual activity is measured using nitrophenyl phosphate (pNPP) provided in the Phosphatase Assay Kit as a chromogenic substrate that is converted via SAP to p-nitrophenol according to the reaction:

Shrimp Alkaline Phosphatase + *p*-Nitrophenyl phosphate >>> *p*-nitrophenol + phosphate (pH 10)

Procedure using cuvettes:

1. Transfer 50 μL samples into 1 cm light path cuvettes.

2. Pipet 950 μL reaction mixture into samples. Addition of reaction mix containing the samples and mixing should be quick. Assays are carried out at 30 °C.
3. Mix briefly and monitor formation of pNP at OD_{630} spectrophotometrically.

Alternative procedure using 96-well microtiter plates:

1. Transfer aliquots of the samples to a microtiter plate with a multichannel pipettor.
2. Transfer 200 μL distilled H_2O and 200 μL into separate wells of a clear bottom 96-well plate.
3. Carefully transfer 5–50 μL of the withdrawn samples into adjacent wells.
4. Pipet 150–195 μL of reaction mixture to sample wells. The final reaction volume in the sample wells should be 200 μL. Tap plate mix.
5. Read OD_{630} ($t = 0$), and again after 4 min ($t = 4$) on a plate reader (*see* **Notes 23** and **24**).

3.3 Fibril Formation

1. Form fibrils by following a specific protocol. Many proteins form fibrils in vitro at low pH and high temperature. Other factors such as, salinity and stirring the sample during incubation can speed up fibrillation.
2. Isolate the fibrils from the remaning soluble protein by centrifugation at 15,682 × *g* for 30 min.
3. Discard or transfer the supernatant if needed for further analysis of the soluble protein. The pellet contains fibrils and possibly other insoluble aggregates, such as oligomers and/or amorphous aggregates.
4. Resuspend the pellet with a fresh solution used for the initial fibrillation proces. Changing the solution properties at this stage, e.g., pH and salinity can result in spontanious refolding of amyloid fibrils from one polymorph to another [22, 26].
5. Dilute samples to the point when individual fibrils are observed on AFM images. For example, for a starting protein concentration of 25 mg/mL, the dilution factor is 1000 × (1 μL sample resuspended in 1000 μL of solution). From our experience, keeping the concentration in the lows μg/mL yield adsorbed fibrils on mica that can be individually imaged.
6. On a petri dish, place a microscope slide. The petri dish will help to protect a mica sample from the environment dust until it is ready to be imaged. The AFM tip could pick up loose particles from the environment, which can affect its performance.

7. On the microslide, place a small drop of nail polish (works well as temporary glue) and put a mica on the drop so that it temporarily attaches to the microslide.
8. Push the mica to make it leveled on the microslide using a micropipette tip.
9. Remove a couple of mica layers with Scotch tape, until a fresh, clean, and even surface is obtained (the tape can remove uneven layers of mica, keep removing layers until a smooth layer is revealed).
10. Cover the petri dish while not working on the mica to avoid dust particles landing on the freshly cleaved surface.
11. Resuspend the diluted fibrils sample right before depositing them on the mica slide.
12. Deposit approximately 150 μL of the sample from a close distance to the center of the mica slide (do not touch the mica with the micropipette tip) and slowly form a big drop of sample on the mica surface (*see* **Note 25**).
13. Wait 2 min for the sample to adsorb on the mica.
14. Then, remove the remaining drop slowly with a micropipette from an edge of the mica avoiding touching the surface which is going to be imaged.
15. To remove excess sample that has not attached to the mica, add approximately 120 μL of solvent slowly to the center of the mica slide (without touching the surface with the pipette tip) and remove it slowly from an edge of the mica avoiding touching the surface which is going to be imaged.
16. Then, completely dry the mica under a slow airflow (1–2 min).
17. Place a small piece of double sided tape on the AFM sample holder.
18. Carefully slide the mica from the microscope slide with clean forceps.
19. Place the mica on the tape on the AFM sample holder, and gently push down on the edges of the mica. The sample is now ready to be imaged by AFM.
20. Mount the cantilever.
21. Select the tapping mode.
22. To localize the tip of the cantilever press “find tip”.
23. Then align the laser beam on the tip of the cantilever by pressing “adjust diode”.
24. Find the resonance frequency of the cantilever (*see* **Note 26**).
25. Approach the mica surface with the cantilever tip.

26. For a quick first image, scan a 10 μm × 10 μm area with 1 Hz scanning speed and 300 pixels.
27. Select an area of interest (where individual fibrils, or their bundles are located) and image a 1 μm × 1 μm or 2 μm × 2 μm area (depending on how big is the area of interest) with 1 Hz scanning speed and 400 pixels (*see* **Note 27**).
28. Image different areas of the mica surface to have a good representation of the sample morphology (*see* **Note 28**).

4 Notes

Notes for Subheading 3.1

1. The choice of 30 kDa cutoff allows a fast processing speed while assuring the retention of the solute (60 × 12 kDa in our case).
2. Glycerol is added to stock solutions to stabilize the native structure of proteins and preventing their aggregation.
3. Each step of the filtering procedure implies a 4-time dilution of the initial solvent. After four steps, the concentration of low molecular weight materials and glycerol will be reduced to about 0.4% of the initial value. Note that this procedure provides a practical alternative to dialysis for buffer exchange.
4. With modern DSC equipment, good heat capacity data can usually be measured down to 0.1 mg/mL.
5. Reference and sample solutions must be degassed prior to be placed in DSC to prevent (or at least minimize) the formation of gas bubbles that will occur with increasing temperature because of the decreased solubility of dissolved gases. Bubble formation during the DSC run is typically reflected into noisy heat capacity data.
6. The fixed-cell power compensation instrument would assure the absolute maximum in baseline repeatability. Although reference and sample cells in the DSC are manufactured to be as identical as possible, unavoidably there will be slight differences in volume, shape, and thermal contact. Thus, a preliminary DSC run with both cells loaded with buffer is necessary to measure the effective baseline. This signal will be subtracted from data collected during a DSC run on samples of interest before proceeding with data analysis. To obtain a reliable baseline, it is advisable to perform the protein/buffer scan without changing the buffer in the reference cell.
7. A constant excess pressure (3–6 atm) can be applied by the DSC manostat to prevent gas bubble formation during the

run and heating the aqueous solutions above 100 °C will accomplish the same.

8. When studying protein unfolding, a scan rate of 60 °C/h is usually adequate for simple, reversible unfolding transitions. However, kinetical processes such as aggregation upon unfolding can affect the shape of the thermogram. To verify the absence of scan rate-dependent processes, it is necessary to repeat the experiments at different scan rates.
9. Several buffer–buffer baselines are typically recorded to ensure baseline repeatability, which is essential for obtaining absolute heat capacity profiles.
10. Even small traces of aggregated protein in the DSC cell may cause irregular baseline. The use of detergents or strong acids/bases is recommended by the manufacturer as essential for optimal DSC operation.
11. Nano-Analyze TA and MicroCal Origin Pro 8.0 Software are used for baseline subtraction and ΔH_{cal} calculation.
12. Thermal unfolding of single-domain proteins generally results in a single peak that is well described by the two-state unfolding model. Thermal unfolding of multidomain proteins reveals several endothermic peaks that may be fully or partially resolved. Structural information from other techniques may be required for assigning each peak to a certain transition. However, DSC experiments at varying protein concentrations are essential to verify the nature of multiple peak thermograms. As a general rule, the total molar calorimetric enthalpy increases with protein concentration if the unfolding is coupled to dissociation of oligomeric complexes. Results in Fig. 2 for Pf-CD1 show that the amplitude of the lower temperature peak increases with protein concentration, while the higher temperature peak remains unchanged, thus validating the hypothesis of unfolding coupled to oligomers dissociation. Moreover, the decrease of T_m with increasing concentration is a clear indication that a larger oligomerization degree [7].
13. When titrating a multimeric protein solution into buffer alone, the choice of the concentration of the protein solution loaded in the injection syringe is not as crucial as in the case of binding experiments. A common practice is to start from concentrations at which the protein is in stable oligomeric state. This can be ascertained with other techniques like as SEC-HPLC chromatography. By subsequent injections of a few microliters (up to a total of 100) into buffer of about 1000 μL, the protein concentration inside the sample cell will span from a few hundredth up to one tenth of the concentration of solution loaded. This should be enough to investigate the stability of protein oligomers against dilution. If there are signals of

further dissociation at the lowest concentration explored, it is worth to repeat the experiment at lower concentration of the injected solution. However, the signal will also become too low as a lower number of molecules are injected in the cell. To correct this signal decrease, one can increase the injection volume.

14. The needle of titrant syringe also functions as the stirrer. The stirring helps the diffusion of the titrant through the volume of the sample cell. Usually, the stirring rate is not a crucial parameter as the diffusion of small titrant molecules takes very short time. It might be important for bulky macromolecules, whose slower diffusion broadens the peak of heat developed by the single injection. However, the peak area (that is the total heat due to injection) will not be affected.
15. The time delay is set to check automatically that the system has reached the thermal equilibrium before starting the experiment. It corresponds to the interval time over which software routinely controls if the amplitude of baseline fluctuations is always inside a very small percent. On the same interval time, the baseline flatness is continuously calculated to check for occurrence of instrumental drift. When both baseline fluctuations amplitude and slope become lower than the chosen limiting values, the experiment starts. The choice of the injection volume results from a compromise between a sufficient number of experimental points and high signal/noise ratio. The optimal condition is when the molar heat of dilution is very high. This allows the use of small injection volumes, thus obtaining a large number of experimental points. When the signal is very low, it is preferable to increase the injection volume and obtain few but good experimental points. Repeated experiments by varying appropriately the protein concentration of the injected solution will allow increasing the number of experimental points in the concentration range of interest. Obviously, in each experiment the number of successive injections has to be consistent with the volume of each injection and the total volume of the injection syringe. The time between injections must be long enough to allow the level of the baseline after each injection be the same as before. Typically, 5 min is a good choice, but longer times could be required in some cases.
16. Running a control experiment with both sample cell and injection syringe loaded with buffer will provide the contribution to the heat signal due to slight mismatch between reference and sample cells.
17. Total heat for each injection of the control experiment should show no dependence on the injection number. In fact, this is a

good test to verify that sample cell is clean and the injection syringe is correctly operating.

18. The total concentration of protein (in a monomer base) in the calorimetric cell after each injection i is given by:

$$[M]_i = P_0\left(1-\left(1-\frac{v_{\text{inj}}}{V_0}\right)^i\right)$$

where P_0 is the protein concentration in the injection syringe, v_{inj} is the injection volume, and V_0 is the cell volume.

19. The choice of the most appropriate model to analyze ITC data would be guided by the already known information on protein structure. Here, we analyze ITC curves by using a hexadecamer–monomer equilibrium model. In this model the equilibrium constant and mass balance equations are:

$$M_{16} \leftrightarrow 16M$$

$$K_{\text{d}} = \frac{[M]^{16}}{[M_{16}]}$$

$$[M]_t = [M] + 16[M_{16}]$$

$$16\frac{[M]^{16}}{K_{\text{d}}} + [M] - [M]_t = 0$$

Here K_{d} is the dissociation equilibrium constant, $[M]$ and $[M_{16}]$ are the equilibrium concentrations of the monomer and hexadecamer, respectively, and $[M]_t$ is the total monomer concentration.

The free monomer concentration after each injection i, $[M]_i$, can be calculated by using the previous equations. The normalized heat associated with each injection, Q_i, is given by:

$$Q_{\text{i}} = \frac{V_0 \Delta H_{\text{d}}}{v_{\text{inj}} P_0}\left([M]_i - [M]_{i-1}\left(1-\frac{v_{\text{inj}}}{V_0}\right) - F_0 P_0 \frac{v_{\text{inj}}}{V_0}\right) + q_{\text{d}}$$

where ΔH_{d} is the enthalpy change due to dissociation, V_0 is the reaction volume, v_{inj} the injected volume, P_0 is the molar concentration in the syringe, and F_0 is the fraction of monomers in the syringe. The heat released or absorbed in each injection is proportional to the increment of the concentration of monomer concentration in the cell after injection i and triggered by protein dilution. In the calculation, we also accounted for the fraction of monomers introduced directly from the syringe and not contributing to the heat effect.

Table 1
Thermodynamic parameters for the hexadecamer–monomer equilibrium of Pf-CD1

k_d (μM)	ΔG_d (kcal/mol)	ΔS_d (kcal/mol K)	ΔH_d (kcal/mol)
0.31 ± 0.06	−8.8 ± 0.1	−1.14 ± 0.01	−350 ± 5

Fitting parameters for Pf-CD1 disassembly are summarized in Table 1. K_d is related to the Gibbs free energy of dissociation via $\Delta G_d = -RT\ln(K_d)$, while ΔS_d is calculated from $\Delta G_d = \Delta H_d - T\Delta S_d$. Pf-CD1 disassembly involves a favorable and considerable enthalpy change. This, in turn, implies that the assembly of the hexadecameric complex is endothermic and is entropically driven. The negative Gibbs free energy indicates that disassembly is a thermodynamically favorable reaction as long as the concentration is less than the critical transition concentration (CTC).

Notes for Subheading 3.2

20. Since ATPase is assayed at a broad range of temperatures, light mineral oil is overlaid on the samples before incubation to prevent evaporation and consequent errors due to volume change.
21. When using denatured MDH or SAP as amendments in the standard Cpn ATPase activity, the assay reaction buffer is composed of 25 mM Hepes–KOH buffer at pH 7.2, 300 mM KCl, 1 mM $MgCl_2$ in the presence or absence of 0.05 mg/mL final concentration of purified Cpn. Addition of denatured proteins activates ATPase activity by allosteric stimulation of the Cpn up to 100-fold.
22. MDH and SAP are stable for 48 h at 4 °C and 2 months at −20 °C. EDTA, oxalate, fluoride, and citrate are known inhibitors of SAP and should be avoided in sample preparation. It is recommended to measure the residual activities of heat treated samples on the same day as the experiment is started, and to maintain them at 4 °C until they are assayed.
23. Reagent preparation for plate reading: Equilibrate reagents to 23 °C. The reaction mix is prepared by mixing for each 96-well assay, 200 μL Assay Buffer, 5 μL Mg Acetate (final 5 mM) and 2 μL pNPP liquid substrate (10 mM). Fresh reconstitution of enzymes is recommended, although the reaction mixture is stable for at least 1 day at 23 °C.
24. If sample SAP activity exceeds 800 IU/L, dilute samples in saline and repeat the assay, multiply the result by the dilution factor. Incubation can be prolonged to 2–4 h for samples with low activity.

Notes for Subheading 3.3

25. If the sample drop breaks during the adsorption process, the sample may not be optimally dispersed through the mica. Tap dry the mica surface and peel a couple of mica layers (*see* **step 9**) to obtain a clean surface. Deposit the sample again.
26. The experimental resonance frequency should be very close to those recommended by the manufacturer. Resonance frequencies, which are significantly different from those recommended by the manufacturer, can indicate certain problems with the tip. For example, a higher resonance frequency could be due to a broken cantilever and a lower resonance frequency could be due to tip contamination with loose sample or dust particles.
27. To increase the image resolution, decrease the scanning speed and increase the number of pixels.
28. After imaging multiple areas consecutively or after a long scanning time, the images can show up distorted, due to thermal drift. Allow sufficient time (approximately 5 min) to equilibrate the temperature between individual scans.

Acknowledgments

Part of the work by PLSB and DB was carried out using instruments funded by PO. FESR 2007/2013 "*Piattaforma regionale di ricerca traslazionale per la salute*" 4.1.2. Asse IV. We thank Drs. A. Provenzano and F. Impallari for technical support.

Part of this work was carried out at the Biomolecular Labeling Laboratory-National Institute of Standards and Technology (NIST), Rockville, USA. FTR, ECdeM, and AJLM thank Zvi Kelman for his hospitality and stimulating discussions.

FTR was supported by NSF USA grants EAR0747394, EAR0747412, and MCB0605301; and Air Force Office of Scientific Research under Grants AFOSR 03-S-28900; and 496 9550-10-1-0272 (IKL and FTR). IKL was supported in part by the National Science Foundation under Grant No. CHE-1152752 (I.K.L.). AJLM and ECdeM were partially supported by IMET. This work was done under the agreement between IEMEST (Italy) and IMET (USA) (IMET contribution number IMET 18-003).

References

1. Macario AJL, Conway de Macario E (2005) Sick chaperones, cellular stress and disease. N Engl J Med 353:1489–1501
2. Macario AJL, Conway de Macario E (2016) The chaperoning and the immune systems with the microbiome integrate a matrix that supports health: when one of them is disturbed the others

suffer and disease ensues. Life Safety Security (LiSS) 4:101–123 http://www.iemest.eu/life-safety-and-security/images/Doc/ARTICOLI/2016/macario_24/Macario04.pdf

3. Bouhouche A, Benomar A, Bouslam N, Chkili T, Yahyaoui M (2006a) Mutation in the epsilon subunit of the cytosolic chaperonin-containing t-complex peptide-1 (Cct5) gene causes autosomal recessive mutilating sensory neuropathy with spastic paraplegia. J Med Genet 43(5):441–443
4. Bouhouche A, Benomar A, Bouslam N, Ouazzani R, Chkili T, Yahyaoui M (2006b) Autosomal recessive mutilating sensory neuropathy with spastic paraplegia maps to chromosome 5p15.31-14.1. Eur J Hum Genet 14(2):249–252
5. Conway de Macario E, Robb FT, Macario AJL (2017) Prokaryotic chaperonins as experimental models for elucidating structure-function abnormalities of human pathogenic mutant counterparts. Front Mol Biosci 3:84. https://doi.org/10.3389/fmolb.2016.00084 https://www.ncbi.nlm.nih.gov/pubmed/28119916 http://journal.frontiersin.org/article/10.3389/fmolb.2016.00084/full
6. Privalov PL, Potekhin SA (1986) Scanning microcalorimetry in studying temperature-induced changes in proteins. In: Hirs CHW, Timasheff SN (eds) Enzyme structure part L. methods enzymology, vol 131. Academic Press, San Diego, pp 4–51
7. Sturtevant JM (1987) Biochemical applications of differential scanning calorimetry. Ann Rev Phys Chem 38:463–488. https://doi.org/10.1146/annurev.pc.38.100187.002335
8. Freire E (1995) Thermal denaturation methods in the study of protein folding. In: Abelson J, Simon M, Johnson M, Ackers G (eds) Methods in enzymology, Energetics of biological macromolecules, vol 259. Academic Press, San Diego, pp 144–168
9. Privalov G, Kavina V, Freire E, Privalov PL (1995) Precise scanning calorimeter for studying thermal properties of biological macromolecules in dilute solution1. Anal Chem 232:79–852. https://doi.org/10.1006/abio.1995.9957
10. Freire E, Mayorga OL, Straume M (1990) Isothermal titration calorimetry. Anal Chem 62:950–959. https://doi.org/10.1021/ac00217a002
11. Velazquez-Campoy A, Leavitt SA, Freire E (2015) Characterization of protein-protein interactions by isothermal titration calorimetry. In: Meyerkord CL, Fu H (eds) Methods in molecular biology, Protein-protein interactions: methods and applications, vol 1278, 2nd edn. Springer, New York, pp 183–204
12. Loh W, Brinatti C, Tam KC (2016) Use of isothermal titration calorimetry to study surfactant aggregation in colloidal systems. BBA-Gen Subjects 1860:999–1016. https://doi.org/10.1016/j.bbagen.2015.10.003
13. Min W, Angileri F, Luo H, Lauria A, Shanmugasundaram M, Almerico AM, Cappello F, Conway de Macario E, Lednev IK, Macario AJL, Robb FT (2014) A human CCT5 gene mutation causing distal neuropathy impairs hexadecamer assembly in an archaeal model. Sci Rep 4:6688. https://doi.org/10.1038/srep06688
14. Luo H, Robb FT (2011) A modulator domain controlling thermal stability in the Group II chaperonins of Archaea. Arch Biochem Biophys 512:111–118. https://doi.org/10.1016/j.abb.2011.04.017
15. Serpell LC (2000) Alzheimer's amyloid fibrils: structure and assembly. Biochim Biophys Acta 1502(1):16–30
16. Conway KA, Lee SJ, Rochet JC, Ding TT, Williamson RE, Lansbury PT (2000) Acceleration of oligomerization, not fibrillization, is a shared property of both α-synuclein mutations linked to early-onset Parkinson's disease: Implications for pathogenesis and therapy. Proc Natl Acad Sci U S A 97(2):571–576
17. Arnaudov LN, de Vries R (2005) Thermally induced fibrillar aggregation of hen egg white lysozyme. Biophys J 88(1):515–526 https://doi.org/10.1529/biophysj.104.048819
18. Uversky V, Fink A (2004) Conformational constraints for amyloid fibrillation: the importance of being unfolded. Biochim Biophys Acta 1698(2):131–153. https://doi.org/10.1016/j.bbapap.2003.12.008
19. Adamcik J, Mezzenga R (2012) Study of amyloid fibrils via atomic force microscopy. Curr Opin Colloid Interface Sci 17(6):369–376 https://doi.org/10.1016/j.cocis.2012.08.001
20. Rosario-Alomar MF, Quiñones-Ruiz T, Kurouski D, Sereda V, Ferreira EB, Jesús-Kim LD, Lednev IK (2015) Hydrogen sulfide inhibits amyloid formation. J Phys Chem B 119(4):1265–1274 https://doi.org/10.1021/jp508471v
21. Quiñones-Ruiz T, Rosario-Alomar MF, Ruiz-Esteves K, Shanmugasundaram M, Grigoryants V, Scholes C, López-Garriga J, Lednev IK (2017) Purple fibrils: a new type of protein chromophore. J Am Chem Soc 139(29):9755–9758 https://doi.org/10.1021/jacs.7b03056
22. Kurouski D, Dukor RK, Lu X, Nafie LA, Lednev IK (2012) Spontaneous inter-

conversion of insulin fibril chirality. Chem Comm 48(23):2837–2839

23. Gasteiger E, Hoogland C, Gattiker A, Duvaud S, Wilkins MR, Appel RD, Bairoch A (2005) Protein identification and analysis tools on the ExPASy server. In: Walker JM (ed) The proteomics protocols handbook. Humana Press Inc, Totowa, NJ, pp 571–607

24. Spigolon D, Gallagher DT, Velazquez-Campoy A, Bulone D, Narang J, San Biagio PL, Cappello F, Macario AJL, Conway de Macario E, Robb FT (2017) Quantitative analysis of the impact of a human pathogenic mutation on the CCT5 chaperonin subunit using a proxy archaeal ortholog. Biochem Biophys Rep 12:66–71, 2017. https://doi.org/10.1016/j.bbrep.2017.07.011

25. Geladopoulos TP, Sotiroudis TG, Evangelopoulos AE (1991) A malachite green colorimetric assay for protein phosphatase activity. Anal Biochem 192:112–116 PMID: 1646572

26. Kurouski D, Lauro W, Lednev IK (2010) Amyloid fibrils are "alive": spontaneous refolding from one polymorph to another. Chem Commun 46:4249–4251. https://doi.org/10.1039/B926758A

Chapter 6

High-Throughput Microplate-Based Fluorescence Assays for Studying Stochastic Aggregation of Superoxide Dismutase-1

Alireza Abdolvahabi, Sanaz Rasouli, Corbin M. Croom, and Devon L. Plewman

Abstract

Investigating in vitro kinetics of superoxide dismutase-1 (SOD1) aggregation with high-throughput microplate-based assays provides valuable information regarding SOD1 pathogenesis in amyotrophic lateral sclerosis (ALS) and opens venues for the development of effective therapies. In this chapter, we first explain the step-by-step purification and demetallation of wild-type (WT) and ALS-variant SOD1 proteins from *Saccharomyces cerevisiae* (baker's yeast). We then describe the methodology for a microplate-based fluorescence assay that is used to study real-time kinetics of metal-free (apo)-SOD1 aggregation. This technique is highly sensitive, semiautomated, requires minimum modifications to protein, and produces a plethora of data in a short period of time. We also describe a new approach for extracting clinically relevant information from SOD1 aggregation data using Kaplan–Meier estimators.

Key words Cu, Zn superoxide dismutase, Fibrillization, Protein purification, Demetallation, Amyloid, Fluorescence spectroscopy, Amyotrophic lateral sclerosis, Thioflavin-T

1 Introduction

Over the past decade, high-throughput microplate-based assays have become the method of choice for studying kinetics of aggregation of amyloidogenic proteins (in vitro) including, but not limited to amyloid-β (Aβ), α-synuclein, and SOD1 [1–4], owing to their ease of use, comparatively low cost of materials, and high yield of data. These assays have the flexibility to be coupled with various detection techniques such as fluorescence spectroscopy, absorption spectroscopy, and light scattering, which provides a broad range of applicability in the areas of neurodegeneration and drug discovery. Specifically, this unique feature of microplate-based assays facilitates drug screening studies where a large library of small molecules can be tested with high efficiency in data production [5].

Cláudio M. Gomes (ed.), *Protein Misfolding Diseases: Methods and Protocols*, Methods in Molecular Biology, vol. 1873, https://doi.org/10.1007/978-1-4939-8820-4_6,

Because protein aggregation is an intrinsically stochastic process in vitro and in vivo [6, 7], performing a large number of replicate assays ($n > 30$) is necessary for extracting statistically significant kinetic parameters and drawing flawless conclusions from protein aggregation experiments. Utilizing high-throughput kinetic assays is crucial for achieving this level of data reproducibility. For example, by utilizing high-throughput microplate-based thioflavin-T (ThT) fluorescence aggregation assays (a total of 825 replicate assays), our group has recently shown that the rate of aggregation of some ALS-variants of apo-SOD1 is significantly lower than rate of aggregation of WT apo-SOD1 [3], which is in contrast to other studies that conclude the rate of aggregation of SOD1 is uniformly increased by ALS variants [7, 8]. However, the results from the latter suffer from sufficient reproducibility for all studied SOD1 proteins ($n = 3$ in one study, and $n = 6–196$ in another), which might affect the robustness of some of their conclusions. This lack of reliability of conclusions in the absence of high data reproducibility further emphasizes the importance of high-throughput assays for studying protein aggregation.

In this chapter, we first outline a detailed protocol for expression, purification, and demetallation of SOD1 from *S. cerevisiae*, because producing pure SOD1 protein is a crucial step for studying kinetics of SOD1 aggregation. We then lay out a systematic and detailed procedure for implementing high-throughput microplate-based fluorescence assays for studying kinetics of aggregation of WT and ALS-variant apo-SOD1 proteins, which are causally linked to the pathogenesis of sporadic and familial cases of ALS. Finally, we present two different approaches for analyzing kinetic data of SOD1 aggregation.

2 Materials

All solutions should be prepared with ultrapure, metal-free water (conductivity = 18.2 MΩ.cm at 25 °C). Filter all buffers using vacuum filtration to remove undissolved particles. All buffer solutions should be stored at 4 °C. Store all other materials at −20 °C, unless stated otherwise. Follow waste disposal regulations for disposing of each solution.

2.1 SOD1 Expression and Purification

1. SD Medium-Leu yeast culture media: pH 5.8. Dissolve 26.7 g Yeast Base and 0.69 g SD Medium-Leu supplement in 1 L of ultrapure water. Adjust pH with HCl.
2. 50 mg/mL yeast extract peptone dextrose (YPD) media, pH 6.8. Prepare 27 L in ultrapure water. This can be done in

eighteen Fernbach flasks, each containing 1.5 L of YPD media (*see* **Note 1**).

3. Three blenders: 2 L blenders are preferred.
4. Lysis buffer: 250 mM Tris base, 150 mM NaCl, 100 μM EDTA, pH 8.0. Dissolve 121.1 g Tris base, 35 g NaCl, and 0.15 g EDTA disodium salt in ultrapure water and bring the total volume to 4 L (*see* **Note 2**).
5. 1 M dithiothreitol (DTT). Prepare 10 mL in lysis buffer.
6. Deoxyribonuclease (DNAse) enzyme: 264 μg/mL (≥5000 U/mg).
7. Glass beads: 0.5 mm in diameter.
8. Magnetic stir plate and stir bars.
9. Ammonium sulfate ($(NH_4)_2SO_4$).
10. Packing material for hydrophobic interaction chromatography (HIC) column: Phenyl Sepharose 6 Fast Flow (High Sub) material, V_t = 300 mL.
11. Automated fraction collector.
12. Buffer A: 150 mM NaCl, 50 mM sodium phosphate, 25 mM EDTA, 400 μM $(NH_4)_2SO_4$, pH 7.4. Mix 150 mL of 3 M NaCl solution, 150 mL of 1 M sodium phosphate solution pH 7.0, 150 mL of 0.5 M disodium EDTA solution pH 8.0, and 0.6 mL of 2 M $(NH_4)_2SO_4$ solution, and bring up to 3 L with ultrapure water.
13. Buffer B: 150 mM NaCl, 50 mM sodium phosphate, 25 mM EDTA, pH 7.4. Mix 150 mL of 3 M NaCl solution, 150 mL of 1 M sodium phosphate solution, pH 7.0, and 150 mL of 0.5 M EDTA solution, pH 8.0, and bring up to 3 L with ultrapure water.
14. 180 mL Amicon® concentrator: 10-kDa cutoff filter.
15. Packing material for ion-exchange column: Diethylaminoethyl (DEAE) cellulose, V_t = 250 mL.
16. 1 M potassium phosphate buffer, pH 7. Make 1 L in ultrapure water.
17. 2.25 mM potassium phosphate buffer, pH 7. Make 1 L in ultrapure water.
18. Packing material for size-exclusion chromatography (SEC) column: Sephadex G-75, V_t = 600 mL.
19. SEC buffer: 20 mM KH_2PO_4, 40 mM K_2HPO_4, 200 mM NaCl, pH 7. Dissolve 28.8 g KH_2PO_4, 65.8 g K_2HPO_4, and 116.8 g NaCl in ultrapure water and bring the total volume to 10 L.

2.2 Demetallation

1. Demetallation buffer 1: 100 mM sodium acetate, 10 mM EDTA, 45 mM acetic acid, pH 3.8. Dissolve 164 g sodium acetate, 74.5 g EDTA disodium salt, and 52 mL glacial acetic acid in ultrapure water and bring the total volume to 20 L.
2. Demetallation buffer 2: 100 mM sodium acetate, 10 mM NaCl, 45 mM acetic acid, pH 3.8. Dissolve 164 g sodium acetate, 11.7 g NaCl, and 52 mL glacial acetic acid in ultrapure water and bring the total volume to 20 L.
3. Demetallation buffer 3: 100 mM sodium acetate, 15 mM acetic acid, pH 5.5. Dissolve 164 g sodium acetate and 17.2 mL glacial acetic acid in ultrapure water and bring the total volume to 20 L.
4. Spin-X® UF centrifugal concentrators: 10-kDa molecular weight cutoff.
5. Dialysis tubing and clamps.

2.3 SOD1 Aggregation Assay

1. Tris(2-carboxyethyl)phosphine (TCEP).
2. ThT and 1-anilino-8-naphthalenesulfonate (ANS) fluorescent dyes.
3. Aggregation assay buffer 1: 10 mM potassium phosphate, 5 mM EDTA, pH 7.4. Make 1 L using ultrapure water.
4. Aggregation assay buffer 2: 10 mM potassium phosphate, 5 mM EDTA, 10 mM TCEP, 20 μM ThT or ANS, pH 7.4. Prepare 1 L using ultrapure water (*see* **Notes 3** and **4**).
5. Purified SOD1 protein.
6. Acrodisc® syringe filters: 200 nm cutoff.
7. Flat-bottom, nonbinding, black polystyrene 96-well microplates.
8. Transparent polypropylene plate seals.
9. Extreme-temperature slippery PTFE Teflon® beads: 1/8″ (3.13 mm) in diameter (McMaster-Carr, Elmhurst, IL, USA).
10. Fluoroskan® Ascent FL Microplate Fluorometer.

3 Methods

3.1 Expression, Purification, and Demetallation of SOD1

3.1.1 Growing the Yeast Cell Culture

1. Make 1 L of SD Medium-Leu yeast culture media the day before purification and sterilize in an autoclave for 15 min.
2. Inoculate 50 μL of *S. cerevisiae* cell stock transfected with human SOD1 plasmid (that carries overexpressed human-SOD1 gene and is a knockout for yeast SOD1) into 3 mL of SD Medium-Leu in a 15-mL falcon tube. Follow regular sterile procedures.

3. Grow the yeast culture with continuous shaking at 200 rpm (30 °C) for 36 h, or until the optical density (OD_{600}) of the culture reaches ~1.5 a.u.
4. From the grown yeast cell culture, inoculate 1–1.5 mL into 150 mL of sterilized SD Medium-Leu (i.e., secondary yeast culture). Follow regular sterile procedures.
5. Grow this secondary culture with continuous shaking at 200 rpm (30 °C) for 36 h.
6. Sterilize 27 L of YPD media (*see* **Note 1**).
7. Aliquot ~12 mL of the secondary SD Medium-Leu yeast culture into each flask containing YPD medium. Incubate in a shaker at 200 rpm (30 °C) for ~7 days. Check the OD_{600} of each culture every day. Cultures should reach an $OD_{600} \approx 1.5$ at day 7 (*see* **Note 5**).
8. Centrifuge all solutions at 2600 × *g* (4 °C) for 20 min, discard the supernatant, and store the pellets at −80 °C.

3.1.2 Purification

1. Thaw the yeast cell pellets at room temperature approximately 4 h before starting the purification process. Prechill three beakers (2 L each) and three blenders at 4 °C.
2. Add chilled lysis buffer to cell pellets (2:1 v/v) in centrifuge bottles to loosen and break up the pellets (*see* **Note 6**). Once the cell pellets are roughly broken apart, pour the contents of the centrifuge bottles into a chilled, 2 L beaker. Immediately add 1 mL of 1 M DTT solution ($[DTT]_{final}$ = 1 mM) and 10 μL of DNase enzyme (per 1 L of total volume) to the suspension and mix thoroughly until the solution is homogenized.
3. Evenly aliquot the pellet/lysis buffer suspension into three prechilled blenders and add 0.5 mm glass beads (2:1 v/v) to each blender. Blend for 2 min at medium speed followed by cooling blenders for 2 min in ice. Repeat this cycle 12 times. After the sixth cycle, chill the suspension for 10 min, and then continue blending.
4. Upon completion of blending cycles, allow glass beads to settle to the bottom of the blenders, then carefully decant the solution into a beaker that is placed on ice. Ensure that foaming is minimized during decanting and that all glass beads remain in the blenders while pouring the homogenate into the beaker.
5. Centrifuge the suspension at 8200 × *g* (4 °C) for 45 min to remove cellular debris from the lysed pellets. Pour the supernatant into a prechilled, clean beaker placed in an ice bath.
6. Place the entire ice bath on a stir plate, add a clean magnetic stir bar, and begin stirring the supernatant at medium speed. Slowly add ammonium sulfate ($[(NH_4)_2SO_4]_{final}$ = 60% w/v)

to supernatant over a time period of 30 min. This is to prevent the formation of high local concentrations of $(NH_4)_2SO_4$ salt that would precipitate (i.e., salt out) SOD1. After all $(NH_4)_2SO_4$ is added, continue stirring for an additional 30 min. Prevent foaming during mixing and/or addition of $(NH_4)_2SO_4$.

7. Centrifuge the mixture at 15,500 × *g* (4 °C) for 30 min, pour the supernatant into a chilled, clean beaker, and add lysis buffer (20% v/v) to the supernatant.
8. Load the mixture onto a prepacked HIC column at a flow rate of 2 mL/min (~ one droplet per 5 s). As a rule of thumb, loading should take approximately 10 h. After the completion of loading, elute the sample from the column with ~200 mL buffer A, followed by elution with a gradient of buffers A/B, as shown in Fig. 1. Program the fraction collector to collect 60 fractions at 7 min (~20 mL) per fraction.
9. Perform SDS-PAGE on all collected fractions from HIC column. Determine the fractions that contain SOD1 protein (~16 kDa) and combine them in a 180 mL Amicon® concentrator, equipped with a 10-kDa cutoff filter. Concentrate to ~30 mL and exchange buffer into 2.25 mM potassium phosphate, pH 7 (with continuous and gentle stirring at 4 °C) using house nitrogen (~20 psi) to help push the solution through the filter. Repeat this cycle at least three times to ensure that the buffer exchange is completed.
10. Load the content of Amicon® concentrator onto a prepacked DEAE cellulose ion-exchange column. Load the solution slowly and evenly on top of DEAE column to avoid disturbing

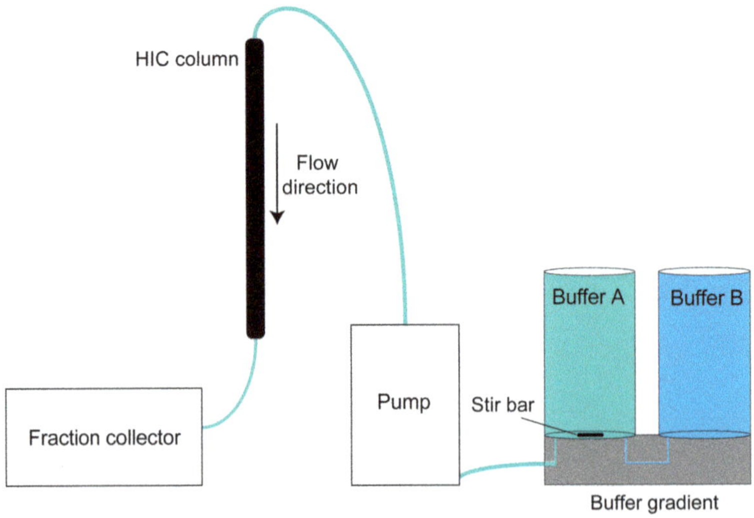

Fig. 1 Layout of the experimental setup for running the HIC column

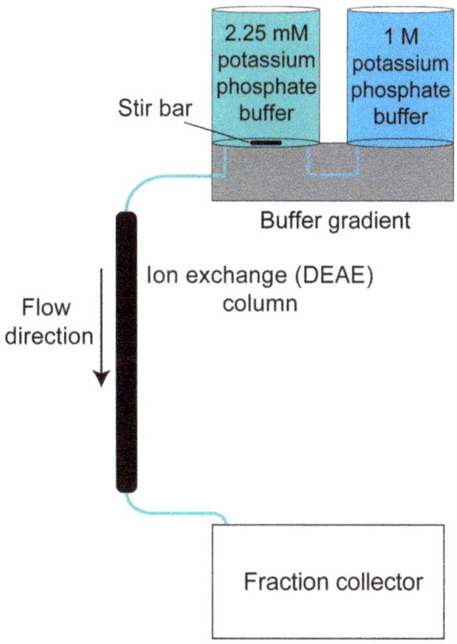

Fig. 2 Layout of the experimental setup for running DEAE cellulose ion-exchange column. Note that buffer containers are placed higher than the column to facilitate the gravity-driven flow of buffers onto the column

the surface of packed media. Setup the gradient as shown in Fig. 2, and elute the sample from the column with 1 M and 2.25 mM potassium phosphate buffers, pH 7, at a flow rate of 3–4 mL/min (Fig. 2). Program the fraction collector to collect ~41 fractions at 4 min per fraction.

11. Prepack a G-75 SEC column. Determine the fractions from DEAE column that contain SOD1 protein (from SDS-PAGE analysis) and combine them in a 180 mL Amicon® concentrator, equipped with a 10-kDa cutoff filter. Concentrate and exchange with SEC buffer. Repeat this cycle at least three times to ensure that buffer exchange is completed. Load ~25 mL of concentrated protein solution onto SEC column and elute with ~800 mL SEC buffer. Program the fraction collector to collect ~60 fractions (8–10 mL per fraction).
12. Combine the fractions and test for the presence of SOD1 protein using SDS-PAGE. After this step, SOD1 should be ~99% pure (*see* **Note 7**).

3.1.3 Demetallation

1. Combine all fractions from SEC column that contain purified SOD1 protein (freshly purified SOD1 is historically called "as-isolated" SOD1 [9], which is partially mismetallated [10]), and aliquot into precut dialysis tubing. Clamp dialysis tubing, place them in a 3 L plastic beaker filled with demetallation

buffer 1, and dialyze against demetallation buffer 1 with gentle stirring (*see* **Note 8**). Cover the beaker with a plastic wrap. Discard demetallation buffer 1 and add fresh buffer every 6–8 h. Repeat this procedure at least six times (*see* **Note 9**).

2. Exchange demetallation buffer 1 with demetallation buffer 2 to remove EDTA. Discard demetallation buffer 2 and add fresh buffer every 6–8 h. Repeat this procedure at least six times.
3. Exchange demetallation buffer 2 with demetallation buffer 3 to drive the refolding of the protein into its native state. Discard demetallation buffer 3 and add fresh buffer every 6–8 h. Repeat this procedure at least six times.
4. After the last step of demetallation buffer exchange, concentrate the protein via centrifugation using 10-kDa cutoff Spin-X® UF centrifugal concentrator tubes (~3200 × *g* at 4 °C; *see* **Note 10**). Determine the content of Cu^{2+} and Zn^{2+} metal ions in purified SOD1 protein using inductively coupled plasma mass spectrometry (ICP-MS) (*see* **Notes 11** and **12**). Determine the concentration of apo-SOD1 solutions with UV-vis spectroscopy (ε_{280nm} = 10,800 M^{-1} cm^{-1} per SOD1 dimer; *see* **Note 13**), aliquot into nonbinding, metal-free 1.5 mL Eppendorf tubes (~300 μL), flash-freeze the aliquots in liquid nitrogen, and store at −80 °C for further use.

3.2 High-Throughput Fluorescence Assays for Studying Kinetics of SOD1 Aggregation

3.2.1 Preparation of SOD1 Solution

1. Take out desired number of tubes containing flash-frozen apo-SOD1 proteins from −80 °C freezer (which are stored in demetallation buffer 3) and flash-thaw under cold running water. Keep all protein solutions on ice throughout the assay preparation. The required number of apo-SOD1 tubes depends on the desired number of replicate assays. As a rule of thumb, one should thaw 7–8 tubes for performing a typical SOD1 aggregation assay ($[\text{apo-SOD1}]_{\text{aggregation assay}}$ = 30 μM, per dimeric protein) with 60 replicates in a 96-well plate (assuming that concentration of apo-SOD1 is 100–200 μM in each tube, and each tube contains ~300 μL apo-SOD1 solution).
2. Pipette all apo-SOD1 solutions into a 20 mL Spin-X® UF centrifugal concentrator tube with 10-kDa cutoff. Transfer of protein should be as gentle as possible to minimize foaming, which is a sign of protein denaturation.
3. Fill the concentrator tube with cold aggregation assay buffer 1. Centrifuge at 4000 × *g* (4 °C) until the volume of solution reaches ~2 mL. Discard the effluent and rinse the effluent tube with ultrapure water. Add fresh aggregation assay buffer 1 to the concentrated protein solution, gently mix with a pipettor, and centrifuge at 4000 × *g*. Repeat this washing/concentration cycle 5–6 times to ensure that the buffer exchange is completed.

Keep the dilution/concentration factor constant for each washing cycle.

4. To remove any possible preformed protein oligomers/aggregates, apo-SOD1 solution must be filtered. Attach a 200-nm low-protein binding Acrodisc® syringe filter to a 6 mL sterile syringe, take the pump out, pour the protein solution inside the syringe, and slowly insert the pump back in. Gently push the pump down to filter the protein solution into a prechilled, nonbinding falcon tube. Store the filtered protein on ice.
5. Measure the concentration of apo-SOD1 using UV-vis spectroscopy (*see* **Note 14**). Dilute apo-SOD1 protein in aggregation assay buffer 1 to achieve an absorbance between 0.3 and 1.0.

3.2.2 Initiating the Aggregation Assay

1. Add aggregation assay buffer 2 to a nonbinding falcon tube (15 mL or 45 mL tube, depending on the total assay volume; however, we recommend using 45 mL tubes to facilitate mixing the solution).
2. Add apo-SOD1 solution ($[\text{apo-SOD1}]_{\text{final}}$ = 30 μM, per dimeric protein) to aggregation assay buffer 2 and mix gently. Do NOT vortex or sonicate the solution, and prevent solution from foaming.
3. Incubate apo-SOD1 with aggregation assay buffer 2 for 30 min at room temperature without agitation to achieve ~5% disulfide reduction (*see* **Note 15**).
4. Thoroughly flush (with house nitrogen) a flat-bottom, black polystyrene 96-well microplate to remove dust particles. Place a 1/8" Teflon® bead inside each well. Do not use wells on the edge of the microplate (i.e., wells on rows A and H, and wells on columns 1 and 12), because of high evaporation rate (*see* **Note 16**).
5. Dispense 200 μL aliquots of the "mother stock" (i.e., apo-SOD1/aggregation assay buffer 2) into each well of microplate (Fig. 3). Prior to dispensing the solution into microplate, make sure that Teflon® beads rotate freely in each well when slowly gyrating the microplate. The total volume of solution in each well should be constant: different solution volumes can largely affect kinetics of protein aggregation due to variations in air–water interface [11].
6. Seal the microplate carefully using a transparent polypropylene plate seal. Prewarm the seal to avoid condensation.
7. Place the well in a Fluoroskan® fluorescence plate-reader equipped with thermal controller and shaker. Turn on the plate-reader at least 30 min before initiating the assay to allow the instrument to warm up and equilibrate at 37 °C. Set the parameters for the aggregation assay as desired, depending on the nature of the experiment (*see* **Note 17**).

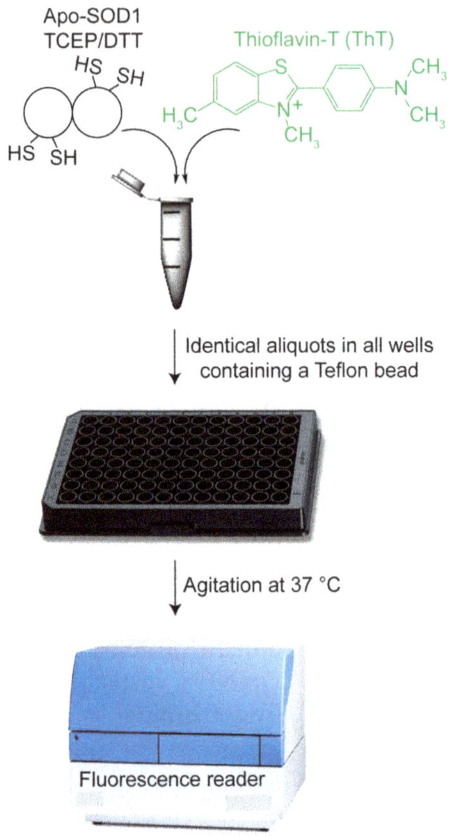

Fig. 3 Schematic showing the setup for microplate-based ThT fluorescence assays for studying kinetics of apo-SOD1 aggregation. ANS can also be used for studying protein aggregation kinetics instead of ThT

3.3 Data Analysis

The main readout from apo-SOD1 aggregation assay is a sigmoidal plot of ThT fluorescence vs. time, which demonstrates a three-step mechanism for apo-SOD1 aggregation (Fig. 4). ANS fluorescence can also be used to study kinetics of SOD1 aggregation (*see* **Note 18**). The nucleation step (also known as "primary nucleation" [12]) involves the formation of a "critical nucleus" composed of a certain number of monomeric SOD1 polypeptides [13]. The elongation (propagation) phase involves the formation of oligomeric protofibrils upon the addition of monomeric protein to the critical nucleus. It has been proposed that the principle mechanism underlying the elongation of protein oligomers/protofibrils is the continuous and dynamic fragmentation of oligomers into smaller species that can seed the oligomerization of free monomeric polypeptides (i.e., "secondary nucleation") [12]. The termination step is accompanied with the formation of mature, micrometer-sized SOD1 fibrils that have reached the thermodynamic equilibrium (the plateau of the sigmoid in Fig. 4).

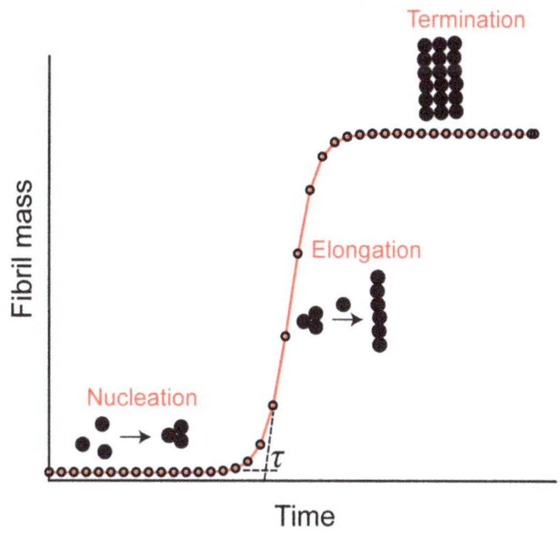

Fig. 4 Output of a typical fluorescence-based apo-SOD1 aggregation assay, depicting three stages of apo-SOD1 fibrillization. The time required for the formation of "critical nucleus" is known as fibrillization lag time (τ). Each black circle illustrates a monomeric apo-SOD1 polypeptide. In most cases, fibril mass correlates with ThT/ANS fluorescence intensity; however, several factors can break this correlation

Data from SOD1 aggregation assays can be analyzed in two ways: (1) classical method, that produces rate of fibril nucleation and elongation, and maximal ThT/ANS fluorescence; and (2) Kaplan–Meier estimators, that calculate the probability of apo-SOD1 fibrillization in each well and produce a single metric (namely Hazard ratio, see below) that can be used to correlate in vitro kinetics of apo-SOD1 aggregation with clinical data from ALS patients (see ref. 3 for more details). Below, we have described the procedures for performing both methods.

3.3.1 Extracting Classical Kinetic Parameters of Apo-SOD1 Aggregation from Longitudinal ThT/ANS Fluorescence Data

1. Import the raw kinetic data (ThT/ANS fluorescence vs. time) into graphing software such as SigmaPlot® or GraphPad Prism®.
2. Fit each set of replicate data (i.e., each sigmoidal trace of ThT/ANS fluorescence vs. time) with Eq. 1 (*see* **Note 19**):

$$F = F_0 + \frac{F_{\max}}{1 + e^{(t - t_0)k}} \quad (1)$$

In Eq. 1, F is ThT/ANS fluorescence intensity at time t; $F_{\max}$ and F_0 are minimal and maximal ThT/ANS fluorescence intensities, respectively; t_0 is the inflection point of sigmoid ($F = ½\, F_{\max}$); and k is the rate of fibril elongation. The rate of fibril nucleation (i.e., lag time, τ; *see* **Note 20**) can then be calculated as $\tau = t_0 - 2\,k$.

3. Collect parameters τ, k, and F_{max} for all replicate aggregation assays for each apo-SOD1 protein (e.g., WT vs. ALS-variant SOD1) and apply appropriate statistical analyses (*see* **Note 21**).

3.3.2 Kaplan–Meier Method for Calculating Probability of Apo-SOD1 Fibrillization

1. Populate all t_0 values obtained in **step 1** (*see* above) for each apo-SOD1 variant in GraphPad Prism® software (under "Survival Curve" mode, in the first column), and assign "1" to all t_0 values (in the second column).
2. Count the number of replicate assays that did not result in any ThT/ANS fluorescence throughout the course of the assay (*see* **Note 22**), and list them below t_0 values with a "0" assigned to each (*see* **Note 23**).
3. Repeat **step 2** for each apo-SOD1 protein, and generate the Kaplan–Meier survivability plots at the end of the process. Kaplan–Meier plots for aggregation of WT and some ALS-variant apo-SOD1 proteins are shown in Fig. 5.
4. Calculate the Hazard ratio for each ALS-variant SOD1 protein with respect to WT SOD1 protein. Values of Hazard ratio are interpreted as follows: HR > 1: higher probability of aggregation of ALS-variant SOD1 compared to WT protein. HR < 1: lower probability of aggregation of ALS-variant SOD1 compared to WT protein. HR = 1: similar aggregation probability in ALS-variant SOD1 compared to WT protein.

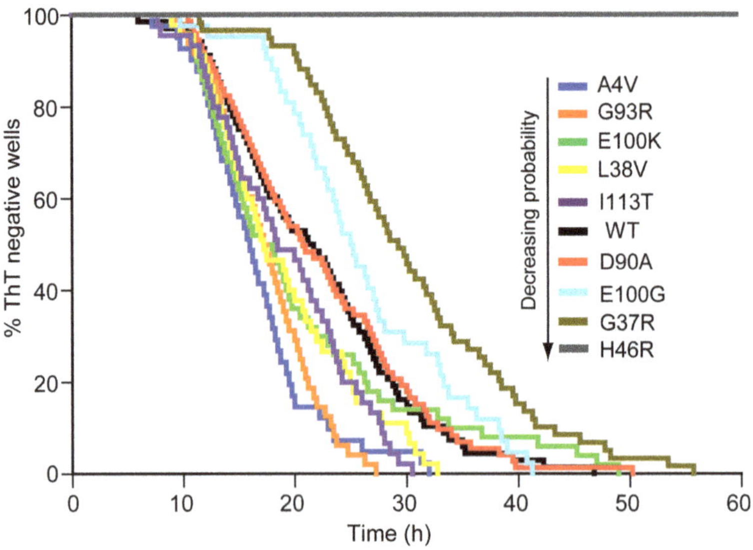

Fig. 5 Kaplan–Meier plots of aggregation of WT and 9 ALS-variant apo-SOD1 proteins. Note that each downward step in each plot corresponds to one aggregation event. The flat line in the case of H46R apo-SOD1 indicates that this apo-SOD1 variant did not aggregate during aggregation assay. Reprinted with permission from (Abdolvahabi et al. (2017) ACS Chem Neurosci 8:1378–1389). Copyright (2017) American Chemical Society

4 Notes

1. The amount of YPD media that is prepared depends on the required protein yield as well as the variant of SOD1 protein that is being purified (some ALS-variant SOD1 proteins, such as A4V SOD1, are very unstable and hard to purify). Starting with 27 L of YPD media for purification of WT SOD1 usually results in ~400–500 mg of protein.
2. Use disodium salt EDTA for all experiments in this chapter. Free-acid EDTA has a low solubility in water at pH 7.
3. Reduction of disulfide bonds in apo-SOD1 can be achieved using different reducing agents such as TCEP and DTT. The equivalent DTT concentration in aggregation assay buffer 2 is 100 mM. Since dissolving TCEP significantly decreases pH of the solution, prepare a 50 mM stock solution of TCEP in aggregation assay buffer 1 and adjust the pH to 7.4 with KOH, prior to addition to aggregation assay buffer 2. Filter TCEP stock solution with a 200-nm Acrodisc® syringe filter if necessary. Prepare fresh TCEP stock solution immediately before starting the aggregation assay, because TCEP is gradually oxidized over time. Stock solutions of TCEP can be stored at −20 °C for up to 1 month; however, we recommend that freshly made solutions be used.
4. Dissolve the fluorescent dye (ThT or ANS) in aggregation assay buffer 1 immediately prior to initiation of the assay. Stock concentration of fluorescent dye is 5 mM, and its final concentration for aggregation assay is 20 μM. Cover the tube containing fluorescent dye solution with aluminum foil to prevent light exposure. Filtering fluorescent dye solution with a 200-nm Acrodisc® syringe filter is also recommended.
5. To calculate the approximate cell density for each sample, assume that an OD_{600} of 1.0 corresponds to ~1.3×10^7 cells/mL.
6. The process of breaking up the pellets can be expedited by gently homogenizing the pellets with a clean, glass spatula.
7. At this stage, perform mass spectrometry on purified SOD1 proteins to ensure that the correct SOD1 variant is purified. Mass spectrometry can also determine the extent of posttranslational modifications (e.g., tryptophan oxidation) on purified SOD1 proteins.
8. All dialysis steps for SOD1 demetallation should be performed at 4 °C.
9. Place the plastic beaker on the lab bench, carefully take the plastic cover off and place it upside down on the bench. Place the dialysis tubing on the plastic cover, refill the beaker with fresh buffer, and put the dialysis tubes back in the beaker, followed by

covering with the plastic cover. This procedure is crucial for preventing exogenous metal ions from contaminating SOD1 proteins.

10. To ensure that no metal contaminants come into contact with SOD1 protein, centrifuge a 10 mM solution of disodium EDTA through the concentrator tubes, followed by three centrifugal runs with ultrapure water prior to concentrating the protein.
11. Demetallated SOD1 must contain <0.1 equivalents of Cu^{2+} and/or Zn^{2+} ions, per dimeric protein, following the demetallation procedure outlined in this Chapter. If, however, higher metal contents (>0.1 equivalents per dimer) were detected, the demetallation procedure should be repeated until the protein is fully demetallated. Presence of metal ions in SOD1 can affect the kinetics of SOD1 aggregation [14].
12. If excess equivalents of Cu^{2+} and/or Zn^{2+} ions remain bound to purified SOD1 proteins (after rerunning demetallation procedure, *see* **Note 11**), pack a 200 mL column with Chelex® 100 chelating ion-exchange resin, equilibrate the column with demetallation buffer 1, and run SOD1 protein solution through Chelex® ion-exchange column. This procedure will almost completely strip the protein off the metals.
13. To ensure that no DNA is present in purified apo-SOD1 solution, measure the UV-vis absorbance of purified apo-SOD1 proteins at 260 nm and 280 nm. The 260 nm:280 nm absorbance ratio should be <0.6 to ensure the absence of DNA fragments. If significantly higher ratios are observed, more DNase enzyme (10 μL of DNase stock per 1 L of total solution volume) should be added to purified apo-SOD1. We also recommend repeating DEAE ion-exchange chromatography step (followed by SEC chromatography) to ensure the complete removal of DNA fragments.
14. Warm up UV lamp for 30 min prior to measuring the protein concentration. Blank the instrument with prewarmed aggregation assay buffer 1. Do not use cold solutions for UV measurements since the condensation leads to blocking of light path which introduces error in measuring protein concentration.
15. If higher extents of disulfide reduction are desired, increase the incubation time of apo-SOD1 in aggregation assay buffer 2. For example, incubation of apo-SOD1 with 10 mM TCEP in aggregation assay buffer 2 for 6 h (with gentle agitation) results in ~30% disulfide reduction [6]. Accordingly, incubation of apo-SOD1 with 100 mM TCEP in aggregation assay buffer 2 for 6 h (with gentle agitation) results in ~90% disulfide reduction [6].
16. Gradual evaporation of aggregation assay solution can affect kinetics of apo-SOD1 aggregation via (1) increasing the con-

centration of apo-SOD1 which increases the rate of aggregation; and (2) increasing the air–water interface that might significantly affect kinetics of protein aggregation [11].

17. Program the Fluoroskan® to scan wells every 15 min (under "Kinetic" mode) for a total of 1000 scans, with intermittent shaking every 15 s (15 s OFF and 15 s ON). Set the shaking speed at 360 rpm (under "orbital shaking" mode) with a gyration diameter of 3 mm. The excitation/emission wavelength pairs for ThT and ANS are 444 nm/485 nm and 350 nm/500 nm, respectively.
18. The ANS fluorescent dye has been historically used to study the dynamics of protein folding, because it shows an increase in fluorescence upon binding to aromatic residues (i.e., tryptophan) on the surface of the protein [15]. ANS can also be utilized to study the kinetics of formation of amorphous protein aggregates (*see* refs. 6, 16 for more details). ANS is more sensitive to light-induced quenching than ThT. Therefore, ANS stock solution must be prepared in a dark room. This process is not necessary for preparing ThT solutions.
19. Occasionally, a drop in fluorescence signal is observed after the sigmoid has plateaued. This artifact occurs because SOD1 fibrils scatter emitted light (as they form and increase in size) and block it from reaching the detector. To fit these plots, simply omit the data points up to where the plateau of the sigmoid ends, followed by fitting with Eq. 1.
20. Some wells show bimodal aggregation kinetics that is characterized by two sequential sigmoidal traces. In this case, fit both sigmoidal traces separately, assign separate sets of kinetic parameters to each trace (e.g., τ_1 and τ_2), and report all extracted parameters.
21. Try to fit all sigmoidal traces, no matter how low the ThT/ANS fluorescence intensity might be. As long as a sigmoidal trace is detectable, it should be fitted and included in the analysis.
22. Occasionally, apo-SOD1 does not aggregate in some wells (stochastically) and no fluorescence can be detected. Do NOT discard data from these wells (i.e., ThT(−) wells [17]). Count the number of ThT(−) wells and report them along with the rest of the results.
23. We consider ThT(−) wells as "censured" subjects in our Kaplan–Meier analysis, and thus assign "0" to each.

Acknowledgment

The authors wish to dedicate this work to Bryan F. Shaw.

References

1. Guo JP, Yu S, McGeer PL (2010) Simple *in vitro* assays to identify amyloid-β aggregation blockers for Alzheimer's disease therapy. J Alzheimers Dis 19:1359–1370. https://doi.org/10.3233/JAD-2010-1331
2. Yedlapudi D, Joshi GS, Luo D, Todi SV, Dutta AK (2016) Inhibition of α-synuclein aggregation by multifunctional dopamine agonists assessed by a novel *in vitro* assay and an *in vivo* drosophila synucleinopathy model. Sci Rep 6:38510. https://doi.org/10.1038/srep38510
3. Abdolvahabi A, Shi Y, Rasouli S, Croom CM, Aliyan A, Marti AA, Shaw BF (2017) Kaplan-Meier meets chemical kinetics: intrinsic rate of SOD1 amyloidogenesis decreased by subset of ALS mutations and cannot fully explain age of disease onset. ACS Chem Neurosci 8:1378–1389. https://doi.org/10.1021/acschemneuro.7b00029
4. Abdolvahabi A, Shi Y, Rasouli S, Croom CM, Chuprin A, Shaw BF (2017) How do gyrating beads accelerate amyloid fibrillization? Biophys J 112:250–264. https://doi.org/10.1016/j.bpj.2016.12.004
5. Varma H, Lo DC, Stockwell BR (2008) High throughput screening for neurodegeneration and complex disease phenotypes. Comb Chem High Throughput Screen 11:238–248
6. Abdolvahabi A, Shi Y, Chuprin A, Rasouli S, Shaw BF (2016) Stochastic formation of fibrillar and amorphous superoxide dismutase oligomers linked to amyotrophic lateral sclerosis. ACS Chem Neurosci 7:799–810. https://doi.org/10.1021/acschemneuro.6b00048
7. Lang L, Zetterstrom P, Brannstrom T, Marklund SL, Danielsson J, Oliveberg M (2015) SOD1 aggregation in ALS mice shows simplistic test tube behavior. Proc Natl Acad Sci U S A 112:9878–9883. https://doi.org/10.1073/pnas.1503328112
8. Vassall KA, Stubbs HR, Primmer HA, Tong MS, Sullivan SM, Sobering R et al (2011) Decreased stability and increased formation of soluble aggregates by immature superoxide dismutase do not account for disease severity in ALS. Proc Natl Acad Sci U S A 108:2210–2215. https://doi.org/10.1073/pnas.0913021108
9. Hayward LJ, Rodriguez JA, Kim JW, Tiwari A, Goto JJ, Cabelli DE et al (2002) Decreased metallation and activity in subsets of mutant superoxide dismutases associated with familial amyotrophic lateral sclerosis. J Biol Chem 277:15923–15931. https://doi.org/10.1074/jbc.M112087200
10. Goto JJ, Zhu H, Sanchez RJ, Nersissian A, Gralla EB, Valentine JS, Cabelli DE (2000) Loss of *in vitro* metal ion binding specificity in mutant copper-zinc superoxide dismutases associated with familial amyotrophic lateral sclerosis. J Biol Chem 275:1007–1014. https://doi.org/10.1074/jbc.275.2.1007
11. Campioni S, Carret G, Jordens S, Nicoud L, Mezzenga R, Riek R (2014) The presence of an air-water interface affects formation and elongation of α-synuclein fibrils. J Am Chem Soc 136:2866–2875. https://doi.org/10.1021/ja412105t
12. Knowles TP, Waudby CA, Devlin GL, Cohen SI, Aguzzi A, Vendruscolo M et al (2009) An analytical solution to the kinetics of breakable filament assembly. Science 326:1533–1537. https://doi.org/10.1126/science.1178250
13. Knowles TP, Vendruscolo M, Dobson CM (2014) The amyloid state and its association with protein misfolding diseases. Nat Rev Mol Cell Biol 15:384–396. https://doi.org/10.1038/nrm3810
14. Oztug Durer ZA, Cohlberg JA, Dinh P, Padua S, Ehrenclou K, Downes S et al (2009) Loss of metal ions, disulfide reduction and mutations related to familial ALS promote formation of amyloid-like aggregates from superoxide dismutase. PLoS One 4:e5004. https://doi.org/10.1371/journal.pone.0005004
15. Royer CA (2006) Probing protein folding and conformational transitions with fluorescence. Chem Rev 106:1769–1784. https://doi.org/10.1021/cr0404390
16. Younan ND, Viles JH (2015) A comparison of three fluorophores for the detection of amyloid fibers and prefibrillar oligomeric assemblies. ThT (thioflavin T); ANS (1-anilinonaphthalene-8-sulfonic acid); and bisANS (4,4′-dianilino-1,1′-binaphthyl-5,5′-disulfonic acid). Biochemistry 54:4297–4306. https://doi.org/10.1021/acs.biochem.5b00309
17. Rasouli S, Abdolvahabi A, Croom CM, Plewman DL, Shi Y, Ayers JI, Shaw BF (2017) Lysine acylation in superoxide dismutase-1 electrostatically inhibits formation of fibrils with prion-like seeding. J Biol Chem 292:19366–19380. https://doi.org/10.1074/jbc.M117.805283

Chapter 7

Methods for Structural Analysis of Amyloid Fibrils in Misfolding Diseases

Devkee M. Vadukul, Youssra K. Al-Hilaly, and Louise C. Serpell

Abstract

Many proteins and peptides are able to self-assemble in solution in vitro and in vivo to form amyloid-like fibrils. These fibrils share common structural characteristics. In order for a fibril to be characterized as amyloid, it is expected to fit certain criteria including the composition of cross-β. Here we describe how the formation of amyloid fibrils can be characterized in vitro using a variety of methods including circular dichroism and intrinsic tyrosine/tryptophan fluoresence to follow conformational changes; Thioflavin and/or ThS assembly to monitor nucleation and growth; transmission electron microscopy to visualize fibrillar morphology and X-ray fiber diffraction to examine cross-β structure.

Key words Self-assembly, Cross-β, Protofilaments, Protofibril, Oligomer, Electron microscopy, Thioflavin T/S fluorescence, Tyrosine/tryptophan fluorescence, Circular dichroism, X-ray fiber diffraction

1 Introduction

Amyloid fibrils can be formed from a very large number of proteins and peptides. Polypeptide sequences may have the propensity to self-assemble to form highly organized β-sheet containing structures, in which the β-sheet core of the protofilaments is made up of β-strands that run perpendicular to the fiber axis and several β-sheets associate via side-chain interactions. Protofilaments can then further associate via surface exposed residue interactions and the resulting morphology of the fibrils is dependent on the chemical composition of these surface side chains.

Amyloid fibrils have been defined as pathological fibrous structures that are usually deposited in the extracellular spaces in the tissues [1]. Structurally, amyloid fibrils appear as long, unbranching fibrils by negative stain transmission electron microscopy (and atomic force microscopy) and show a major contribution of β-strand secondary structure by circular dichroism (and Fourier transform infrared spectroscopy) and a cross-β pattern by X-ray

Cláudio M. Gomes (ed.), *Protein Misfolding Diseases: Methods and Protocols*, Methods in Molecular Biology, vol. 1873, https://doi.org/10.1007/978-1-4939-8820-4_7, © Springer Science+Business Media, LLC, part of Springer Nature 2019

fiber diffraction. Traditionally, amyloid fibrils have been stained using Congo red and give a green birefringence under cross-polarized light [2]. Amyloid self-assembly can also be followed using Thioflavin T (ThT) fluorescence [3]. However, Congo red and ThT can also bind to other fibrillar structures and so these assays are not a definitive test for amyloid alone.

Many different peptides and proteins self-assemble to form structures that fit the above criteria and these are often referred to as amyloid-like fibrils. These include those pathological fibrils that are deposited intracellularly [1]. For the purposes of this article, we will refer to all fibrils that fit the "amyloid" structural criteria as amyloid.

During the assembly process, amyloidogenic proteins and peptides can (in many cases) form intermediate species. These small, spherical assemblies are referred to as oligomers and specific conformation antibodies have been used to define these oligomeric assemblies formed from many different proteins [4, 5]. The oligomers elongate to form protofibrils, which often form less ordered, curved and immature fibrils before maturing to form amyloid fibrils which are long and straight. Occasionally, fibrils may mature further to form fibrous crystals [6, 7].

In more recent studies, solid state NMR and X-ray crystallography, as well as cryo-electron microscopy have been utilized to yield highly detailed structural information about the composition of fibrils and their molecular architecture [7–11]. Atomic force microscopy is frequently used in place of electron microscopy to visualize amyloid morphology [12, 13]. The aim of this manuscript is to cover general methods for amyloid structural analysis and these more detailed methods are outside the scope of this manuscript.

2 Materials

All buffers should be passed through a 0.2 μm Millipore filter prior to use to reduce bacterial growth and remove contaminants. 0.01% Sodium Azide can be added to reduce any potential bacterial growth, particularly if incubation will be conducted at above room temperature. Low retention tips and eppendorf tubes should be used throughout preparation.

2.1 Prepreparation of Amyloidogenic Peptides for Assembly Assays (e.g., Aβ)

1. Aliquot of peptide of desired concentration.
2. 1,1,1,3,3,3-Hexafluoro-2-propanol (HFIP).
3. Vortex.
4. Water bath sonicator at maximal setting, kept at room temperature.

5. Sterile needle and syringe with 2 μm filter attached to nitrogen supply.
6. Precooled centrifuge (4 °C) with RAD 99 mm falcon tube rotor.
7. Zeba 7 K MWCO (ThermoScientific) 2 mL columns.
8. HEPES buffer pH 7.4: 10 mM HEPES, 50 mM NaCl, 1.6 mM KCl, 2 mM $MgCl_2$, 3.5 mM $CaCl_2$, or other suitable buffer (e.g., 20 mM phosphate buffer pH 7.4).
9. Anhydrous dimethyl sulphoxide (DMSO).
10. Eppendorf UVette (Sigma-Aldrich) cuvettes.
11. Spectrofluorometer.

2.2 Negative Stain Electron Microscopy

1. Formvar/carbon coated 400 mesh copper grids.
2. Whatman filter paper.
3. 0.2 μm filtered MilliQ water
4. 2% (w/v) Uranyl acetate.
5. Fine, antimagnetic tweezer.

2.3 Thioflavin T/S Fluorescence

1. 1 mM Thioflavin T (ThT) in water/Thioflavin S (ThS) in 20 mM 3-(N-morpholino)propanesulfonic acid (MOPS) buffer pH 6.8.
2. 20 mM MOPS buffer; dissolve 4.2 g MOPS in 1 L milli-Q filtered water.
3. 10 mm quartz cuvette.
4. Spectrofluorometer.

2.4 Tyrosine Fluorescence

1. 10 mm quartz cuvette.
2. Spectrofluorometer.

2.5 Circular Dichroism

1. Nitrogen prepurged circular dichroism spectropolorimeter.
2. Peptide/protein in appropriate buffer.
3. Quartz cuvette of appropriate path-length.

2.6 SDS-PAGE Electrophoresis and Immunoblotting

1. 4–20% gradient gels.
2. 4× Laemmlli sample buffer.
3. β-Mercaptoethanol (BME) or dithiothreitol (DTT),
4. 10× running buffer pH 8.5. Dilute to 1× before use: 0.025 M Tris, 0.192 M glycine, 0.1% sodium dodecyl sulphate (SDS).
5. 10× transfer buffer pH 8.5. Dilute to 1× before use: 0.025 M Tris, 0.192 M glycine.
6. 0.45 μm nitrocellulose membrane.

7. 10× Tris buffered saline (TBS) pH 7.4: 1.5 M NaCl, 0.1 M Tris–HCl. Dilute to 1× TBS before use.
8. Washing solution: TBS containing 0.1% tween (TBS-T).
9. Blocking solution and antibody dilutent: 10% milk in 0.1% TBST stored at 4°.
10. ECL substrate.
11. X-ray film.

2.7 X-Ray Fiber Diffraction

2.7.1 X-Ray Fiber Diffraction Sample Preparation

1. Protein/peptide fiber at high concentration in water or low-salt buffer.
2. Filtered Milli-Q water.
3. Refrigerated centrifuge.
4. Glass capillaries (Harvard apparatus ltd., Fircroft way, Edenbridge, Kent, UK), (1.5 mm OD × 1.17 ID borosilicate).
5. Candle wax.
6. Hot plate.
7. Diamond glass cutter pen.
8. Plasticine.
9. Petri dish.

2.7.2 X-Ray Fiber Diffraction Data Collection

1. Goniometer head (Hampton).
2. X-ray source and detector.

2.7.3 X-Ray Fiber Diffraction Data Analysis and Modeling

1. Computer workstation running Mosflm [14]/fit2d [15]/ FibreFix/CLEARER [16] (latest version of java).

2.8 Immunogold Labeling Negative Stain TEM

1. A modified phosphate-buffered saline, pH 8.2, containing 1% bovine serum albumin (BSA): 500 μL/L Tween-20, 10 mM Na EDTA, and 0.2 g/L NaN_3 (henceforward termed PBS+).
2. Formvar/carbon coated 400 mesh copper TEM support grids.
3. Filter paper.
4. 0.2 μm filtered MilliQ water.
5. 0.5% (w/v) uranyl acetate.
6. Petri dish.
7. Parafilm.
8. Fine, antimagnetic tweezer.
9. Blocking solution: Normal goat serum (1:10 dilution).
10. Primary antibody.
11. Suitable secondary antibody gold conjugated (5, 10, or 15 nm) antibody.

3 Methods

3.1 Prepreparation of Amyloidogenic Peptides for Assembly Assays (e.g., Aβ)

The method below is based on a method by Broerssen et al. [17] for preparing Aβ to ensure removal of preaggregated species and removal of solvents. This method can be used for other peptides, but some peptides/proteins may be more suitable for dissolution in water or buffer at the desired concentration. Other methodologies are available. For example *see* [18, 19].

Before beginning to prepare the peptide/protein, keep two low retention eppendorf tubes on ice, one with the lid cut off.

1. In order to disaggregate any preformed aggregates, add HFIP (*see* **Note 1**) to the peptide/protein to obtain the concentration needed (e.g., 1 mg/mL).
2. Vortex the solution for 1 min.
3. Sonicate for 5 min in a water bath sonicator. This will help dissolve the peptide as well as further disaggregate any preformed aggregates.
4. In a fume hood, dry off the HFIP with a low, steady flow of nitrogen using a sterile syringe and needle using a 0.2 μm filter until you can see a film of peptide (*see* **Note 2**).
5. Allow the peptide to continue to dry in a sterile hood with the lid off while equilibrating the buffer- exchange Zeba spin columns.
6. (*See* **Note 3**) Loosen the lid of the spin column, snap off tab at the bottom and place in a 15 mL falcon tube. Centrifuge at 1000 × *g* for 2 min at 4 °C and discard the sodium azide. Add 1 mL of buffer dropwise into the column and centrifuge again. Repeat three times, discarding the buffer after each spin (*see* **Note 4**).
7. Resuspend peptide in anhydrous DMSO using a long needle and syringe, inserting through the rubber bung of the cap (again, to desired concentration). Vortex for 1 min followed by a 1 min sonication.
8. Place the low retention, capless eppendorf at the end of the spin column and reinsert into the falcon tube. Add the peptide into the spin column with the buffer as stack and centrifuge.
9. Remove the peptide containing eppendorf from the bottom of the spin column and immediately keep on ice. This will slow down the assembly process and allow enough time to obtain an accurate concentration reading at time zero. Transfer the peptide into a new eppendorf.
10. Measure the A280 of the peptide using a spectrofluorometer and calculate the concentration using the molecular coefficient (concentration = absorbance/molecular coefficient), or use suitable protein concentration assay.

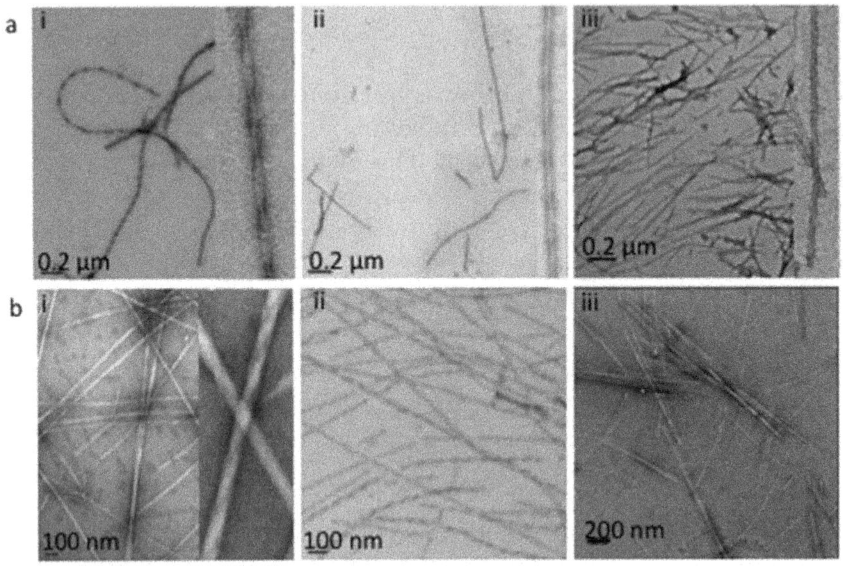

Fig. 1 Electron micrographs of amyloid fibrils. Amyloid fibrils formed from: (**a**) pathogenic amyloidogenic proteins (i) 20 μM Aβ42 (incubated in 50 mM PB at pH 7.4 for 7 days at 37 °C). (ii) 100 μM α-synuclein (incubated in 20 mM HEPES+2.5 mM EDTA at pH 7.4 for 2 weeks at 37 °C with agitation). (iii) 400 μM truncated tau ($tau_{297-391}$) in 10 mM PB with 10 mM DTT at pH 7.4 for 48 h with agitation). (**b**) short amyloidogenic peptides; (i) 1 mg/mL VIYKI incubated in H_2O for 7 days at 22 ° C. (ii) 1 mg/mL HYFNIF in H_2O, for 7 days at 22 °C. (iii) 1 mg/mL GNNQQNY in H_2O for 48 h at 22 °C

11. Incubate the amyloidogenic protein/peptide in a suitable buffer under the desired temperature, pH and concentration for the required number of hour/days, with or without shaking.

3.2 Negative Stain Electron Microscopy

1. Place 4 μL peptide solution on Formvar/carbon film coated, 400 mesh copper grids and incubate for 1 min. Blot excess using filter paper from the edge of the grid.
2. Wash the grid with 4 μL milliQ-filtered water and blot dry.
3. Place 4 μL 2% (w/v) uranyl acetate onto the grid for 1 min. Blot any excess dye and allow to air-dry. Repeat once more.
4. Examine and image the grids on a transmission electron microscope (TEM) and compatible camera (Fig. 1).

3.3 Thioflavin T/S Fluorescence

1. Mix peptide (for example, 100 μM) and dye to get a final concentration of between 10 and 20 μM ThT/5 μM ThS, vortex and allow to incubate for 2 min. The amount of dye added can be optimized for individual peptides.
2. Prepare a buffer blank incubated with either ThT/ThS in the same way.
3. Place in 10 mm quartz cuvette and insert into spectrofluorometer.

4. For ThT set an emission scan between 460 and 600 nm. For ThS, set an emission scan between 450 and 600 nm.
5. Set excitation and emission slits to 5 and 10 nm respectively. Set the sample compartment at a temperature of 21 °C, scan rate to 600 nm/min and use the average three spectra for each measurement.
6. Subtract the blank reading from the sample reading.
7. *See* **Note 5**.

3.4 Tyrosine Fluorescence

1. Place sample in 10 mm quartz cuvette and insert into spectrofluorometer remembering to take a reading for the blank first.
2. Set the excitation wavelength to 280 nm and collect an emission scan between wavelength 290–500 nm. Set the sample compartment to 21 °C with a scan rate of 600 nm/min. Use the average of three spectra for each measurement.
3. Subtract the blank reading from the sample reading.
4. *See* **Note 5**.

3.5 Circular Dichroism

1. Prepare the peptide in an appropriate buffer. Low chloride concentrations and low detergents for optimum spectra [20]. If the HT signal reaches above 600 mV, the reading is no longer reliable. 10–20 mM phosphate buffer is recommended but further information can be found in Kelly et al. [20].
2. Take a reading of the blank first. The signal should be neglible.
3. Place sample in appropriate cuvette; as a general rule, the higher the concentration, the lower the path-length.
4. Use a water bath to equilibrate samples at 20 °C (or suitable temperature).
5. Take scans between 180 and 320 nm using a spectropolarimeter (Fig. 2).
6. Samples can be centrifuged to separate out fibrils from a heterogeneous population. Centrifugation at max speed for 30 min results in a pellet which can then be resuspended in a small volume of buffer (20–100 μL).
7. Dense fibers may "fall out" of solution and can be sonicated between 10 and 20 min to resuspend them.
8. Linear dichroism (LD) can be observed from some fibril samples that align within the cuvette [21, 22]. To check whether there is a contribution from LD, the cuvette can be rotated 90°. If the spectrum is CD, then the two spectra (0 and 90°) will be identical. If they differ, then the spectrum should be interpreted as LD. (*See* [23, 24] for more detail).

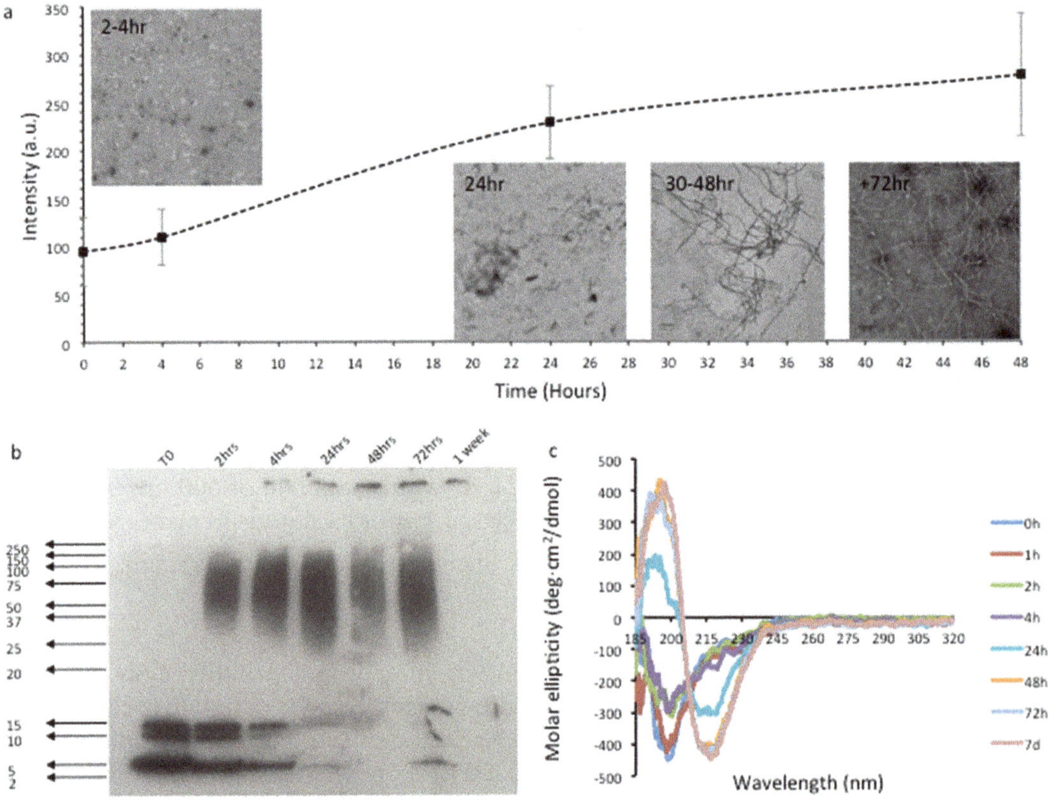

Fig. 2 Assembly time course. (**a**) Shows an example of a Thioflavin T spectra during the self-assembly process with inserted electron micrographs showing the morphology of Aβ1–42 over time. (**b**) Shows the separation of proteins by SDS gel electrophoresis at different Aβ42 assembly time points from monomer to high molecular weight species. (**c**) An example of a CD spectra showing the changing spectra to monitor the conformational change that Aβ1-42 adopts during its self-assembly process; oligomeric Aβ1–42 shows a random coil structure where as Aβ1-42 fibrils display a strong β-sheet secondary structure

9. Use an average of three spectra for each measurement and subtract the blank readings before replotting data.
10. Spectral data can be converted to molar ellipticity using the following equation: Mdeg × molecular weight/(10 × mg/mL × pathlength of cuvette (cm) × number of amino acids).

3.6 SDS Page/Western Blotting

1. Prepare samples by adding sample buffer plus BME (25% of total sample volume). Vortex and centrifuge samples and keep on ice until ready to load into gel.
2. Load the protein standard marker in the first well and samples into subsequence wells taking extra care not to let any spill over into neighboring wells.
3. Run the gel at 120 kV until the marker has reached the bottom of the gel (*see* **Note 5**).
4. Break open the glass plates using either a spatula or a knife.

5. Assemble the transfer equipment on the black side of the cassette as follows; sponge, two pieces of filter paper, gel, nitrocellulose membrane (precut to the size of the gel), two pieces of filter paper and sponge.
6. Transfer the gel onto the nitrocellulose membrane at 25 kV for 2 h.
7. After transferring is complete, block the membrane in 10% milk/TBST at room temperature for 1 h.
8. Apply primary antibody for 1 h at room temperature or overnight at 4 °C.
9. Wash the membrane for 10 min three times with 0.1% TBST.
10. Apply secondary antibody for 1 h at room temperature.
11. Wash the membrane for 10 min three times with 0.1% TBST.
12. Apply the ECL substrate for detection between 1 and 5 min before exposing the membrane to film and developing.

3.7 X-Ray Fiber Diffraction (XRFD)

3.7.1 X-Ray Fiber Diffraction Sample Preparation

1. For XRFD, sample with relatively high concentration of peptide/protein may often be needed to produce a suitable partially aligned fiber sample (*see* Fig. 3). Approximately 5–10 mg/mL of peptide is often sufficient. Alternatively, fibrils can be collected from the pellet by centrifugation 20,000 × *g* for 30–45 min.
2. If the sample contains buffer salts wash the pelleted fibrils three times with plenty of milli-Q water (400–600 μL) by centrifugation 20,000 × *g* for 15 min.
3. Resuspend the pellet with 10–15 μL of milli-Q water.
4. Cut the glass capillaries into 20–30 mm lengths and block one end with wax [25].
5. Put two wax-tipped capillary tubes ~10 mm away in petri dish using plasticine.
6. Place 10 μL of the sample between wax-tipped capillary tubes, ensuring that the droplet is suspended and does not run along the glass. Allow to dry overnight at room temperature, or, more slowly at 4 °C. Alternative alignment methods and further detail are described by [25].
7. Carefully separate the glass capillaries to yield a fiber sample on the end of one or both of the capillaries.

3.7.2 X-Ray Fiber Diffraction Data Collection

1. Mount the partially aligned fiber sample, attached to the capillary, on a goniometer head using plasticine.
2. Collect a diffraction pattern using a suitable X-ray source (e.g., CuKα wavelength 1.5419 Å or 1 Å for synchrotron) and detector (e.g., CCD) using an oscillation of 0.5° or still. Exposure times will vary depending on X-ray source. Specimen

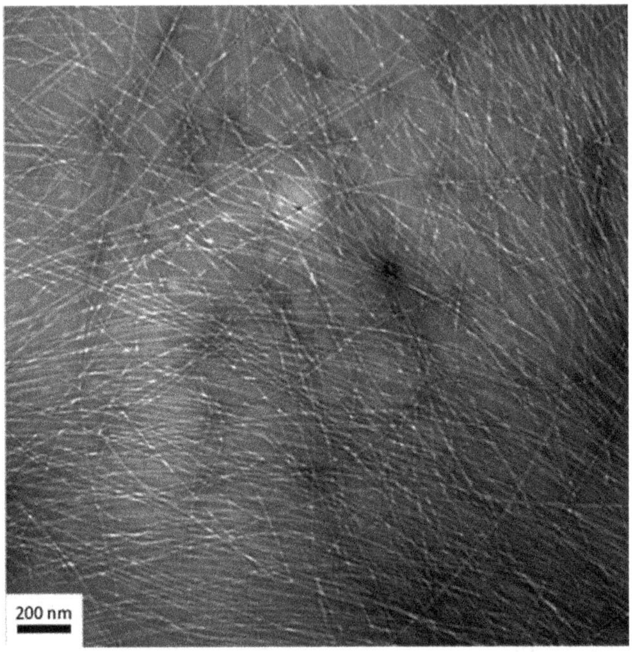

Fig. 3 An electron micrograph showing a high density of amyloid fibrils suitable for X-ray fiber diffraction alignment

to detector distance should be set to allow a maximum of 2 Å at the detector edge and can be moved further away to allow better separation of reflections. The sample should be rotated and data collected at different angles (e.g., 0° and 90°) to ensure the best data is collected from the sample.

3.7.3 X-Ray Fiber Diffraction Data Analysis

1. Examine the diffraction pattern using the software associated with the X-ray source and detector, or using another data display programme (e.g. Imosflm [14, 26]).
2. Diffraction signal measurement can be performed using the display program or the image can be converted to TIFF and loaded into CLEARER [16].
3. Measurement of the diffraction signal positions should be conducted along the fibrer axis (meridian) and perpendicular to it (equator).
4. To confirm the presence of amyloid fibrils, cross-β, reflections are expected at 4.7 Å on the meridian and 9–11 Å on the equator (Fig. 4).
5. Further reflections may be measured on meridian and equator to aid interpretation of structural architecture and arrangements of the peptide/protein within the fiber. Meridional reflections provide information about the repeat distance(s) along the

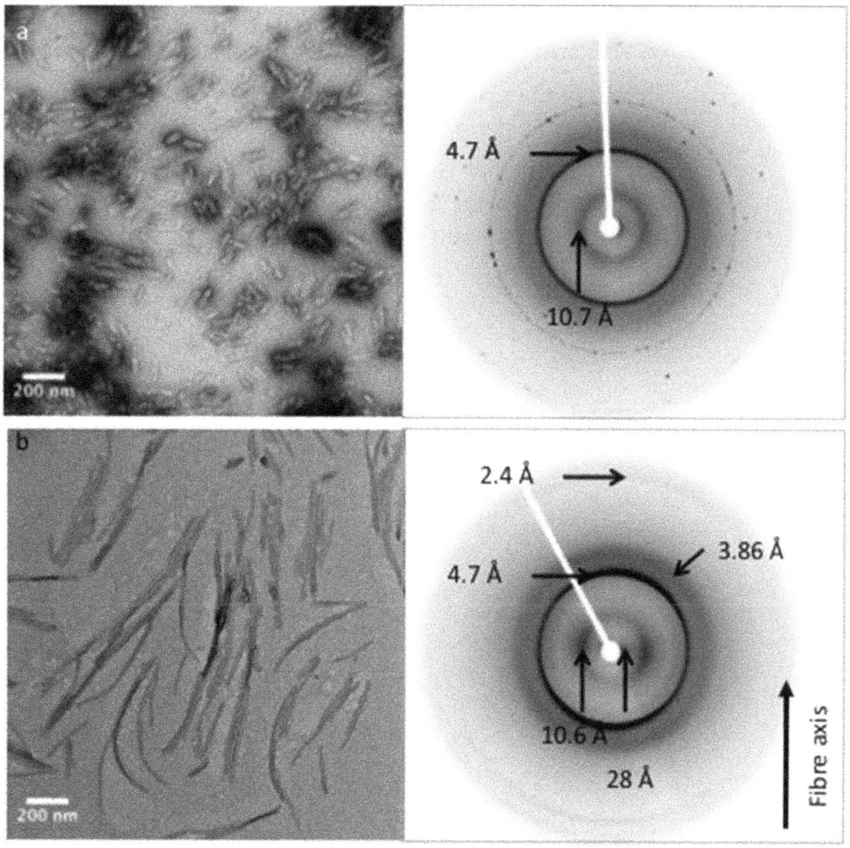

Fig. 4 X-ray fiber diffraction patterns. Collected from (**a**) truncated tau and (**b**) a variant C322A. Short filaments formed by truncated tau are hard to align and yield a poorly oriented pattern where the meridional and equatorial axes cannot be distinguished. C322A forms very long filaments which align well and produce an aligned diffraction pattern. The fiber axis is vertical (meridional) and the horizontal axis is the equator. This provides the required information to conclude that the fibers are cross-β and therefore considered to be amyloid

fiber axis, while equatorial reflections provide information about chain length, protofilament width and packing.

6. Further details regarding interpretation of amyloid fibril diffraction data can be found in [25, 27].

3.8 Immunogold Labeling Negative Stain TEM

1. Pipette 4 μL aliquots of the protein/peptide fibrils onto Formvar/ carbon coated 400 mesh copper TEM support grids (Agar Scientific, Essex, UK) and incubate for 1 min.
2. Blot excess using filter paper.
3. Leave the grid to dry for 2–3 min. Do not over dry, as this will affect the binding of the antibody with the fibrils.
4. Block the grid using normal goat serum (1:10 in PBS+) for 15 min. To do this, place a 20 μL drop of normal goat serum on to parafilm fixed in petri dish. Then, put the grid carefully on the top of the drop.

5. Incubate the grids with primary antibody (5–10 μg/mL) for 2 h at room temperature using the method in **step 4**.
6. Rinse the grid three times for 2 min using PBS+ by placing the grid on a drop of water as described in **step 4**.
7. Immunolabel the grid with a suitable 5, 10 or 15 nm gold particle-conjugated secondary probe for 1 h at room temperature, (as in point 4).
8. Rinse the grid five times, 2 min each with PBS+ and then five times, 2 min each with distilled water, (as in **step 4**).
9. Finally, stain the grid with (centrifuged and filtered) 0.5% aqueous uranyl acetate for 30 s (Fig. 5).

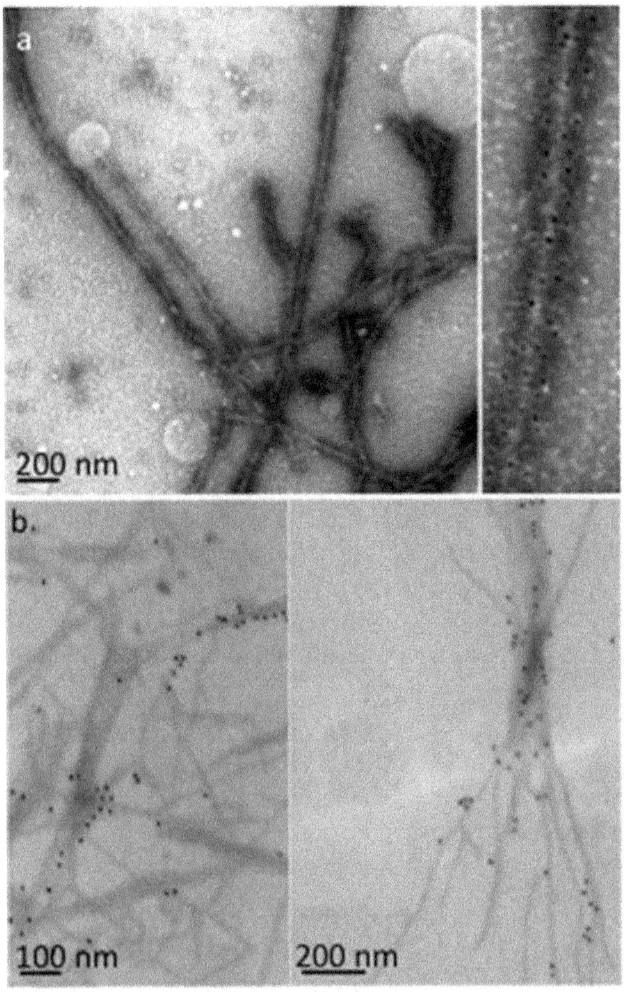

Fig. 5 Electron micrographs showing immunogold labeling. (**a**) Aβ42 fibrils using the oligomer-specific NU1 antibody [5], revealing oligomers bound to the fibrils. (**b**) β-synuclein fibrils using the dityrosine antibody (JaICA) [28]

4 Notes

1. When adding HFIP, use a needle and glass syringe. As it is extremely volatile, it is also recommended that the cap is kept closed and instead insert the needle through the rubber bung in the cap.
2. If the nitrogen flow is too high, the peptide solution will begin to "spray." It is helpful to first begin drying the HFIP by keeping the needle near the rim of the vial.
3. If you are working with short peptides that have the propensity to aggregate into amyloid fibrils immediately, do not use the spin columns. The fibrils will remain in the column and very little peptide will be recovered at the end. Instead, after drying off the HFIP, the peptide film can be immediately resuspended in buffer.
4. The spin column can be kept in the centrifuge until the peptide is ready to be added. If, however, the waiting time is greater than 30 min, do one more buffer spin.
5. The presence of substrates, metals, small molecules, or inhibitors in the sample can lead to fluorescence quenching. This can hamper conclusions from the data, so care should be taken to check this carefully.

References

1. Sipe JD, Benson MD, Buxbaum JN, Ikeda S, Merlini G, Saraiva MJ, Westermark P (2014) Nomenclature 2014: amyloid fibril proteins and clinical classification of the amyloidosis. Amyloid 21(4):221–224. https://doi.org/10.3109/13506129.2014.964858
2. Puchtler H, Sweat F (1965) Congo red as a stain for fluorescence microscopy of amyloid. J Histochem Cytochem 13(8):693–694
3. LeVine H 3rd (1999) Quantification of beta-sheet amyloid fibril structures with thioflavin T. Methods Enzymol 309:274–284
4. Glabe CG (2004) Conformation-dependent antibodies target diseases of protein misfolding. Trends Biochem Sci 29(10):542–547
5. Lambert MP, Velasco PT, Chang L, Viola KL, Fernandez S, Lacor PN, Khuon D, Gong Y, Bigio EH, Shaw P, De Felice FG, Krafft GA, Klein WL (2007) Monoclonal antibodies that target pathological assemblies of Abeta. J Neurochem 100(1):23–35. https://doi.org/10.1111/j.1471-4159.2006.04157.x JNC4157 [pii]
6. Marshall KE, Hicks MR, Williams TL, Hoffmann SV, Rodger A, Dafforn TR, Serpell LC (2010) Characterizing the assembly of the Sup35 yeast prion fragment, GNNQQNY: structural changes accompany a fiber-to-crystal switch. Biophys J 98(2):330–338. https://doi.org/10.1016/j.bpj.2009.10.020 S0006-3495(09)01627-0 [pii]
7. Sawaya MR, Sambashivan S, Nelson R, Ivanova MI, Sievers SA, Apostol MI, Thompson MJ, Balbirnie M, Wiltzius JJ, McFarlane HT, Madsen AO, Riekel C, Eisenberg D (2007) Atomic structures of amyloid cross-beta spines reveal varied steric zippers. Nature 447(7143):453–457
8. Tycko R (2001) Biomolecular solid state NMR: advances in structural methodology and applications to peptide and protein fibrils. Annu Rev Phys Chem 52:575–606
9. Fitzpatrick AWP, Falcon B, He S, Murzin AG, Murshudov G, Garringer HJ, Crowther RA, Ghetti B, Goedert M, Scheres SHW (2017) Cryo-EM structures of tau filaments from Alzheimer's disease. Nature 547(7662):185–190. https://doi.org/10.1038/nature23002
10. van der Wel PC, Lewandowski JR, Griffin RG (2007) Solid-state NMR study of amyloid

nanocrystals and fibrils formed by the peptide GNNQQNY from yeast prion protein Sup35p. J Am Chem Soc 129(16):5117–5130

11. White HE, Hodgkinson JL, Jahn TR, Cohen-Krausz S, Gosal WS, Muller S, Orlova EV, Radford SE, Saibil HR (2009) Globular tetramers of beta(2)-microglobulin assemble into elaborate amyloid fibrils. J Mol Biol 389(1):48–57
12. Xue WF, Radford SE (2013) An imaging and systems modeling approach to fibril breakage enables prediction of amyloid behavior. Biophys J 105(12):2811–2819. https://doi.org/10.1016/j.bpj.2013.10.034
13. Goldsbury C, Kistler J, Aebi U, Arvinte T, Cooper GJ (1999) Watching amyloid fibrils grow by time-lapse atomic force microscopy. J Mol Biol 285(1):33–39
14. CCP4 (1994) The CCP4 suite programs for crystallography. Acta Cryst D50:760–763
15. Hammersley AP (1998) FIT2D V9.129 Reference Manual V3.1. ESRF Internal Report ESRF98HA01T
16. Makin OS, Sikorski P, Serpell LC (2007) CLEARER: a new tool for the analysis of X-ray fibre diffraction patterns and diffraction simulation from atomic structural models. Appl Cryst 40(5):966–972
17. Broersen K, Jonckheere W, Rozenski J, Vandersteen A, Pauwels K, Pastore A, Rousseau F, Schymkowitz J (2011) A standardized and biocompatible preparation of aggregate-free amyloid beta peptide for biophysical and biological studies of Alzheimer's disease. Protein Eng Des Sel 24(9):743–750. https://doi.org/10.1093/protein/gzr020
18. Bitan G, Frandinger EA, Spring S, Teplow D (2005) Neurotoxic protein oligomers- what you see is not always what you get. Amyloid 2(Jun):88–95
19. Fezoui Y, Hartley DM, Harper JD, Khurana R, Walsh DM, Condron MM, Selkoe DJ, Lansbury PT Jr, Fink AL, Teplow DB (2000) An improved method of preparing the amyloid beta-protein for fibrillogenesis and neurotoxicity experiments. Amyloid 7(3):166–178
20. Kelly SM, Jess TJ, Price NC (2005) How to study proteins by circular dichroism. Biochim Biophys Acta 1751(2):119–139. https://doi.org/10.1016/j.bbapap.2005.06.005
21. Morris KL, Rodger A, Hicks MR, Debulpaep M, Schymkowitz J, Rousseau F, Serpell LC (2013) Exploring the sequence-structure relationship for amyloid peptides. Biochem J 450(2):275–283. https://doi.org/10.1042/BJ20121773
22. Dafforn TR, Rajendra J, Halsall DJ, Serpell LC, Rodger A (2004) Protein fiber linear dichroism for structure determination and kinetics in a low-volume, low-wavelength couette flow cell. Biophys J 86(1 Pt 1):404–410
23. Rodger A, Marrington R, Geeves MA, Hicks M, de Alwis L, Halsall DJ, Dafforn TR (2006) Looking at long molecules in solution: what happens when they are subjected to Couette flow? Phys Chem Chem Phys 8(27):3161–3171
24. Rodger A, Norden B (1997) Circular Dichroism and linear Dichroism. Oxford University Press, Oxford
25. Morris KL, Serpell LC (2012) X-ray fibre diffraction studies of amyloid fibrils. Methods Mol Biol 849:121–135. https://doi.org/10.1007/978-1-61779-551-0_9
26. Winn MD (2003) An overview of the CCP4 project in protein crystallography: an example of a collaborative project. J Synchrotron Radiat 10. (Pt 1:23–25
27. Makin OS, Serpell LC (2005) X-ray diffraction studies of amyloid structure. In: Sigurdsson EM (ed) Amyloid proteins: methods and protocols. Humana press, Totowa, NJ, pp 67–80
28. Al-Hilaly YK, Biasetti L, Blakeman BJ, Pollack SJ, Zibaee S, Abdul-Sada A, Thorpe JR, Xue WF, Serpell LC (2016) The involvement of dityrosine crosslinking in alpha-synuclein assembly and deposition in Lewy bodies in Parkinson's disease. Sci Rep 6:39171. https://doi.org/10.1038/srep39171

Chapter 8

Assays for Light Chain Amyloidosis Formation and Cytotoxicity

Luis M. Blancas-Mejia, Pinaki Misra, Christopher J. Dick, Marta Marin-Argany, Keely R. Redhage, Shawna A. Cooper, and Marina Ramirez-Alvarado

Abstract

Common biophysical techniques like absorption and fluorescence spectroscopy, microscopy, and light scattering studies have been in use to investigate fibril assembly for a long time. However, there is sometimes a lack of consensus from the findings of an individual technique when compared in parallel with the other techniques. In this chapter, we aim to provide a concise compilation of techniques that can effectively be used to obtain a comprehensive representation of the structural, aggregation, and toxicity determinants in immunoglobulin light chain amyloidosis. We start by giving a brief introduction on amyloid assembly and the advantages of using simple and readily available techniques to study aggregation. After an overview on preparation of protein to set up parallel experiments, we provide a systematic description of the in vitro techniques used to study aggregation in AL protein. Additionally, we thoroughly discuss the steps needed in our experience during the individual experiments for better reproducibility and data analysis.

Key words Immunoglobulin light chain, Amyloidosis, Amyloid formation, Thioflavin T, Electron microscopy, Sedimentation assay, Cell viability, Caspase assay

1 Introduction

Amyloid disease initially starts from soluble proteins that polymerize through multiple stages into insoluble amyloid fibrils which get deposited in various organs, leading to organ dysfunction and tissue degeneration [1–3]. The aggregation pathway involves multiple species that self-interact in a variety of ways to produce insoluble amyloid fibrils.

Luis M. Blancas-Mejia and Pinaki Misra contributed equally to this work.

Cláudio M. Gomes (ed.), *Protein Misfolding Diseases: Methods and Protocols*, Methods in Molecular Biology, vol. 1873, https://doi.org/10.1007/978-1-4939-8820-4_8, © Springer Science+Business Media, LLC, part of Springer Nature 2019

Simple and readily available techniques like thioflavin T (ThT) fluorescence assay and electron microscopy in conjunction with complementary techniques as high performance liquid chromatography (HPLC) sedimentation assay, dynamic light scattering (DLS), and size exclusion chromatography (SEC) can provide a global insight of the whole aggregation process. Experiments with these techniques are easy to set up, accurate and due to minimal technical variability are highly reproducible. Composite analysis of the data obtained from these complementary techniques at periodic intervals of time can provide substantial quantitative information concerning transformation of different species into amyloid assembly. We will focus on the experience we have had with immunoglobulin light chain (AL) amyloidosis. In the following section, we describe our laboratory techniques developed to study in vitro aggregation of AL proteins (Fig. 1) and assays to analyze cytotoxicity and apoptosis in AL amyloidosis. The various subsections in the chapter include preparation of AL proteins and general setup for aggregation kinetic experiments, use of thioflavin T (ThT) to study fibril formation in vitro, electron microscopy to study the morphology of fibrils and intermediate species, use of HPLC-based sedimentation assay for quantitative determination of soluble proteins present in aggregate suspension, particle size analysis by dynamic light scattering (DLS) to explore the involvement of low molecular weight oligomers (LMW) in the aggregation pathway, and size-exclusion chromatography (SEC) analysis of the soluble aggregate suspension to characterize the gradual transformation of soluble proteins and LMW oligomers into protofibrillar and fibrillar forms, nanoparticle tracking analysis (NTA) to detect species ranging from 30 to 1000 nm in size, and in vitro assays to analyze cytotoxicity and apoptosis induced by fibrils/soluble proteins in light chain amyloidosis.

2 Materials

2.1 Preparation of AL Variable Domain Proteins for In Vitro Assays

Use ultrapure water (prepared by purifying deionized water, to attain a conductivity of 18 MΩ-cm at 25 °C) and analytical grade reagents for all solutions. Prepare and store all reagents at room temperature (unless indicated otherwise). Diligently follow all waste disposal regulations when disposing of waste materials. We do not add sodium azide to reagents.

1. Frozen aliquots of the protein of interest.
2. Bucket of ice.
3. Vortex.
4. 0.22 μM membrane syringe filters.

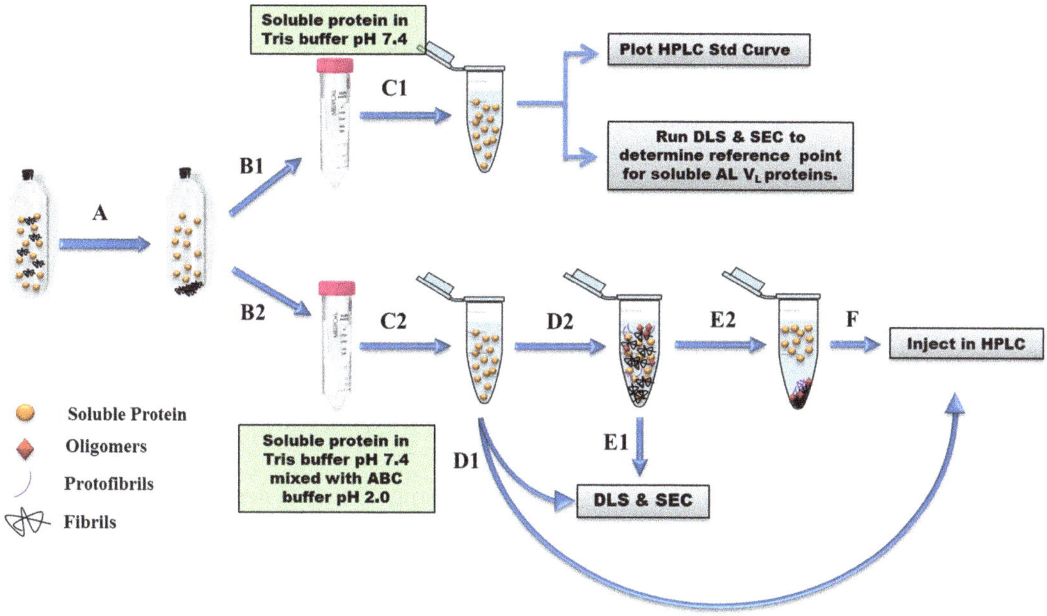

Fig. 1 Schematic representation of the laboratory techniques we used to study in vitro aggregation of AL proteins. A: After pooling thawed AL protein (from −80 °C) in the centrifuge tubes, spin at 643,867 × *g* for 3 h 20 min. B1: Decant three-fourth parts of the centrifuged solution into 10,000 MWCO centrifugal filter and concentrate to 350–400 μM. B2: Decant three-fourth parts of the centrifuged solution into 10,000 MWCO centrifugal filter mix ABC buffer pH 2.0 and concentrate to 350–400 μM. C1: Serially dilute the concentrated AL protein with Tris buffer pH 7.4 and plot the HPLC standard curve and determine the reference point (in terms of particle size and MW) for individual soluble AL proteins from DLS and SEC. C2: Dilute the concentrated AL protein with ABC buffer pH 2.0 to the desired concentration and add 150 mM NaCl. This is the starting reaction mixture for the aggregation reaction kinetics. D1: As soon as the aggregation reaction mixture is prepared, parallel reading in DLS, SEC, and HPLC is obtained to test the quantity and quality of soluble protein. D2: After taking the first measurements, setup the reaction mixture in an incubator shaker (37 °C; 300 rpm). E1: At predetermined time intervals take required aliquots from the reaction mixture in the incubator shaker for DLS and SEC experiments. E2: At predetermined time intervals take aliquot ~80 μL and centrifuge at 160,000 × *g* for 45 min. F: Mix 40 μL of the supernatant from step E2 with 20 μL of 1% formic acid and inject in the analytical reverse phase HPLC to determine the concentration of unreacted monomers

5. 10 mL conical tubes.
6. 10 mL conical centrifugal filter concentrators.
7. 5 mL polyallomer centrifuge tubes, or equivalent.
8. Ultracentrifuge.
9. UV-vis spectrophotometer.

2.2 Thioflavin T (ThT) to Study Amyloid Fibril Formation In Vitro

1. 10 mM ThT stock solution: 32 mg ThT to10 mL of ultrapure water and filter through a 0.2 μm syringe filter (*see* **Note 1**). Determine the real ThT concentration after filtration by UV absorbance spectroscopy using a molar extinction coefficient of 26,620 M^{-1} cm^{-1} at 416 nm [4, 5]. Store the solution protected from light (*see* **Note 2**).

2. 10 mM ABC buffer pH 2–11: 10 mM of sodium acetate, boric acid, and sodium citrate buffer at the desired pH. 0.68 g of sodium acetate, 0.32 g of boric acid, and 1.48 g of sodium citrate, in 500 mL of ultrapure water. Filter the buffer solution using a 0.22 μm filter and aliquot into 10 mL conical tubes. Measure the pH of each conical tube and adjust the pH with either 1 M HCl or 1 M NaOH accordingly (*see* **Note 3**).
3. 5.0 M NaCl: 14.61 g of NaCl in 50 mL of ultrapure water. Filter it using a 0.22 μm filter. Store it at room temperature.

2.3 Kinetic Data Analysis

1. Microsoft Excel and/or any scientific graphing and data analysis software, i.e., Origin software package (Origin lab, USA), GraphPad Prism (GraphPad Software Inc., USA).

2.4 Negative Staining of Amyloid Fibrils for Transmission Electron Microscopy (TEM)

1. 0 4% uranyl acetate: 400 mg of uranyl acetate in 10 mL of filtered ultrapure water.
2. Formvar/carbon 300 mesh copper TEM grids.
3. Filter paper.
4. TEM grid tweezers.

2.5 Sedimentation Assay by HPLC

1. HPLC grade water.
2. HPLC grade acetonitrile.
3. HPLC grade formic acid.
4. C8 RP- HPLC analytic column.
5. HPLC chromatographic system.

2.6 Particle Size Analysis by Dynamic Light Scattering (DLS)

1. Cuvette-based DLS molecular sizing instrument.
2. DLS-cuvette (12 μL square quartz microcuvette).
3. Long round gel loading pipette tips.
4. Ultrapure filtered water.

2.7 Particle Size Analysis by Size Exclusion Chromatography (SEC)

1. Analytical chromatographic system.
2. Analytical size exclusion column with a broad separation range, we recommend agarose-dextran based matrix (we used Superdex 200 size exclusion column with a separation range of 10,000–600,000 Da).
3. Molecular weight protein markers.
4. Phosphate buffer at pH 6.8 as the mobile phase for the size exclusion experiments: 7.09 g of sodium phosphate dibasic, 5.99 g of sodium phosphate monobasic, and 8.76 g of sodium chloride. Adjust the pH of the solution using 1 M NaOH. Bring the volume to 1 L using ultrapure water, mix well and filter through a 0.22 μM membrane filter.

2.8 Nanoparticle Tracking Analysis

1. 1 mL disposable Luer lock syringes.
2. Phosphate buffered saline (PBS), pH 7.4. (filtered, free of all particulates).
3. Ultrapure H_2O.
4. 10% ethanol for cleaning.
5. Clean tubes and pipets for making sample dilutions.
6. External hard drive disk for large files.
7. Lens paper.
8. Can of compressed air.
9. NanoSight NS300 with software package NTA 2.3.5.

2.9 In Vitro Assays for Light Chain Amyloidosis Cytotoxicity

This protocol is ideal for AC16 human cardiomyocytes cells and mesenchymal stromal cells (MSCs) (*see* **Note 4**).

1. Cell Culture Media—DMEM:F12 (1:1) (Dulbecco's modified Eagle medium and Ham's F-12 medium), L-glutamine, 15 mM HEPES, 12.5% fetal bovine serum (FBS), 1% penicillin–streptomycin (P/S). Remove and discard 67.5 mL of media from bottle, add 62.5 mL of FBS and 5 mL of P/S to DMEM/F12 media bottle. Filter using a 500 mL, 0.22 μm vacuum driven sterile filter bottle. Store at 4 °C.
2. 0.25% trypsin–EDTA.
3. Fetal bovine serum (FBS).
4. Sterile phosphate buffered saline (PBS), pH 7.4.
5. 3-(4,5-Dimethylthiazol-2-yl)-2,5-diphenyltetrazolium Bromide (MTT) reagent: 10 mg of MTT reagent in 5 mL of sterile PBS buffer, pH 7.4. Filter solution using a sterile syringe filter with a 0.22 μm pore size hydrophilic polyvinylidene difluoride (PVDF) membrane. Store at 4 °C, protected from light (*see* **Note 5**).
6. Dimethyl sulfoxide (DMSO).
7. Trypan blue stain 0.04% solution.
8. Sterile syringe filter with a 0.22 μm pore size hydrophilic PVDF membrane.
9. CO_2 incubator.
10. Sterile pipette tips.
11. Centrifuge with rotor for 96-well plates.
12. T75 cell flasks.
13. Flat-bottom sterile 96-well polystyrene plates (cell culture quality).
14. Visible/UV plate reader.

15. LDH cytotoxicity detection kit (according to manufacturer's instructions, Clontech).
16. Homogeneous Caspases Assay, fluorimetric kit (according to manufacturer's instructions, Roche).

2.10 In Vitro Assays to Analyze/Study Apoptosis in Light Chain Amyloidosis

1. Cells–chosen and treated based on the experimental model (*see* **Note 6**).
2. Complete cell culture medium of choice (varies by cell type).
3. Phosphate buffered saline (PBS), pH 7.4.
4. Methanol–acetic acid (3:1) solution (MeOH–AcOH).
5. 50% glycerol–50% 0.1 M Tris–HCl (pH 7.4) containing 1 μg/mL Hoechst 33258 (*see* **Note 7**).
6. Glass microscope slides and coverslips.
7. Fluorescence microscope possessing the appropriate excitation and emissions filters (*see* **Note 8**).

3 Methods

3.1 Preparation of AL Variable Domain Proteins for In Vitro Assays

After purification and confirmation of the biophysical properties of your protein or peptide of interest [6–10], it is important to take into account the storage conditions of the protein and peptide prior to the amyloid formation reaction. We particularly suggest aliquoting 1 mL of the proteins at ~100 μM, in 1.5 mL cryogenic tubes, flash-freezing in liquid nitrogen, and storing at −80 °C. In the case of synthetic peptides, it is a common practice to store them as lyophilized peptides on a desiccator, and/or under anaerobic conditions to prevent oxidation of the peptides. These recommendations work well with the light chains proteins in our laboratory, but other proteins may vary in their optimal storage conditions.

1. Thaw frozen aliquots accordingly to the total amount of protein estimated for the experiment, on ice (*see* **Note 9**).
2. Once thawed, gently mix the content of the tubes by a gentle vortexing. Pool the protein from all the tubes in 5 mL polyallomer centrifuge tubes and spin at in an ultracentrifuge at 645,000 × *g* (*see* **Note 10**) (Fig. 1; A). This step removes oligomers, protofibrils and large aggregates (if any) formed during the storage process.
3. Gently remove the top three-fourth part of the supernatant and filter through 0.2 μM low protein binding membrane filter. Additionally you can check the filtered protein solution by SEC and dynamic light scattering (described in Subheadings **2.6** and **2.7**) for aggregates, and to obtain a reference point for the soluble protein before the assays (Figs. 1C1 and 2).

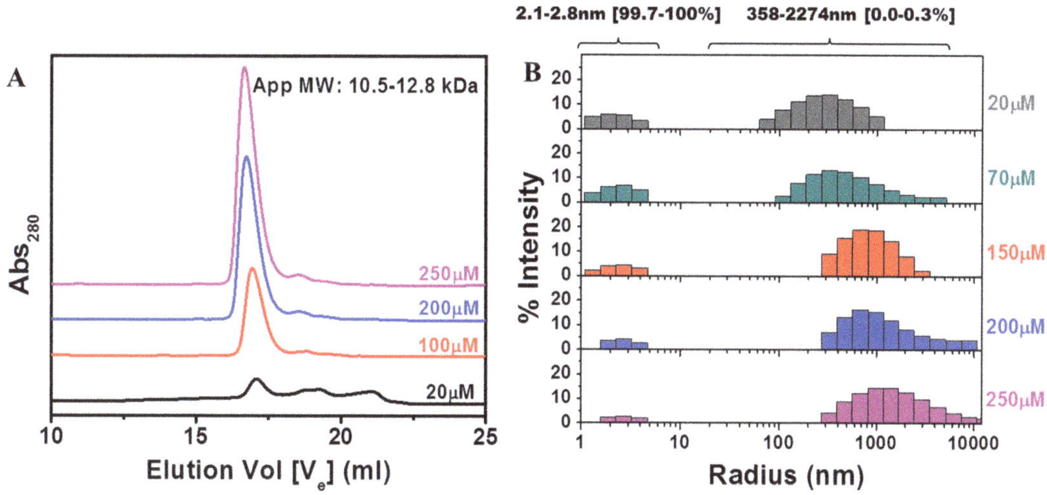

Fig. 2 Example of a serial dilution of AL protein (κI 018/08 V_L) with Tris buffer pH 7.4 to obtain a reference point for soluble protein (see step C1 of Fig. 1). (**a**) Size exclusion chromatogram representing apparent MW observed in κI 018/08 V_L over a range of concentrations. (**b**) Hydrodynamic radii of particles observed for κI 018/08 V_L over a range of concentrations. The hydrodynamic radii of the particles are mentioned on top of the graph. Values in parentheses represent mass percentage (population density) of individual particles

4. Mix the filtered protein solution with the buffer of election at desired pH (*see* **Note 11**) and concentrate (we recommend 350–400 μM, but it has to be tested for each protein) using conical centrifugal filters with the appropriated molecular weight cutoff (MWCO) (*see* **Note 12**). This concentrated protein (stock protein) will be used to get the desired concentration for all the experiments reported below (Fig. 1C2).
5. Determine the concentration of the solution by UV absorbance.
6. Prepare the aggregation reaction mixture in 1.5 mL microcentrifuge tube by addition of the 150 mM NaCl [11] and diluting it with an appropriate volume of buffer to the desired concentrations and volume.
7. This reaction mixture would subsequently act as a parallel starting point for ThT binding assay, electron microscopy, HPLC sedimentation assay, DLS, and SEC (Fig. 1; D1) (*see* **Note 13**).
8. Always keep the protein stock on ice during the entire preparation process.

3.2 Use of Thioflavin T (ThT) to Study Amyloid Fibril Formation In Vitro

Thioflavin T (ThT), a benzothiazole dye (Fig. 3) [12–14] has been widely used for the staining of amyloid fibrils, and is considered the gold standard to quantify and determine the presence of amyloid fibrils in both histological samples, and in vitro assays since its first

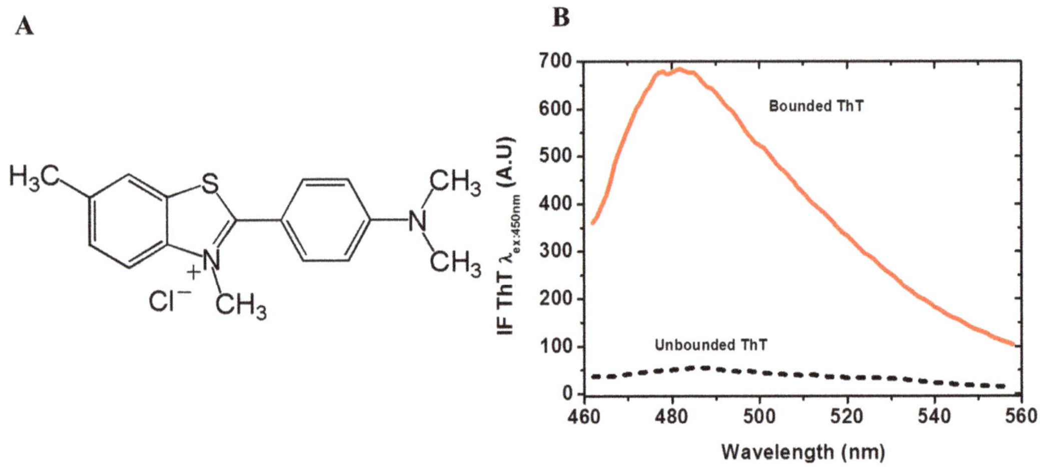

Fig. 3 (**a**) Chemical structure of Thioflavin T. (**b**) Fluorescence emission of unbounded ThT (black line) and bounded ThT to fibrils (red line), $\lambda_{excitation} = 440$ nm and $\lambda_{emission} = 480$ nm

use as a fluorescent amyloid reporter by Vassar and Culling in 1959 [15–17].

ThT binds equally well to amyloid fibrils and cross β-sheet-rich structures of biological (proteins) and synthetic origin (peptides) [17, 18], displaying a characteristic red shift in the emission spectrum [15, 17, 19], and an enhanced fluorescence [20] when bonded, which makes it a cheap, fast, and reliable amyloid probe [15, 21, 22].

The in vitro spectroscopic assay is commonly used to monitor fibril formation over time, but the assay is not strictly quantitative and some deviations/interferences can be observed. Hence, to rule out interferences in the in vitro assay due particular requirements of the experimental setup, some considerations have to be taken into account, like ensuring the quality and conformational homogeneity of the protein sample prior to the experiment, a correct selection of buffer systems compatible with the ThT dye, protein stability at the desired temperature for the experiment, and the correct preparation of the stock solutions, in particular the ThT solution. In our experience, the ThT assay to monitor kinetics of fibril formation in proteins may be set in two different ways depending on the purpose of the planned experiments:

(a) Low volume ThT assays: 260 μL of the desired concentration of soluble protein, buffer, NaCl, and ThT are incubated in a black 96-well polystyrene plate (Greiner 96 well black 6550) in an incubator shaker (*see* **Note 14**).

(b) High volume ThT assays: 1 mL of the desired concentration of soluble protein, buffer, NaCl, and ThT are incubated in low protein binding microcentrifuge tube in an incubator shaker.

300 μL of the reaction mixture is read on a 96-well microplate thereafter (*see* **Note 15**).

To study the aggregation pattern of soluble proteins, we generally prepare an aggregation mixture with the desired concentration of soluble protein, selected buffer of election, 150 mM NaCl at the desired pH (*see* **Note 16**). Typically the aggregation mixture is incubated at 37 °C in low protein binding microcentrifuge tubes with constant shaking at 300 rpm in a temperature controlled incubator with orbital shaking (*see* **Note 17**). Depending on the number of assays and number of time points to be studied, the volume of the aggregation mixture is kept constant for soluble proteins across individual experiments (*see* **Note 18**). At the end of the aggregation experiment (when most of the soluble proteins have transformed into fibrillar form) (*see* **Note 19**), the microcentrifuge tube with the aggregate suspension is frozen at −20 °C. The thawed, washed aggregate suspension is used to study fibril cytotoxicity and apoptosis in light chain amyloidosis cellular model systems.

For the amyloid fibril formation assay in presence of ThT, the protein of interest is incubated with ThT and other relevant buffer components and the ThT signal from each sample is read, often in a plate reader, which allows multiple samples to be measured at once as a function of time [10, 23, 24].

1. Prepare the ThT (*see* **Notes 1** and **2**), 10 mM ABC buffer (*see* **Note 3**), and the protein solution stocks at high concentration to allow you pipetting small volumes with confidence (if it is possible, we suggest ~400 μM protein stock), to reach the final concentration desired (*see* **Note 20**). Keep the protein stock on ice during the entire preparation process.
2. Add the required volume of ABC buffer to a microcentrifuge tube.
3. Add the required volume of 5.0 M NaCl stock to get a final concentration of 150 mM.
4. Add the required volume of protein stock to get a final concentration of 20 μM.
5. Add 10 μM equivalent of ThT to each tube, and mix well by vortexing (keep it on ice until this time) (*see* **Note 21**).
6. Check the pH of the sample and adjust if necessary (*see* **Note 22**).
7. Aliquot the samples in black 96-well polystyrene plate (Greiner 96-well black 6550), seal the wells with sealing film, cover the plate with a black lid, and seal it with tape, to decrease evaporation (*see* **Note 23**).
8. Record how much time it takes you from the time you start adding protein until you take the initial reading in fluorescence plate reader for time corrections.

9. Take an initial reading of the fluorescence emission at time zero, on a plate reader using an excitation wavelength of 440 nm, and 480 nm for the emission (*see* **Note 24**).
10. Shake plate at 300 rpm, 37 °C in a temperature-controlled incubator. We prefer orbital shaking over linear shaking.
11. Take subsequent readings at regular intervals. In cases that the kinetic is really slow, take daily readings at specific time for up to 1 month (~750 h). For fast kinetics, we recommend use a plate reader in kinetic mode. For other reading settings in your reader, please refer to the manual of your equipment, and adjust the settings accordingly.

3.3 Kinetic Data Analysis

Analyzing in vitro amyloid fibril formation kinetics is not trivial; misleading conclusions can be drawn due to incorrect analysis. A thorough discussion of the models to analyze fibril formation kinetics is beyond the scope of this chapter (for a good review on the topic, *see* [25, 26]). We will present an empirical approach to analyze and extract kinetic parameters from experimental data, as well as the use of an empirical sigmoidal function (logistic function) to fit kinetic data. The logistic function is not directed to any specific molecular process, but it is a good entry point to study the molecular mechanism of amyloid aggregation.

1. Prepare the kinetic data by plotting the ThT fluorescence values as a function of time. It is preferable to analyze the kinetic traces of each replicate as independent data because averaging fibrillation curves can introduce bias (*see* **Note 25**). Export the ThT fluorescence data as a function of time into a spreadsheet. We advise you to use the first row to name the columns with descriptive headers, especially if your data consist of many experiments.
2. We do not recommend normalizing the kinetic traces (*see* **Note 26**).
3. Verify that all the data is in the same units, as well and check for any outliner data. ThT fluorescence data could be noisy.
4. Next, choose/delimit the region that which will be to fit to obtain the lag, elongation phase, and plateau value. To determine the lag and plateau phases there should be enough data points for the fitting, but not to include any signal before the plateau is reached.
5. A classic example of a ThT-monitored kinetic curve is shown in Fig. 4. Such a curve may be analyzed using the sigmoidal logistic function [23]:

$$Y = y_i + m_i t + \frac{y_f + m_f t}{1 + e^{-[(t - t_{50})/\tau]}} \quad (1)$$

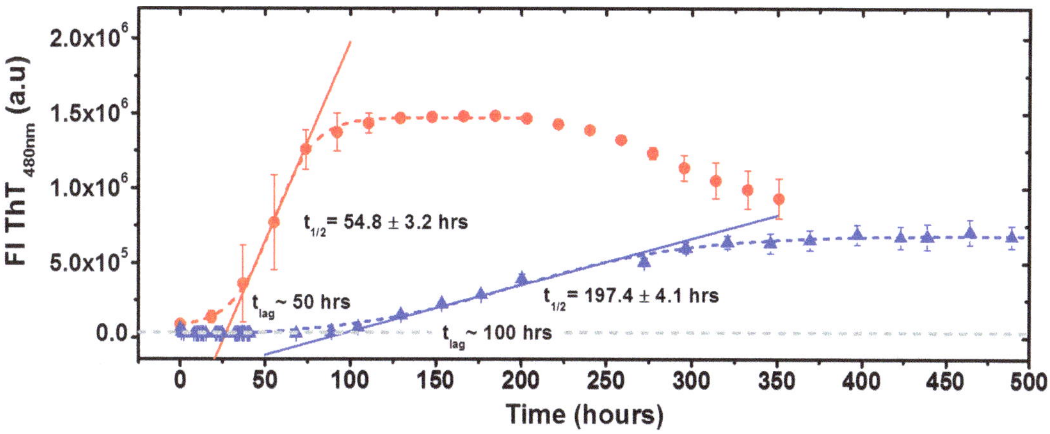

Fig. 4 Typical kinetic traces of fibril formation obtained by in situ *ThT assay*. In this example, reactions were performed with 20 μM protein, 10 mM ABC, 150 mM NaCl, at pH 2.0 in the presence of 10 μM ThT. Dashed lines represent the line of best fit. Continuous lines represent the linear extrapolations of the elongation phase; the intersect with the lag phase baseline is the t_{lag}. t_{50} values were obtained by nonlinear fitting to the Boltzman function (Eq. 2) of each independent kinetics, and were averaged from three independent traces. In this example, red circles show data of the amyloidogenic AL-09 V_L protein, and blue triangles represent data from the control κI 018/08 V_L protein.

where Y is the fluorescence intensity as a function of time t, y_i and y_f are the intercept of the initial and final baselines with the y-axis, m_i and m_f are slopes of the initial and final baselines, $t_{1/2}$ is the time needed to reach halfway through the elongation phase and τ is the elongation time constant. The apparent rate constant, k_{app}, for the growth of fibrils is given by $1/\tau$, and the lag time is usually defined as $t_{lag} = t_{50} - 2\tau$.

Or using the most general sigmoidal function, the Boltzmann equation (defined as a Boltzmann function by the Origin software package http://www.originlab.com/www/helponline/Origin/en/UserGuide/Boltzmann.html).

$$y = \frac{A_1 - A_2}{1 + e^{(x - x_0)/dx}} + A_2 \tag{2}$$

Where A_1 is the initial fluorescence value, A_2 is the final fluorescence value, x_0 is the center ($t_{1/2}$ value), dx is defined as the time constant. In both cases, a larger t_{50} value indicates a longer time required to form fibrils (*see* **Note 27**).

6. Once all the data has been preprocessed, data fitting can be performed using your favorite scientific graphics and data analysis software (*see* **Note 28**).
7. In the case of noisy kinetics data, with deviations from the sigmoidal shape, it is a common procedure to determine the t_{lag} and/or t_{50} of your reaction by fitting the elongation phase using a linear function. The t_{lag} is the intercept between linear extrap-

olations of the lag phase (Fig. 4, dashed lines) and the elongation phase respectively (Fig. 4, continuous lines) (*see* **Note 29**).

8. The ThT endpoint intensity could be proportional to the amount of amyloid formed. This correlation differs among proteins and peptides; therefore it has to be determined/standardized for each protein by an independent assay, such as measuring the remaining soluble protein at the end of the reaction.

3.4 Negative Staining of Amyloid Fibrils for TEM

Amyloid fibrils are insoluble particles in the nanometer-micrometer-scale in length. This wide size range makes TEM an excellent method to visualize and validate the presence of amyloid structures [27], gain knowledge of the morphological properties of the aggregates and discriminate between protofibrils, mature fibrils and amorphous aggregates [28].

The most common TEM techniques is the negative staining with 2% uranyl acetate [29], using 200–400 mesh spacing copper that are carbon coated and covered in formvar.

1. Spin down the 4% uranyl acetate solution at 13,000 × *g* for 3 min to pellet any insolubilized uranyl acetate.
2. Filter the uranyl solution using a 0.22 μM. This solution can be stored for several weeks at 4 °C. Be aware that uranyl acetate may contain some traces of radioactive uranium and therefore the solution and the powder should be stored appropriately.
3. Place 3 μL of protein sample on the grid. After 3 min, apply a torn edge of hardened, ashless filter paper at the edge of the grid to absorb the excess liquid. (This process is known as wicking.)
4. Immediately place 3 μL of 2% uranyl acetate solution on the grid, wait 1 min, wick away excess solution, and air-dry (*see* **Note 30**).
5. The grids can be immediately examined or stored in an EM grid case prior to examination.
6. The grids can be imaged using an electron microscope operating at 80 keV (*see* **Note 31**).

3.5 Sedimentation Assay by HPLC

Sedimentation assay followed by HPLC for tracing amyloid transformation was developed by the Wetzel's laboratory [30] to study aggregation in Aβ and polyglutamine peptides. HPLC sedimentation assay is a quick and ingenious way to quantitate the concentration of the unreacted monomer in the aggregate suspension. In this section we describe how to use this method effectively to study the aggregation of globular proteins with minor modification of the original method described by the Wetzel laboratory.

The HPLC sedimentation assay consist of two parts: (1) creation of a standard curve; which requires plotting the integrated area obtained from HPLC peak against the mass of the soluble AL protein in the sample and (2) determining the concentration of the soluble protein present in the aggregation reaction mixture from the integrated peaks during the course of aggregation. It is important prior to running the experimental samples, to standardize the instrument and to characterize the reverse phase HPLC separation properties of your protein (including the choice of reverse phase column, flow rate, injection volume, gradient of the solvents, and wavelength of the detector). Once the set of parameters are standardized, the method works well for most of the protein/peptides studied. To study AL proteins, we used C8 (2.5 × 50 mm) analytical reverse phase column and a gradient of acetonitrile in 0.1% TFA maintained at 300 μL/min flow rate. The wavelength of the UV detector is fixed at 280 nm to selectively absorb the aromatic residues in proteins.

1. Prepare the AL proteins as described in Subheading 3.1 (Fig. 1A and B1). Make serial dilutions of the protein with the highest possible accuracy, using the protein stock and reestimate the concentration of the serially diluted samples by UV absorbance (Fig. 1C1). These samples are used to plot the standard curve (Fig. 5).
2. Inject the different samples into an RP-HPLC column (being consistent with the volume injected every time) in triplicates to ensure the quality of the calibration curve.
3. Construct the standard curve by plotting the integrated peak areas (obtained from reverse phase HPLC) as a function of protein concentration (calculated from UV-absorbance spectroscopy) and fit the curve to a liner regression. The slope of the curve obtained represents area/μg (Fig. 5).
4. The averaged value of the slope obtained from the triplicates is used to determine the concentration of unreacted soluble protein during the course of aggregation reaction (*see* **Note 32**).
5. After preparing the samples for aggregation reaction as described in Subheading 3.1, inject immediately ~20 μL aliquot (in triplicates) in the RP-HPLC column and note the concentration for the zero time (total soluble protein at the beginning of the reaction) (Fig. 1D1).
6. After injecting protein samples in the HPLC, place microcentrifuge tubes in the orbital shaker at 37 °C and 300 rpm and/or quiescent conditions as per the requirements of the aggregation experiment (*see* **Note 33**).
7. At predetermined time intervals, transfer an aliquot (typically 80 μL) of the suspension into an ultracentrifuge tube, spin at 160,000 × *g* for 45 min in an ultracentrifuge at 4 °C (*see* **Note**

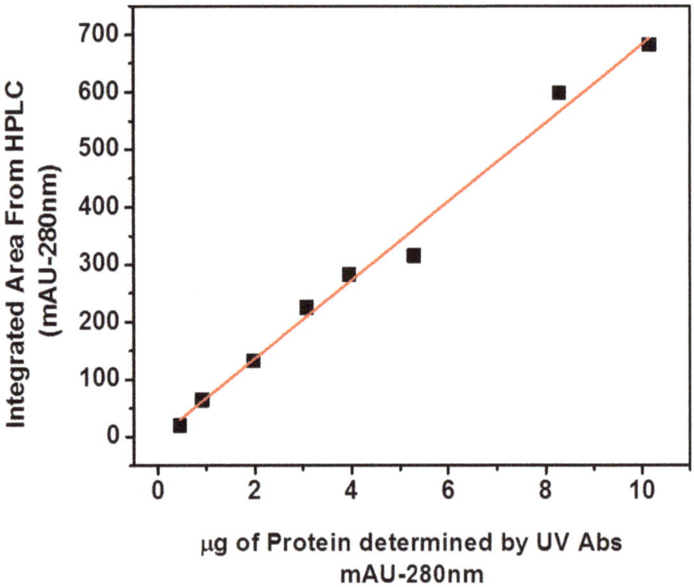

Fig. 5 Example of a standard curve of protein κI 018/08 V_L obtained after plotting the integrated area of the peak from reverse phase HPLC against mass in μg of serially diluted (in Tris buffer pH 7.4) protein

34). Repeat the process till the concentration of soluble protein plateaus at minima.

8. Carefully transfer 40 μL of the centrifuged solution and mix it with 20 μL of 1% formic acid and inject it in the reverse phase HPLC (*see* **Note 35**).
9. After every sample injection, blank injection (with the similar gradient as sample injection) was interspersed to ensure proper flushing of any residual protein from the column before the next run.
10. To determine the aggregation kinetics, plot the protein concentration (calculated from the integrated area and the standard curve for individual protein sample) as a function of the time elapsed.

3.6 Particle Size Analysis by DLS During Aggregation Reactions

Analysis of particle size by dynamic light scattering (DLS) is a versatile and accurate way of determining the monodisperse or polydisperse distribution of species during protein assembly. DLS computes the autocorrelation function by measuring the intensity of light scattered by the particles undergoing Brownian motion. Fitting of the autocorrelation function thus obtained provides diffusion coefficients of particles which can then be used to calculate the hydrodynamic radius from Stokes–Einstein equation [31].

DLS can be accurately employed to study the early events during the course of aggregation process. It suffers, however, from major drawbacks while studying the entire aggregation spectrum as during the elongation phase scattering of light might predominantly be contributed from a fraction of amyloid assembly in the suspension. Fibril length, flexibility and self-interaction in the amyloid assembly during the elongation phase might also lead to complexity in the mutual diffusion [31]. Although these complexities can be expressed by multiexponential correlation function and the apparent hydrodynamic radius calculated, the data is not reliable and leads to overestimation of particle size in a polydisperse sample.

1. For this protocol, we are using the cuvette-based DLS with 1 μM upper size limit for detection of hydrodynamic radius, as reference.
2. Prepare the samples of aggregation reaction for measurement as described in Subheading 3.1.
3. Turn on the DLS instrument, and start the software.
4. In the software interface of the instrument, set the viscosity and refractive index of the solvent for water before the start of scattering measurement. Also select the power for the laser beam.
5. Immediately take an aliquot of ~30 μL, and gently pipette it into cuvette using a gel tip (*see* **Note 36**).
6. Insert the cuvette into the DLS instrument cell holder and tightly close the lid to start the measurement. Repeat the process at predetermined time intervals.
7. Check the average intensity autocorrelation function, from at least 30 independent acquisitions from samples at early time point (*see* **Note 37**).
8. As the light scattered from a solution is extremely sensitive to sample impurities and dust, after every sample measurement the cuvette is cleaned several times in soap solution and ultrapure water.
9. Dry any excess of water sticking to the cuvette by compressed air.

3.7 Size Exclusion Chromatography (SEC)

SEC is a popular and wide spread analytical tool to study the in vitro amyloid aggregation, given its ability for quick interpretation of data and separation based on size/molecular weight (MW) of various species coexisting at any given time. Further, the variations resulting from shape and equilibrium condition among the heterogeneous population of aggregates is largely minimized due to significant difference in molecular weight between aggregates and prefibrillar species. The chromatogram obtained has optimal resolution and can be used to demonstrate a concomitant appearance or disappearance of species over a period of time. Furthermore the species detected during the course of aggregation can easily be isolated and processed for structural analysis and toxicity assays.

1. Prior to sample injection, filter all the solvents used as mobile phase (phosphate buffer at pH 6.8) using 0.2 μM membrane filters (*see* **Note 38**).
2. Inject molecular weight protein markers to obtain a standard curve for the column, by plotting the elution volume against MW of the protein standards (*see* **Note 39**).
3. Inject approximately ~150 μL of the aggregation reaction mixture (preparation described in Subheading 3.1).
4. Determine the apparent molecular weight from the elution volume of individual chromatographic peaks using the standard curve.
5. After a few injections, intercalate blank samples to verify the baseline of the column.

3.8 Nanoparticle Tracking Analysis

Similar to DLS, Nanoparticle Tracking Analysis (NTA) utilizes light scattering intensity from particles undergoing Brownian motion to determine the hydrodynamic radius of individual particles [32]. Due to the low refractive index of protein, particles ranging from 30 nm to 1 μm are ideal for NTA analysis [33]. NTA differs from DLS in that particles flow and are tracked using video and tracking software to simultaneously determine the values for a multitude of individual particles per frame. The characteristics/ properties of the sample population are thus available and overcome the problem of large aggregates within polydisperse samples obscuring the majority sample signal in DLS. With NTA, it is also possible to visualize sample particles in real time and results are available quickly. This is an advantage over electron microscopy and other methods prone to preparation artifact [34]. Calculations assume a spherical shape, so irregularly shaped particles are difficult to measure reliably. Thus, it is ideal to complement NTA analyses with methods such as DLS and electron microscopy to confirm measurements.

1. Assemble sample chamber, optical flat, and attach tubing. Insert chamber into the Nanosight NS300 device and power it on.
2. Launch NTA 2.3.5 Software Stage and Temperature Comms.
3. Launch software interface and within the main menu, **Connect Syringe Pump**.
4. Select **Capture** to activate camera display (*see* **Note 40**).
5. Load 1 mL disposable Luer lock syringe with ddH_2O. Expel a small amount of liquid against the sample syringe port to ensure there are no bubbles and attach. Slowly advance the

plunger to advance approximately 900 μL of liquid through the chamber (*see* **Note 41**).

6. Watch the camera display to ensure there are no air bubbles and the chamber is filling with fluid. Once the chamber is filled with fluid, adjust the **Camera Level** to the highest level that does not produce excessive background.
7. Repeat syringe flush at least twice to ensure that there are no contaminating particles left in the lines from the last user or air bubbles (*see* **Note 42**).
8. Load one syringe filled with PBS dilution buffer into the chamber, and check that it is indeed particle free.
9. Prepare diluted samples of particles of interest in PBS buffer. You will need at least 1 mL volumes to fill the sample syringe for measurement (*see* **Note 43**).
10. Fill a clean 1 mL syringe with sample, attach to syringe port, and now load syringe into the syringe pump. Run at Infusion rate of '250' until sample particles begin to appear on the screen. Slow the infusion rate down to an infusion rate of '25' (*see* **Note 44**).
11. Adjust **Focus** and **Detection Threshold** until you can clearly see particles with small light halos around them. Record these settings and keep them consistent for samples you want to compare between.
12. Select **Capture Duration** based on polydispersity of samples. For 10–40 particles per frame and medium polydispersity, the recommended capture is 60 s. For highly polydisperse samples, 90 s is preferable.
13. Click **Record.** Complete sample description screens and make folders for the data to be exported to. Be sure to record the dilution factor used (1:1000, etc.), the camera level, and detection threshold. Save when prompted.
14. **Process Sequence** to analyze the video and obtain data. Excel and PDF files will be exported and include a graphical representation of particle size vs. concentration, a numerical list of particles of particular bin sizes with concentrations, and finally a sample population mean, mode, standard deviation, and overall sample concentration. For disperse samples, the mode tends to be more descriptive than the mean values.
15. Multiple videos may be recorded from the same syringe, but consider sample settling in the syringe when evaluating values. Sometimes running the infusion pump rate at '250' for several seconds will sufficiently agitate the sample that you can get another good measurement (*see* **Note 45**).
16. Flush the lines with sufficient ddH_2O to ensure no particles are seen between syringes of sample.

17. Finally, the chamber must be cleaned with ddH_2O, 10% ethanol, and air. Chamber must be disassembled and dried thoroughly with compressed air and lens paper before storage (*see* **Note 46**).

3.9 In Vitro Assays for Light Chain Amyloidosis Cytotoxicity

MTT and LDH Cytotoxicity Assays are colorimetric assays to quantitatively measure cellular cytotoxicity (*see* **Note 47**). MTT assay quantifies cell proliferation by measuring the mitochondrial activity of the succinate dehydrogenase enzyme. LDH assay measures lactate dehydrogenase enzyme (LDH) released into the extracellular culture media from damaged cells. Both methods result in a formazan product that can be measured spectrophotometrically.

Always warm the culture media to 37 °C in incubator or water bath, 15 min before use. The age of the cultures, number of passages, and details of the growth medium are important factors for accurate cell viability results. Be consistent between experiments and avoid cells reaching confluent growth and high passages (< passage #25).

1. Thaw the cells by gently warming the vial tube containing the cells for 2 min at room temperature and 2 min at 37 °C.
2. Transfer contents of vial (approximately 1 mL) to 50 mL conical tube containing 24 mL of prewarmed media and mix gently.
3. Transfer all contents into a T75 cell culture flask.
4. Place flask in incubator overnight.
5. Next day, check for cell growth and change media.
6. Cell passage is kept at 20 before disposal.
7. To count the cells, aspirate cell media with a glass pipette.
8. Wash the cells with 25 mL PBS solution twice.
9. Add 3 mL 0.25% trypsin–EDTA to cells.
10. Put flask in incubator for no more than 2 min.
11. Gently rock the flask to help cell detachment.
12. Add 10 mL PBS to flask to inhibit the trypsin effect.
13. Transfer all contents of the cell flask into 50 mL conical tube.
14. Centrifuge the 50 mL conical tube at 225 × *g* for 5 min, at 22 °C.
15. Resuspend the pellet with 2–4 mL of prewarmed media until solution looks homogeneous.
16. Transfer 100 μL of cell/media solution to a 1.5 mL microcentrifuge tube and vortex.
17. Gently mix 10 μL cells with 10 μL of trypan blue stain 0.04% and apply 10 μL into a Countess Cell Chamber slide.
18. Count viable cell concentration using a Cell Countess.

19. Calculate the volume of cell/media solution that you need for splitting 3 × 10^5 cells in a new T75 flask. (*3 × 10^5 AC16 cells usually take 3 days to reach overconfluence.*)
20. Transfer the cells to a 50-mL conical tube containing 25 mL of prewarmed media and mix gently.
21. Transfer all contents into a T75 cell culture flask.
22. Place flask in incubator overnight.
23. Next day, check for cell growth and change media.
24. Calculate the volume of cell/media solution that you need for preparing 25 mL of prewarmed media containing 10,000 cells per mL.
25. Transfer the cells to 50-mL conical tube containing media and mix gently.
26. Plate 200 μL of cells at a concentration of 10,000 cells per mL (2000 cells/well) in a 96-well polystyrene plate.
27. Incubate the cells at 37 °C, 5% of CO_2, and allow them to reach 70% of confluence (within 24–48 h) before addition of the soluble protein.
28. To prepare the amyloid fibrils for the cell culture, if frozen, thaw the amyloid fibrils at room temperature from your stock, and gently vortex to resuspend the fibrils (*see* **Note 48**).
29. Dilute the samples to approximately 2 μM (substock) with cell culture media, such that the final concentration of fibrils in 96-well plate is around 1 μM.
30. Resuspend 68 μL from the substock in 1.6 mL of AC16 stock media.
31. For the co-culture experiments for cell growth effects, seed AC16 cardiomyocytes in a 96-well plate (*see* previous protocol).
32. Place the 96-well plate in the IncuCyte Incubator (*see* IncuCyte Protocol) and allow the cells to adhere to the bottom of the plate for 24 h.
33. Remove 96-well plate from incubator and remove media.
34. Add 100 μL of fresh DMEM:F12 and 100 μL of 2 μM fibril substock solution to each well of the plate. This concentration is a 1:1 preparation; dilutions may be created to achieve the desired concentration of plated cells, fibrils, and other components such as soluble proteins (*see* **Note 49**).
35. Place 96-well plate back in incubator. Plates can be scanned every 4 h or as needed.
36. Follow the incubation for 6 days using IncuCyte. IncuCyte ZOOM software was used to count cells by measuring number of cells with red fluorescence protein (RFP).

37. For the caspase activity assay, seed the cells as described above with either 1 μM fibrils or 1/12 μM soluble protein. Cell viability can/must be measured for each experimental condition (fibril/soluble protein) by 80 h using the IncuCyte ZOOM (*see* **Note 50**).
38. After 80 h, 100 μL of substrate working solution was added to each well of a 96-well plate. Incubate the plates for 1 h, at 37 °C, 5% CO_2.
39. Following incubation, read the plate using a fluorescence plate reader with the following setup: 480 nm on the excitation wavelength and 520 nm for the emission wavelength, using medium attenuator mode with a continuous lamp.
40. The collected data must be reported as averaged fluorescence units/well, from triplicates of each well and each condition.
41. For the assay using soluble protein, remove the media by aspiration and replace it with 100 μL of fresh media (*see* **Note 47** *for LDH assays, see* **Note 51** *for MTT assays*).
42. Add 100 μL of soluble protein to the cells at a different final protein concentration (we recommend concentrations within the range of 1–20 μM). Dilute the soluble proteins to the appropriated concentration using fresh cell culture media.
43. Incubate cells with soluble protein for various time points (24, 48, and 72 h).
44. Include a negative control made of 200 μL media alone (blank), along a positive control of cells without proteins (control).
45. For the mitochondrial viability assay, replace the cell media with 100 μL of fresh media without norepinephrine and phenol red (*see* **Note 51**).
46. Add 10 μL of 5 mg/mL MTT stock solution to each well (final concentration of 0.5 mg/mL per well) (*see* MTT reagent preparation in Subheading 2.9), and incubate the plate at 37 °C, 5% of CO_2, for 4 h.
47. Add 100 μL of DMSO to each well and incubate them overnight to solubilize the precipitates.
48. Read the absorbance in a plate reader device at 570 and 690 nm.
49. Subtract the 690 nm absorbance reading (reference) from the 570 nm absorbance values for each well.
50. Calculate the percentage of living cells using the following formula:

    % Living cells = (sample absorbance − blank absorbance) × 100/(control absorbance + blank absorbance).

51. For the Lactose Dehydrogenase Assay, transfer the 200 μL of media from each well into a new 96-well polystyrene plate.
52. Centrifuge the plate at 300 × *g* for 10 min, at 22 °C.
53. Transfer 100 μL of supernatant to a new plate carefully, without disturbing the cell pellet.
54. Add 100 μL of freshly prepared reaction mixture to each well, and incubate it at room temperature for 30 min, protected from light.
55. Read the absorbance in a plate reader device at 492 and 690 nm.
56. Subtract the 690 nm absorbance reading (reference) from the 492 nm values for each well.
57. Calculate the cytotoxic effect using the following formula:
    % Cytotoxicity = (sample absorbance - spontaneous absorbance) × 100 / (maximum absorbance - spontaneous absorbance).

3.10 In Vitro Assays to Analyze/Study Apoptosis in Light Chain Amyloidosis

Multiple studies implicate AL amyloid light chains in the induction of apoptotic events in a variety of target cells [35–40]. Consequently, it can be advantageous to utilize methodologies that detect characteristic apoptotic changes [41, 42] resulting from amyloidogenic cytotoxicity. Many indicators of apoptosis can be quantified via biochemical analyses. Common examples include: Caspase assays, which detect the activation of cysteine aspartate-specific proteases [43]; dUTP nick end labeling (TUNEL) assay to detect DNA fragmentation [44]; and Annexin V assay for flow cytometric detection of translocated cell-surface phosphatidylserine [45]. Alternatively, analyses of characteristic morphologic changes to apoptotic cells can be utilized. Among these, DNA staining of viable cells with fluorescent bisbenzimide (Hoechst) dyes in conjunction with fluorescence microscopy allows for ready visualization of chromatin condensation and related apoptotic events [46, 47].

Hoechst dyes offer some of the most frequently used reagents for dsDNA staining and characterization. The Hoechst dyes, in particular Hoechst 33258, permeable to intact cell membranes, bind to the minor groove of DNA at regions rich in AT sequences, and fluorescence bright blue emission upon excitation by ultraviolet light when in the bound state. This fluorescence intensifies with the staining of condensed chromatin, characteristic of apoptotic cells, allowing for an observable distinction from non-apoptotic cells [48]. Additional indicators of apoptosis, such as nuclear shrinkage and apoptotic body formation, can also be visualized with this method.

1. For staining DNA with Hoechst 33258, transfer an aliquot of treated cells from a suspension or prepared adherent culture to a 15 mL conical tube (*see* **Note 52**).
2. Sediment the cells by centrifugation at 225 × *g* for 10 min.
3. Aspirate the supernatant medium, reserving a nominal amount (~100 μL) of fluid above the cell pellet. Resuspend the cells by gently flicking the tube.
4. Wash the cells by adding 10 mL PBS and sediment by centrifugation at 225 × *g* for 10 min.
5. Aspirate the supernatant PBS, again reserving ~100 μL of fluid. Gently resuspend the cell pellet by flicking the tube (*see* **Note 53**).
6. Fix the cells by slowly adding 200 μL 3:1 MeOH:AcOH and gently mix with the cell suspension.
7. Add an additional 2 mL 3:1 MeOH:AcOH solution to the cells, mix gently, and incubate for 1 h at room temperature.
8. Sediment the cells by centrifugation at 225 × *g* for 10 min.
9. Aspirate the MeOH:AcOH, leaving ~400 μL in the tube.
10. Resuspend the fixed cells in the remaining liquid.
11. Deposit fixed cells onto a glass microscope slide by dispensing 100 μL of the cell suspension dropwise using a micropipette with a fine tip.
12. Allow the slide to air-dry.
13. Dispense a drop of 50% glycerol–50% 0.1 M Tris–HCl (pH 7.4) containing 1 μg/mL Hoechst 33258 to cover the cells.
14. Overlay a glass coverslip onto the slide. Affix the coverslip by sealing the edges with a thin layer of clear nail polish. Allow the sealant to air-dry (*see* **Note 54**).
15. Observe the samples using a fluorescence microscope with ultraviolet excitation.

4 Notes

1. The solubility of ThT on ethanol (100 g/L) is higher than on water (5 g/L), however water is preferred.
2. A stock solution (~10 mM ThT) protected from light and stored at 4 °C could last for several weeks without apparent damage, as long as they are carefully protected from light [49]. The concentration range between 10 and 60 μM ThT provides a good working concentration [23, 50].

3. The 10 mM sodium acetate, boric acid, and sodium citrate buffer (ABC) presents buffering capacity through a broad range of pH values (from pH 2 to 11). ABC buffer stored at room temperature last for up to 1 year.
4. The cell line has been immortalized by fusion with SV40 transformed fibroblast cell line devoid of mitochondrial DNA.
5. MTT stock can be stored at 4 °C for several months. It is recommendable to filter it before using it.
6. Suspension and/or adherent cell lines should be maintained using standard but cell-specific laboratory practices. Protocols for utilizing AC16 cardiomyocytes, which were analyzed in this staining method, are detailed in the "In Vitro ***Assays for Light Chain Amyloidosis Cytotoxicity***" section of this chapter.
7. There are three related primary Hoechst dyes: Hoechst 34580, Hoechst 33258, and Hoechst 33342, the latter two being the most commonly used for DNA staining [48]. Hoechst staining solutions can be stored at 4 °C for several weeks when protected from light [51].
8. Hoechst 33258 and Hoechst 33342 exhibit similar excitation/emission spectra (~350 nm/460 nm) [52]. This protocol utilized a microscope equipped with an excitation filter of 365 nm and a long pass emission filter of 420 nm.
9. It is important to estimate the amount of protein needed, prior to the experiment, in order to reduce variability associated with the use of different lots of protein. We recommend using an excess of protein in approximately the volume needed for three samples to avoid being short on protein.
10. We determine the correct speed (645,000 × *g*) needed to sediment preformed aggregates, using the following equation:

$$\log s = \log k + \frac{2}{3} \log M$$

$$t = \frac{k}{s}$$

where s = sedimentation coefficient.
k = Clearing factor (specific for the rotor used).
M = Molecular wt (of the species to be removed)

11. Mixing the centrifuged solution with your selected buffer at the desired pH before concentration is recommended as this step works as a buffer exchange, which eliminates the need to readjust the pH while setting up aggregation experiment.
12. It is desirable to use a concentrated stock of protein whenever it is possible (~400 μM). However this depends on the solubility

and stability of the protein under investigation. We suggest performing an experiment testing the dependence of aggregation as a function of the protein concentration.

13. We recommend taking aliquots for all the assays from the same reaction mixture at periodic intervals of time. The rationale behind this is to have a complete set of data that can be extrapolated over each other, if required, to accurately determine any variations in the structure and assembly of AL proteins during the course of aggregation.
14. The ThT assay on black 96-well polystyrene plate is mainly done to explore the changes in kinetics of fibril formation in soluble protein and ensure reproducibility with reported results. The reactions in microcentrifuge tubes display different kinetics, possibly due to the differences in the surface composition and the different air–water interfaces.
15. In order to monitor the stepwise conversion of soluble protein to mature fibrils and correlate it to kinetics of fibril formation, we prepare two sets of aggregation mixture, with and without ThT in low protein binding microcentrifuge tubes. Since subtle changes in the vessel geometry and material of the container affects the kinetics of fibril formation, we recommend using identical low protein binding microcentrifuge tubes to prepare the aggregation mixture with and without ThT. Aggregation mixture without ThT can be used to observe structural changes in the soluble protein and formation of intermediate species during the aggregation pathway which can later be used to correlate with the kinetics of fibril formation as determined by ThT assay.
16. We typically use 10 mM ABC buffer to study aggregation in AL proteins as it has a broad pH range (pH 2–11) which provides an opportunity to monitor AL proteins aggregation kinetics at different pH values. Most of the AL proteins aggregate at pH 2.0 and 4.0.
17. Care should be taken to set the incubator shaker at 37 °C at least an hour prior to the start of the experiment.
18. Maintaining a constant volume of aggregation mixture across individual experiments minimizes error in reproducibility.
19. The end of the aggregation reaction is precisely determined by either by ThT assay (described Subheading 3.2) or HPLC sedimentation assay (described in Subheading 3.5).
20. This allows for flexibility with amyloid formation assays as a function of protein concentration. In case you are using a different buffer, adjust the buffer concentration accordingly to keep the desired buffer concentration unaltered.

21. The use of a master mix prepared in the same way for every experiment is the best way to achieve consistency and good initial ThT values.
22. Reactions that occur at pH 8 or higher pH values and/or elevated temperatures can result in hydroxylation of ThT, resulting in photo bleaching [53, 54].
23. You can add water/buffer to all the unused wells, as well as seal the lids with lab tape to reduce the evaporation.
24. The initial readings may be unusually high, especially at some pH values; this is related to ThT reequilibration at the experimental temperature—Take a new measurement after 5 min and discard the high value from the analysis.
25. In general, fibril formation reaction data should be acquired until the reaction is completed (i.e., until a clear plateau value is reached). However, it is possible that the plateau region will present some fluorescence decay possibly due to photo bleaching.
26. Analyzing the kinetics of each replicate independently helps to avoid misinterpretation of the results, e.g., biphasic traces or poor reproducibility or the strong stochastic amyloidogenic character of some proteins.
27. When preparing the samples, the protein should be kept on ice until the beginning of the measurement, and the measurement is usually taken above room temperature (e.g., at 37 °C). Thus, because there will be thermal reequilibration of the sample as the experiment is being recorded and incubated (usually not more than 5 min) the signal may vary unexpectedly—for example, it may decrease because of the dependence of fluorescence emission with temperatures, and make sure not to include these initial data points into the analysis.
28. Microsoft Excel software provides a lot of flexibility to processing data; however nonlinear fitting and statistical analysis can be easier using more specialized software.
29. The concentration of ThT positive species in the lag phase is far below the detection limit, and therefore there will be only baseline ThT signal before a critical concentration of ThT positive species are formed [2]. The t_{50} value describes the time required for the ThT fluorescence intensity to reach the half point of the reaction; along with an apparent elongation rate (slope of the transition) that is associated to the rate of recruitment of soluble precursors to the growing fibrils. These parameters allow us to characterize the amyloid formation reactions in terms of the lag time and the elongation rate. The plateau at the final phase indicates the end of the reaction.

30. Careful handling of the grids is required to avoid artifacts that can be mistaken for protein aggregates. Some imperfections on the formvar by a poor coating can appear to be interesting shaped protein aggregates. In addition, a high amount of protein or aggregates can result in a "black grid" (over staining). However with practice and experience these artifacts can be easily detected and fixed.
31. Scanning the grids at low magnification (10,000–12,000×) can give an idea of the overall sample composition and then proceed to examine the grids at higher magnification (25,000×). Amyloid fibrils are typically linear, unbranched, and 5–10 nm in width; however it is not uncommon to find big tangles of fibril. Also, it is not uncommon to observe different aggregate morphologies on a single grid [9, 24], coexisting at any given time point. TEM is complimentary to the ThT assay.
32. The use of triplicates potentially minimizes the pipetting errors and errors associated with the HPLC autoinjection.
33. Depending on your needs, you can perform the fibril formation assay under quiescent or shaking conditions. We suggest performing a pilot experiment to determine which condition is the best for your specific protein/system. In general shaking of samples results in a more efficient amyloid fibril formation compared to quiescent conditions, however the amyloid fibrils obtained under quiescent conditions are longer and less tangled.
34. Most of the amyloid like aggregates and protofibrils can be pelleted at 160,000 × *g*. Elimination of preformed aggregates can be critical on the determination of the kinetics of amyloid formation, especially if you have interest in the kinetic parameters of the reaction.
35. Formic acid is known to prevent aggregation in some peptides in the HPLC vials. We have found that in case of AL protein, in the presence of 1% formic acid in the HPLC vials, aggregation is negligible for up to 6 h as determined by dynamic light scattering.
36. Using a gel tip to dispense sample significantly reduces introduction of air bubble inside the cuvette.
37. The hydrodynamic radii and percentage mass (density measure of the population density) of individual particles in the aggregation reaction mixture is measured from the average intensity autocorrelation function computed by the software provided with the instrument.
38. The filtration of the mobile phase is mandatory, especially if the solvent remain unused for at least a week. Avoid filtering the samples before injection as it could eliminate some low

molecular weight oligomers that could be interesting to characterize.

39. The election of the correct SEC column is critical for the determination of the apparent molecular weight of the species present/populated during early events of the fibril formation reaction.
40. Camera sensitivity (camera gain and shutter speed) and software settings will dramatically affect the ability to detect particles. This is defined within the NTA software package as Camera Level CL. To distinguish particles from background, it is possible to specify a Detection Threshold (a minimal light intensity obtained from a particle to identify it as a particle) and a Minimum Track Distance (the area of light must move against background to qualify as a particle).
41. High pressure will damage the system, so always introduce liquid slowly into the port.
42. An air bubble can be detected by the sample particles running at a deflected angle across the screen rather than horizontally. To remedy this, slowly flush out a syringe filled with air into the sample port, and then reflush with fluid until sample runs horizontally again.
43. Optimal concentrations will provide 10–40 particles per frame, which are approximately 10^7–10^9 particles per mL. This is critical to obtain clear images of individual particles without overlap. The Capture Time/Duration (long capture times result in extremely large data files) and whether the sample is monodisperse vs polydisperse will influence the ideal dilution factor and may require multiple attempts to optimize. If you discover that the sample you have introduced is excessively concentrated, stop advancing and immediately flush with copious amount of ddH_2O. Highly concentrated samples can dirty the lens and contribute to background.
44. For good reproducibility between experiments, it is critical to be consistent with the settings for image capture and run comparable samples together whenever possible. Chamber components are sensitive to damage by caustic solutions and solution at extreme pH values. The need to neutralize samples prior to NTA measurement may adversely affect labile protein aggregates and require use of an alternate method.
45. For our studies, triplicate samples are typically taken from each syringe with 10 s "pushes" in between.
46. Day to day changes in chamber age/cleanliness/background can dramatically affect measurements by requiring adjustments to Camera Level and Detection Threshold. The chamber component sensitivity is the primary reason why chamber

cleanliness/background is a factor. This makes the chamber difficult to clean well, so they are meant to be disposable when excessively dirty. However, they are not inexpensive, so it is ideal to get as many measurements from them as possible.

47. Both human and animal sera contain LDH, which may increase the background absorbance of the assay. Therefore, for LDH assays, replace the 12.5% FBS media for 1% FBS fresh media before adding proteins to the cells. Follow instructions from LDH Cytotoxicity Detection Kit User Manual.
48. For fibril formation procedures see reference [40].
49. Fibrillar substocks must be vortexed frequently to ensure a good homogeneity when adding to the 96-well plate. Amyloid fibrils are heavy and will collect at the bottom of tube, thereby skewing the total amount of fibrils plated per well.
50. Apoptotic reagent CellPlayer kinetic caspase-3/7 (Essen Bioscience) was initially used as it allowed for kinetic quantification of cell proliferation over time. However, this reagent became bound to the external amyloid aggregates and emitted a green fluorescence to give off a false positive signal. In order to avoid false positives, the caspase assay kit can be used (Roche) following the instructions from Homogeneous Caspases Assay, fluorimetric kit User Manual.
51. The presence of phenol red in the cell culture media may affect the results of the mitochondrial viability assay. Therefore, for MTT assays, it is recommended to replace the phenol red-containing media for phenol-red free media before adding proteins to the cells.
52. The treatment groups for these experiments included AC16 cardiomyocytes incubated with 12 μM soluble amyloidogenic light chains. Additionally, control groups for comparison included negative (media only), apoptosis positive (camptothecin induced), as well as controls for any specific solvents in use (e.g., DMSO).
53. Depending on cell type, the cell pellet may not always suspend readily with simple flicking of the tube. Should this be the case, a p1000 pipette may be used to gently draw the cells up and down to create a uniform suspension. In any instance, vigorous agitation of the cells should be avoided.
54. Prepared sample slides can be stored in the dark at 4 °C.

Acknowledgments

The methods described in this chapter have been published for the most part in research articles, supported by the National

Institutes of Health (R01 GM 071514), the Mayo Foundation, the Seidler Professorship and Dr. Morie Gertz, and the generous donations of amyloidosis patients and their families. Luis M. Blancas-Mejia and Pinaki Misra contributed equally to this work.

References

1. Eisenberg D, Jucker M (2012) The amyloid state of proteins in human diseases. Cell 148(6): 1188–1203. https://doi.org/10.1016/j.cell.2012.02.022
2. Ferrone F (1999) Analysis of protein aggregation kinetics. Methods Enzymol 309:256–274
3. Knowles TP, Vendruscolo M, Dobson CM (2014) The amyloid state and its association with protein misfolding diseases. Nat Rev Mol Cell Biol 15(6):384–396. https://doi.org/10.1038/nrm3810
4. Khurana R, Coleman C, Ionescu-Zanetti C, Carter SA, Krishna V, Grover RK, Roy R, Singh S (2005) Mechanism of thioflavin T binding to amyloid fibrils. J Struct Biol 151(3):229–238. https://doi.org/10.1016/j.jsb.2005.06.006
5. Wall J, Murphy CL, Solomon A (1999) In vitro immunoglobulin light chain fibrillogenesis. Methods Enzymol 309:204–217
6. Randles EG, Thompson JR, Martin DJ, Ramirez-Alvarado M (2009) Structural alterations within native amyloidogenic immunoglobulin light chains. J Mol Biol 389(1):199–210. https://doi.org/10.1016/j.jmb.2009.04.010
7. Baden EM, Owen BA, Peterson FC, Volkman BF, Ramirez-Alvarado M, Thompson JR (2008) Altered dimer interface decreases stability in an amyloidogenic protein. J Biol Chem 283(23): 15853–15860. https://doi.org/10.1074/jbc.M705347200
8. DiCostanzo AC, Thompson JR, Peterson FC, Volkman BF, Ramirez-Alvarado M (2012) Tyrosine residues mediate fibril formation in a dynamic light chain dimer interface. J Biol Chem 287(33):27997–28006. https://doi.org/10.1074/jbc.M112.362921
9. Blancas-Mejia LM, Tischer A, Thompson JR, Tai J, Wang L, Auton M, Ramirez-Alvarado M (2014) Kinetic control in protein folding for light chain amyloidosis and the differential effects of somatic mutations. J Mol Biol 426(2):347–361. https://doi.org/10.1016/j.jmb.2013.10.016
10. Blancas-Mejia LM, Ramirez-Alvarado M (2016) Recruitment of light chains by homologous and heterologous fibrils shows distinctive kinetic and conformational specificity. Biochemistry 55(21): 2967–2978. https://doi.org/10.1021/acs.biochem.6b00090
11. Sikkink LA, Ramirez-Alvarado M (2008) Salts enhance both protein stability and amyloid formation of an immunoglobulin light chain. Biophys Chem 135(1–3):25–31. https://doi.org/10.1016/j.bpc.2008.02.019
12. Nietzki R (1901) Chemie der organischen Farbstoffe, 4th edn. Springer, Berlin
13. A palette of fluorescent thiophene-based ligands for the identification of protein aggregates. http://doi.wiley.com/10.1002/chem.201502999
14. dye|World dye variety 2017. http://www.worlddyevariety.com/basic-dyes/basic-yellow-1.html
15. Biancalana M, Koide S (2010) Molecular mechanism of Thioflavin-T binding to amyloid fibrils. Biochim Biophys Acta 1804(7): 1405–1412. https://doi.org/10.1016/j.bbapap.2010.04.001
16. Vassar PS, Culling CF (1959) Fluorescent stains, with special reference to amyloid and connective tissues. Arch Pathol 68:487–498
17. Naiki H, Higuchi K, Hosokawa M, Takeda T (1989) Fluorometric determination of amyloid fibrils in vitro using the fluorescent dye, thioflavin T1. Anal Biochem 177(2):244–249
18. LeVine H 3rd (1993) Thioflavine T interaction with synthetic Alzheimer's disease beta-amyloid peptides: detection of amyloid aggregation in solution. Protein Sci 2(3):404–410. https://doi.org/10.1002/pro.5560020312
19. LeVine Iii H (1999) [18] quantification of β-sheet amyloid fibril structures with thioflavin T. Methods Enzymol 309:274–284. https://doi.org/10.1016/S0076-6879(99)09020-5 Academic Press
20. Singh PK, Kumbhakar M, Pal H, Nath S (2010) Ultrafast bond twisting dynamics in amyloid fibril sensor. J Phys Chem B 114(7):2541–2546. https://doi.org/10.1021/jp911544r
21. Groenning M (2010) Binding mode of Thioflavin T and other molecular probes in the context of amyloid fibrils-current status.

J Chem Biol 3(1):1–18. https://doi.org/10.1007/s12154-009-0027-5

22. Gade Malmos K, Blancas-Mejia LM, Weber B, Buchner J, Ramirez-Alvarado M, Naiki H, Otzen D (2017) ThT 101: a primer on the use of thioflavin T to investigate amyloid formation. Amyloid 24(1):1–16. https://doi.org/10.1080/13506129.2017.1304905
23. Malmos KG, Bjerring M, Jessen CM, Nielsen EH, Poulsen ET, Christiansen G, Vosegaard T, Skrydstrup T, Enghild JJ, Pedersen JS, Otzen DE (2016) How Glycosaminoglycans promote fibrillation of Salmon calcitonin. J Biol Chem 291(32):16849–16862. https://doi.org/10.1074/jbc.M116.715466
24. Blancas-Mejia LM, Hammernik J, Marin-Argany M, Ramirez-Alvarado M (2015) Differential effects on light chain amyloid formation depend on mutations and type of glycosaminoglycans. J Biol Chem 290(8):4953–4965. https://doi.org/10.1074/jbc.M114.615401
25. Meisl G, Kirkegaard JB, Arosio P, Michaels TC, Vendruscolo M, Dobson CM, Linse S, Knowles TP (2016) Molecular mechanisms of protein aggregation from global fitting of kinetic models. Nat Protoc 11(2):252–272. https://doi.org/10.1038/nprot.2016.010
26. Arosio P, Michaels TC, Linse S, Mansson C, Emanuelsson C, Presto J, Johansson J, Vendruscolo M, Dobson CM, Knowles TP (2016) Kinetic analysis reveals the diversity of microscopic mechanisms through which molecular chaperones suppress amyloid formation. Nat Commun 7:10948. https://doi.org/10.1038/ncomms10948
27. Serpell LC, Sunde M, Benson MD, Tennent GA, Pepys MB, Fraser PE (2000) The protofilament substructure of amyloid fibrils. J Mol Biol 300(5):1033–1039. https://doi.org/10.1006/jmbi.2000.3908
28. Shirahama T, Cohen AS (1967) High-resolution electron microscopic analysis of the amyloid fibril. J Cell Biol 33(3):679–708
29. Fandrich M (2007) On the structural definition of amyloid fibrils and other polypeptide aggregates. Cell Mol Life Sci 64(16):2066–2078. https://doi.org/10.1007/s00018-007-7110-2
30. O'Nuallain B, Thakur AK, Williams AD, Bhattacharyya AM, Chen S, Thiagarajan G, Wetzel R (2006) Kinetics and thermodynamics of amyloid assembly using a high-performance liquid chromatography-based sedimentation assay. Methods Enzymol 413:34–74. https://doi.org/10.1016/S0076-6879(06)13003-7
31. Lomakin A, Benedek GB, Teplow DB (1999) [27] monitoring protein assembly using quasi-elastic light scattering spectroscopy. Methods Enzymol 309:429–459. https://doi.org/10.1016/S0076-6879(99)09029-1
32. Wright M (2012) Nanoparticle tracking analysis for the multiparameter characterization and counting of nanoparticle suspensions. Methods Mol Biol 906:511–524. https://doi.org/10.1007/978-1-61779-953-2_41
33. (2017) High resolution sizing and concentration measurements of sub-visible protein aggregates using NTA. Technical note
34. Gross J, Sayle S, Karow AR, Bakowsky U, Garidel P (2016) Nanoparticle tracking analysis of particle size and concentration detection in suspensions of polymer and protein samples: influence of experimental and data evaluation parameters. Eur J Pharm Biopharm 104:30–41. https://doi.org/10.1016/j.ejpb.2016.04.013
35. Shi J, Guan J, Jiang B, Brenner DA, Del Monte F, Ward JE, Connors LH, Sawyer DB, Semigran MJ, Macgillivray TE, Seldin DC, Falk R, Liao R (2010) Amyloidogenic light chains induce cardiomyocyte contractile dysfunction and apoptosis via a non-canonical p38alpha MAPK pathway. Proc Natl Acad Sci U S A 107(9):4188–4193. https://doi.org/10.1073/pnas.0912263107
36. Sikkink LA, Ramirez-Alvarado M (2010) Cytotoxicity of amyloidogenic immunoglobulin light chains in cell culture. Cell Death Dis 1:e98. https://doi.org/10.1038/cddis.2010.75
37. Migrino RQ, Truran S, Gutterman DD, Franco DA, Bright M, Schlundt B, Timmons M, Motta A, Phillips SA, Hari P (2011) Human microvascular dysfunction and apoptotic injury induced by AL amyloidosis light chain proteins. Am J Physiol Heart Circ Physiol 301(6):H2305–H2312. https://doi.org/10.1152/ajpheart.00503.2011
38. Shi J, Guan J, Jiang B, Brenner DA, del Monte F, Ward JE, Connors LH, Sawyer DB, Semigran MJ, Macgillivray TE, Seldin DC, Falk R, Liao R (2010) Amyloidogenic light chains induce cardiomyocyte contractile dysfunction and apoptosis via a non-canonical p38α MAPK pathway. Proc Natl Acad Sci U S A 107(9):4188–4193. https://doi.org/10.1073/pnas.0912263107
39. Mishra S, Guan J, Plovie E, Seldin DC, Connors LH, Merlini G, Falk RH, MacRae CA, Liao R (2013) Human amyloidogenic light chain proteins result in cardiac dysfunction, cell death, and early mortality in zebrafish. Am J Physiol Heart Circ Physiol 305(1):H95–H103. https://doi.org/10.1152/ajpheart.00186.2013
40. Marin-Argany M, Lin Y, Misra P, Williams A, Wall JS, Howell KG, Elsbernd LR, McClure M, Ramirez-Alvarado M (2016) Cell damage in light chain amyloidosis: fibril internalization,

toxicity and cell-mediated seeding. J Biol Chem 291(38):19813–19825. https://doi.org/10.1074/jbc.M116.736736

41. Kerr JFR, Wyllie AH, Currie AR (1972) Apoptosis: a basic biological phenomenon with wide-ranging implications in tissue kinetics. Br J Cancer 26(4):239–257
42. Allen RT, Hunter WJ, Agrawal DK (1997) Morphological and biochemical characterization and analysis of apoptosis. J Pharmacol Toxicol Methods 37(4):215–228. https://doi.org/10.1016/S1056-8719(97)00033-6
43. Gurtu V, Kain SR, Zhang G (1997) Fluorometric and colorimetric detection of caspase activity associated with apoptosis. Anal Biochem 251(1):98–102. https://doi.org/10.1006/abio.1997.2220
44. Gavrieli Y, Sherman Y, Ben-Sasson S (1992) Identification of programmed cell death in situ via specific labeling of nuclear DNA fragmentation. J Cell Biol 119(3):493–501
45. Koopman G, Reutelingsperger CP, Kuijten GA, Keehnen RM, Pals ST, van Oers MH (1994) Annexin V for flow cytometric detection of phosphatidylserine expression on B cells undergoing apoptosis. Blood 84(5): 1415–1420
46. Wyllie AH, Kerr JFR, Currie AR (1980) Cell death: the significance of apoptosis. Int Rev Cytol 68:251–306. https://doi.org/10.1016/S0074-7696(08)62312-8
47. Kaufmann SH, Desnoyers S, Ottaviano Y, Davidson NE, Poirier GG (1993) Specific proteolytic cleavage of poly(ADP-ribose) polymerase: an early marker of chemotherapy-induced apoptosis. Cancer Res 53(17): 3976–3985
48. Spence IJMTZ (2010) Molecular probes handbook: a guide to fluorescent probes and labeling technologies. 11th edn. Invitrogen
49. abcam Thioflavin T (ab120751) Product datasheet. http://www.abcam.com/thioflavin-t-ab120751.html. Accessed 16 Dec 2016
50. Kroes-Nijboer A, Lubbersen YS, Venema P, van der Linden E (2009) Thioflavin T fluorescence assay for beta-lactoglobulin fibrils hindered by DAPH. J Struct Biol 165(3):140–145. https://doi.org/10.1016/j.jsb.2008.11.003
51. Tinari A, Giammarioli AM, Manganelli V, Ciarlo L, Malorni W (2008) Chapter one analyzing morphological and ultrastructural features in cell death. Methods Enzymol 442:1–26. https://doi.org/10.1016/S0076-6879(08)01401-8 Academic Press
52. Latt SA, Stetten G (1976) Spectral studies on 33258 Hoechst and related bisbenzimidazole dyes useful for fluorescent detection of deoxyribonucleic acid synthesis. J Histochem Cytochem 24(1):24–33. https://doi.org/10.1177/24.1.943439
53. Hackl EV, Darkwah J, Smith G, Ermolina I (2015) Effect of acidic and basic pH on Thioflavin T absorbance and fluorescence. Eur Biophys J 44(4):249–261. https://doi.org/10.1007/s00249-015-1019-8
54. Fodera V, Groenning M, Vetri V, Librizzi F, Spagnolo S, Cornett C, Olsen L, van de Weert M, Leone M (2008) Thioflavin T hydroxylation at basic pH and its effect on amyloid fibril detection. J Phys Chem B 112(47):15174–15181. https://doi.org/10.1021/jp805560c

Part II

Cellular and Proteostasis Assays

Chapter 9

Monitoring Aggregate Clearance and Formation in Cell-Based Assays

Evelien Eenjes, Young Joo Yang-Klingler, and Ai Yamamoto

Abstract

Understanding the fundamental mechanism underlying the accumulation and clearance of misfolded proteins can lead to insights into the synthetic and degradative pathways that maintain the homeostasis of proteins in all cells. Given the interconnection between protein homeostasis and cell health, as well as the complexity of aggregate formation and the degradation pathways with which it is intertwined, the design of the tools that are used to examine protein aggregation and accumulation can have a profound impact on the interpretation of results. We rely on two previously published stable cell lines that use conditional expression and the ligand-receptor tag known as HaloTag, to temporally distinguish distinct pools of aggregates, and use a combination of biochemical- and imaging-based methods to measure aggregation of a canonical aggregation-prone protein. We measure aggregate load biochemically using Filter Trap Analysis, which combines a filter trap retardation assay and immunoblotting to measure detergent soluble and insoluble protein levels, and visually, using confocal microscopy to monitor simultaneously aggregate formation and growth events in the background of aggregate clearance. As a secondary screen to more simplistic screen based approaches, this method permits further insight into how aggregate load is affected.

Key words Aggregation, Aggregate clearance, Autophagy, Filter trap assay, Cell-based systems

1 Introduction

Maintaining protein homeostasis is fundamental to cell health, and thus it is perhaps unsurprising that cells are equipped with multiple mechanisms to handle problems when they arise [1]. From protein refolding events to different means of degradation, elucidating the players involved in these often well-conserved pathways not only provides fundamental insights, but may also shed light onto pathogenic mechanisms underlying disease. A notable phenomenon in adult disorders of the brain is the abnormal accumulation of ubiquitinated proteins that culminate into intracellular proteinaceous inclusions or aggregates, a neuropathological hallmark that is so prevalent it has diagnostic value. To what extent these structures contribute to disease is unclear, but given that in healthy cells, mis-

Cláudio M. Gomes (ed.), *Protein Misfolding Diseases: Methods and Protocols*, Methods in Molecular Biology, vol. 1873, https://doi.org/10.1007/978-1-4939-8820-4_9, © Springer Science+Business Media, LLC, part of Springer Nature 2019

folded proteins that fail to be refolded are quickly trafficked for degradation, the appearance of these aggregated structures clearly signal something has gone awry.

The prevalence of protein aggregation in diseases has placed particular emphasis on the study of protein aggregation and clearance, which is often studied using cell-based approaches designed around canonical aggregation-prone proteins such as the expanded polyglutamine proteins. Given the strong potential for cytotoxicity and the complexity of aggregate-formation, the design of these assays is critical to provide interpretable information, especially when probing or screening for modifiers of protein accumulation. For example, if a modifier leads to a reduction in the total number of aggregates, using transient or constitutive overexpression based approaches makes it difficult to distinguish if the modifier impacts the formation of aggregates or the clearance of preexisting ones. Moreover, given the diverse number of structures that are represented by the term, "protein aggregate," and the variety of events represented by fluorescent puncta, how a given modifier exerts its effect can be additionally clouded.

To identify modifiers of protein aggregation and clearance, we use two, distinct cell-based assays in a complementary, biochemistry- and imaging-based approach [2–4]. Fundamental for both of these lines is the use of conditional expression or labeling of the canonical aggregation prone protein. Conditional expression is achieved using tetracycline-regulated expression [5], and conditional labeling is achieved using fluorescent dyes that covalently bind to tags such as HaloTag [6]. Using these previously published cell lines as an example, we describe the process through which we screen proteins for their role in protein accumulation versus clearance.

2 Materials

2.1 Cell Lines

The aggregation prone protein central to the two stable cell lines is an expanded polyglutamine (polyQ) protein, which upon a certain range becomes highly aggregation-prone, even in vitro [7, 8]. The lines express a construct encoding an NH_3-terminus fragment of the protein huntingtin (Htt) with an expanded polyQ stretch of 103 glutamines (103Q), which is an expansion length far greater than the expansions typically associated with disease [9]. The first cell line expresses a 17-amino acid Htt fragment with 103Q fused to a monomeric mutant of eCFP (17aaHtt103QmCFP) [4], whereas the second expresses a fragment of Htt encoded by exon1 with 103Q fused to HaloTag (HT)(Exon1Htt103Q-HT) [2]. Both constructs are conditionally expressed in HeLa cells using the tTA-tetO expression system designed by Bujard and colleagues [10]. These cell lines can easily be created in-house should it be necessary for other aggregation-prone proteins. This system permits constitutive expression that is abolished upon the administration of tetracycline (tet) or its analogs. We use doxycycline (dox) at 1 μg/mL to abolish expression, although lesser amounts are also sufficient. Although the tet-regulatable systems have been redesigned to also permit conditional induction of expression, we specifically avoid this approach for several reasons. First, dox is not inert, and thus its presence can potentially interfere unknowingly to outcomes [11–27]. Second, dox is lipophilic, and thus washout times can be problematic. The chemical profile of dox is particularly important given that we selected for, and thereby maintain, the cell lines in a gene-on condition. This is to ensure that the cell lines we use demonstrate no basal toxicity. Although transient overexpression can be quite toxic to cells, steady low-level of expression does not cause outright toxicity. Given that outright toxicity can confound results, establishing cell lines that have achieved a state in which the aggregation event exerts sufficient protection permits better analytic clarity.

1. Maintenance of the stable cell lines: 17aaHtt103Q-mCFP and Exon1htt103Q-HT stable cell lines were created as previously described [2, 4]. Cells are maintained in DMEM + 10% FBS with 100 μg/mL hygromycin and 50 μg/mL neomycin to maintain the presence of the polyQ protein and tTA, respectively. Cells are maintained in the "gene-on" state (i.e., in the absence of doxycycline). Upon thawing, the stable cell lines are maintained up to ten passages. After ten passages, newly thawed cells are highly recommended. When plating cells for experiments cell culture media without hygromycin and neomycin is used. We do not use penicillin and streptomycin regularly in the culture media. This is to prevent the unknowing propagation of mycoplasm and other contaminants.
2. Doxycycline (Dox): Dissolve 1 mg of Dox in 1 mL DMEM. Filter the solution with a 0.2 μm filter and store as aliquots in a −20 °C freezer. Dox is light sensitive.

2.2 Filter Trap Analysis (FTA)

The FTA combines the filter trap retardation assay (FT) to measure levels of detergent-insoluble proteins and western blotting (WB) using SDS-PAGE gels to measure levels of detergent-soluble proteins. Unlike SDS-PAGE which is transferred onto membranes which a high affinity for protein, the filter trap assay is captures proteins onto cellulose acetate, which has a low affinity for protein. The protein that is captured would therefore have to exceed the size of the membrane pores, which in this case is 0.2 μm.

1. Vacuum trap system: Bio-Dot and Bio-Dot SF microfiltration unit. The slot format provides more even loading and easier quantification than the dot format. The unit should be prepared

as recommended by the manufacturer. Given that the assay requires vacuum filtration, ensuring that the unit is well sealed is key, and thus the suggested use of the thicker filter paper is strongly recommended.

2. Membrane for FT: Cellulose acetate 0.2 μm.
3. Buffer for FT: 1× PBS, 0.1% SDS.
4. Multichannel and single channel pipettes: 8-channel pipettes that hold at least 200 μL is particularly useful for washes and loading empty slots, but is optional.
5. WB: A standard SDS-PAGE-based approach is sufficient. We use the precast mini NuPage gels for the Novex gel running systems. 4–12% Bis-Tris gels run in the appropriate running buffer is used. Gels are transferred onto PVDF membranes.
6. Signal detection for WB and FT: Any detection method that is optimized for your system is sufficient. The detection listed in Subheading 3.1, **step 9** is optimized for our system, which is the Vesadoc imaging system by Bio-Rad. Blot developing is optimized for Pierce HRP-conjugated secondaries (Thermo Fisher) and ECL kits we use are SuperSignal WestDura Substrates (ThermoFisher Scientific) or Clarity ECL (Bio-Rad).
7. 8 M Urea lysis buffer: PBS, 1% Triton X-100, 8 M urea, 5 mM EDTA, 100 mM NaCl and 50 mM Tris–HCl. Make fresh and add 1× protease/phosphatase inhibitor cocktail freshly to the lysis buffer before use.
8. 6 well plates.
9. 1× PBS.
10. DMEM + 10% FBS.
11. Trypsin (0.25%).
12. 15 mL falcon tubes.
13. Refrigerated centrifuge.
14. Eppendorf tube.
15. Ice.
16. Sample dilution buffer: 8 M urea lysis buffer, 1× LDS sample buffer, 50 mM 1,4-dithiothreitol (DTT).
17. Protein quantification: Biorad DC protein assay. 96-well format. Protein standard used is gamma globulin.
18. PBS-Tween: 1× PBS, 1% Tween-20.
19. PBS +/+: 1× PBS, 100 μM $CaCl_2$ and 1 mM $MgCl_2$.

2.3 Preparation of Coverslips for Imaging

In parallel to FTA, confocal or epi-fluorescent based imaging approaches are used to count the number of aggregates per number of nuclei. For the 17aaHtt103Q-mCFP line the approach is fairly standard, and will not be described in depth here. Instead we will provide more indepth description using the HT system.

1. Cover glass: 12 mm circles for 24 well culture plates.
2. HaloTag dyes (Promega): Oregon Green (OG) and tetramethylrhodamine (TMR) are aliquoted and stored at −20 °C.
3. 4% paraformaldehyde (PFA) made into PBS(+/+): Make fresh from stock, such as 16% PFA ampoules purchased from Electron Microscopy Science.
4. Hoechst 33342: 1 μg/mL (or another marker of nuclei).
5. Prolong Gold antifade reagent or any other mounting medium.
6. Microscope slides.

3 Methods

3.1 Sample Preparation for FTA

The combined analyses of FTA allows for the simultaneous monitoring of detergent insoluble and soluble protein from a given sample, and the additional ability to run a "loading control" which is not possible with filter trap alone (Fig. 1). Rather than the simple 1% Tx-100 lysis buffer recommended under Subheading 2.2 **step 7**, additional lysis buffers can be used, but we have found that the addition of 8 M urea is important for this step, to maximize recovery of the aggregated proteins which are often lost when creating a postnuclear supernatant (PNS). The incubation also helps denature DNA and RNA and prevents the "goopiness" of lysates, often encountered when exposing total lysates to detergents. This approach can be used for either of the two cell lines.

1. Plate 2.4×10^5 cells/6-well in a 6-well plate. To collect enough protein, typically two 6-wells per condition are required (*see* **Note 1**).
2. The next day, treat cells as desired. For example, a gene of interest can be transfected, or a chemical compound can be administered. For this example, we will add Dox to shut off gene expression and monitor aggregate clearance: To determine the amount of aggregate clearance that occurs 0, 24, 48 or 72 h after suppression of gene expression, Dox at a final concentration of 1 μg/mL to the wells at 72, 48, 24, or 0 h prior to collection. For the 0 h time point, add DMEM alone to serve as vehicle control.
3. On sample collection day, precool the centrifuge to 4 °C.

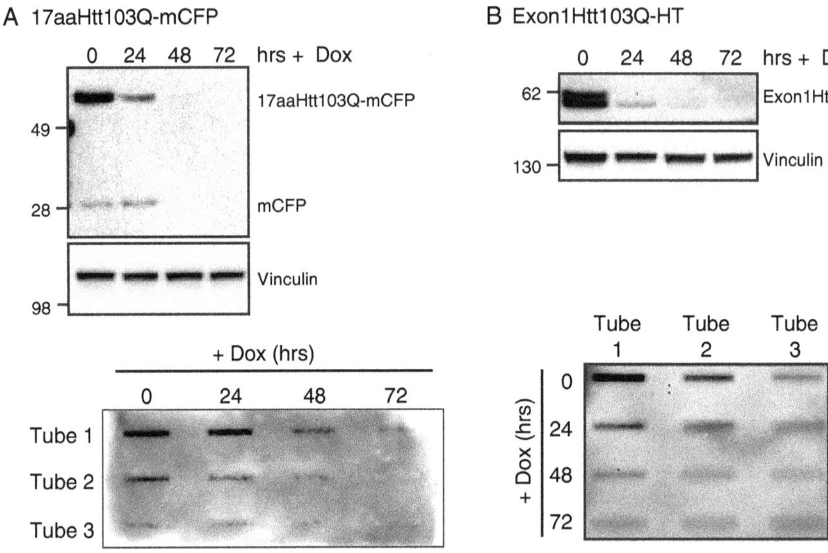

Fig. 1 Sample FTA outputs. Cell lines were treated with 0, 24, 48, or 72 h of Dox to abolish gene expression. (**a**) 17aaHtt103Q-mCFP cell line. (top) WB of the 20 μg of sample. Blots were probed with mouse anti-GFP (Roche) to probe for the expanded polyQ protein and rabbit anti-vinculin as a loading control. A large reduction of soluble 17aaHtt103Q-mCFP protein can already be observed after 24 h. The lower band around 30 kDa most like represents mCFP that has been cleaved from the protein, given that the construct is designed to eliminate the putative start codon before mCFP. (btm) FT of the serial dilution probed with an anti-GFP antibody. Tubes 1, 2, and 3 represent the tubes generated in Subheading 2.2, **items 16** and **17**, which is the threefold serial dilution. Detergent insoluble protein begins to decline at 24 h. (**b**) Exon1Htt103Q-HT cell line. WB and FT probed with the anti-HaloTag antibody. Images are modified from ref. 2

4. Wash the wells once with 2 mL PBS +/+, then trypsinize cells with 0.5 mL trypsin solution to detach cells from the plate (incubate for 10 min at 37 °C). Prepare 15 mL falcon tubes for **step 5** (*see* **Note 2**).
5. To halt the trypsinization, add 2 mL DMEM + 10% FBS, and transfer cells to a 15 mL falcon tube. Add an additional 3 mL DMEM + 10% FBS.
6. Pellet cells at 1000 × *g* in a precooled centrifuge. Prepare eppendorf tubes for **step 8**.
7. Remove media–trypsin supernatant and place tube containing the pellet on ice. Although pellets can be frozen at −80 °C, it is better to continue through **step 11** before stopping.
8. Resuspend the cell pellet in 100 μL 8 M urea lysis buffer with phosphatase/protease inhibitor, and transfer to an eppendorf tube.
9. Incubate on ice for 20 min (*see* **Note 3**). Ready eppendorf tubes to collect the supernatant for **step 11**.

10. Centrifuge the samples on a cooled benchtop centrifuge at 14,000 rpm (20,817 × *g*) for 5 min at 4 °C.
11. Collect supernatant and discard pellet immediately (*see* **Note 4**). The supernatant can then be flash-frozen and stored at −80 °C to permit analyses later.
12. Measure the protein concentration of the samples. The DC protein assay is amenable to the presence of some detergents and urea. The expected amount of total protein is approximately 150 μg.
13. Create equal concentrations of the samples in the sample dilution buffer. Label tubes for **steps 14** and **16**.
14. Set aside 20 μg of protein for WB and leave on ice. WB should be run the same day. Western blotting instructions can be found in Subheading 3.2.
15. Prepare the cellulose acetate membrane and the two pieces of filter paper for FT by soaking it in Buffer for FT (PBS-0.1% SDS). This will be used for FT. Leave this membrane in the buffer while preparing the samples.
16. Dilute 120 μg of the sample in sample dilution buffer to a volume of 300 μL (Tube 1) (*see* **Note 5**).
17. For each sample, prepare two additional eppendorf tubes labeled Tube 2 and Tube 3. In each tube, place 200 μL of sample dilution buffer.
18. Make a threefold serial dilution by taking 100 μL of Tube 1 and adding it to Tube 2. Mix well, then take 100 μL from Tube 2 and add it to Tube 3. The serial dilution permits a cleaner and more reliable readout for FT.

3.2 FTA

FT is a very simple approach, but requires some nuance to ensure clean, reliable results. Critical for this assay if the preparation of the vacuum unit. If the unit is not correctly sealed, the samples run across the membrane prior to being suctioned through the slots giving highly variable results. In addition, the strength of the vacuum is important, such that too strong of a suction can cause the membrane to dimple leading to "ghost" slots upon immunoblotting. The flow valve of the Bio-Dot SF unit is a three-way valve, which permits simultaneous exposure to air and the vacuum to permit gentle suction. In addition to this setting, it is recommended that a 1 mL pipette tip be placed on the air valve and then using one's finger over the tip opening to regulate flow. Practicing with lysates generated also from transiently transfected cells is strongly recommended prior to using stable cell lines that achieve lower levels of expression.

1. Prepare the western blotting apparatus for running samples. This will be performed immediately after FT. Assemble the Bio-

Dot unit as recommended using the cellulose acetate membrane and filter paper soaking in the PBS-0.1% SDS. Ensure to tighten the apparatus to prevent any leaks.

2. Close the suction and fill all of the slots with 200 μL of PBS-0.1% SDS. If the solution begins to drain from the slots, this will indicate that there is leakage. Retighten the Bio-Dot until there is no more leakage. Once tightened, apply gentle suction to allow the buffer to go through the membrane. This washes the membrane.
3. Close the suction and add 200 μL of Tube 1, Tube 2 and Tube 3 for each sample into the slots. Centering the samples is recommended to permit even suction across samples. Fill the remaining empty slots with 200 μL of PBS-0.1% SDS.
4. Apply gentle suction until the slots empty. Given that the slots lacking sample will empty more quickly, continue to fill the slots with PBS-0.1% SDS to prevent the membrane from drying. Wash the sample slots three times with 200 μL of PBS-0.1% SDS (*see* **Note 6**).
5. Remove the membrane from the Bio-Dot and block in PBS-Tween containing 3% BSA. Block for 2 h at RT. Block for too long causes the quality of the immunoblot to significantly be diminished.
6. Begin western blotting the 20 μg samples. Boiling the samples for 2 min is optional (*see* **Note 7**).
7. Run the samples on a 4–12% NuPAGE Bis-Tris gel and transfer the gel onto a PVDF membrane. Block the membrane for 1 h in PBS-Tween containing 3% BSA at RT.
8. Prepare both FT and WB membranes for overnight incubation at 4 °C in primary antibody. FT membranes can be cut to encompass only the sample area, then probed with anti-GFP or anti-HaloTag to detect the trapped, insoluble 17aaHtt103Q-mCFP protein or Exon1Htt103Q protein, respectively. WB membranes can be cut at around 80 kDa to permit detection of the detergent soluble polyQ proteins (smaller than 80 kDa) with anti-GFP or anti-HaloTag by probing the lower portion and vinculin that can serve as a loading control (130 kDa) by probing the upper portion.
9. All membranes can be processed by standard immunoblotting methods the next day. Briefly, membranes are washed three times 10 min with PBS-Tween at RT. Membranes are then incubated from 2 h at RT in HRP-conjugated secondary antibodies at a dilution of 1:5000 to 1:10,000, and we detect signal using with the VersaDoc Imaging system by Bio-Rad (*see* **Note 8**).

3.3 Visualizing the Formation and Clearance of PolyQ Aggregates

As mentioned above, monitoring aggregates by microscopy is fairly straightforward, and will not be addressed in depth here for the 17aaHtt103Q-mCFP cell line. We do suggest, however, that the replicates be prepared simultaneously with the samples for biochemistry to provide more power for analyses. To do so, one 12 mm cover glass can be placed in each six well prepared for FT and WB (Subheading 2.2, **item 1**). Prior to trypsinization, cover glasses can be removed, stained with Hoeschst 33342 to label nuclei, fixed, mounted and imaged. Below, we will describe how we take advantage of the HaloTag system to examine aggregate formation versus clearance (Fig. 2).

1. Place 12 mm cover glasses into a 24-well plate and plate 5×10^4 Exon1Htt103Q-HaloTag cells per coverslip. We typically plate two cover glasses per condition. In this example we will perform a four day time course, and therefore including time 0 we will require ten cover glasses.
2. The next day, thaw an aliquot and prepare a fresh solution of 0.5 μM TMR diluted into DMEM+10% FBS within a dark cell culture hood. Add 200 μL of this mixture to each well. Incubate for 30 min in the cell culture incubator (37 °C, 5% CO_2). This will label all of the existing protein and aggregates with TMR.
3. Wash the wells gently but thoroughly three times with DMEM+10% FBS.
4. Thaw an aliquot and prepare a fresh solution of 1.5 mM OG in DMEM+10% FBS and add 200 μL of this mixture at 0, 24, 48, 72, and 96 h after adding TMR. Incubate for 30 min in the cell culture incubator. This will label the protein and aggregates that were created within the respective hour after TMR-labeling. Given that the TMR-labeling is complete and HaloTag labeling is maintained via covalent attachment of the fluorophore, the 0 h time point serves as a control.
5. Wash the well three times with DMEM+10% FBS and one time with PBS +/+.
6. Fix the cells for 20 min with 4% PFA at RT.
7. Wash the coverslip three times with PBS and stain the nuclei with Hoechst in PBS for 10 min at RT.
8. Wash the coverslip three times with PBS for 10 min, and mount the coverslip to image. For imaging Halotag, confocal microscopy is recommended.
9. For quantification, a minimum of 500 cells per time point is collected. Two different types of analyses can be conducted. The number of structures positive for TMR or OG can be monitored independently or the staining can be monitored simultaneously (TMR-positive only (red), OG-positive only (green) and TMR/OG-positive (yellow)) (Fig. 2).

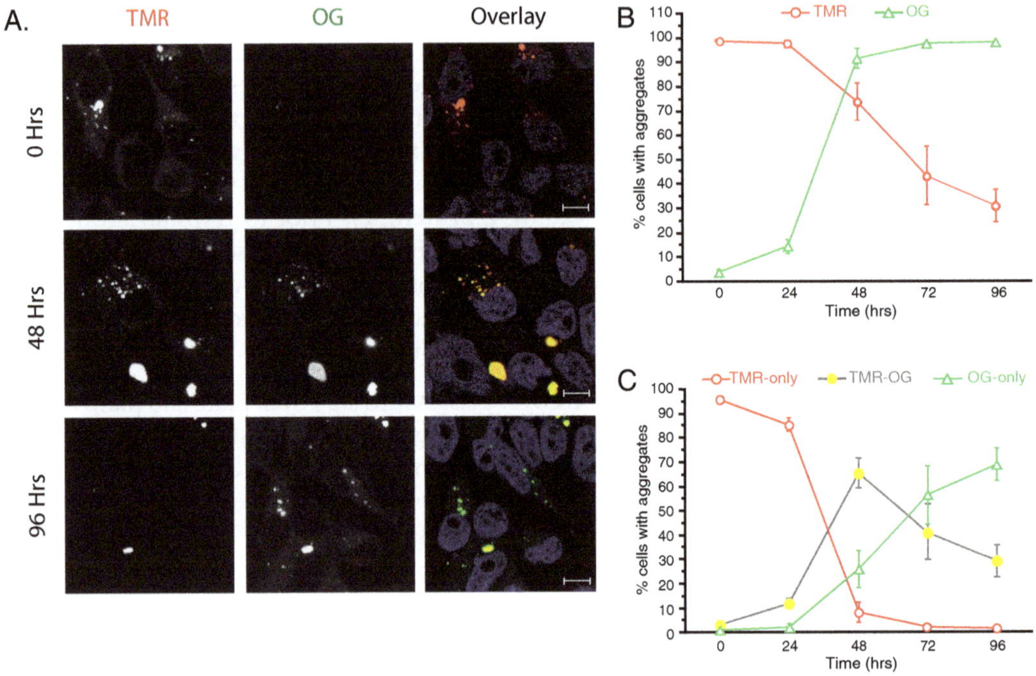

Fig. 2 Using HaloTag to monitor aggregate formation and clearance. (**a**) Confocal images of Exon1Htt103Q-HT cells after labeling with TMR (red) and OG (green) as described in Subheading 3.3. Nuclei are labeled with Hoechst 33342 (blue). The TMR labeling diminishes over time whereas OG labeling increases. (**b**) Quantification of TMR-labeling versus OG-labeling. (**c**) Quantification of TMR-only, OG-only and TMR- and OG-positive labeling. The overlay reveals aggregates that continue to incorporate protein over time. Images are modified from ref. 2

4 Notes

1. The number of cells depends upon the growing condition achieved. This number is based on what is required to achieve 90–95% confluence of our two cell lines on the day of sample collection. To permit the best signal to noise ratio, a relatively large amount of protein is required given that the stable cell lines express less of the aggregation prone protein and thus exhibit less aggregation than under transient conditions. For the two cell lines used as an example here, it is necessary to acquire about 150 μg of protein.
2. We find that trypsinization of HeLa cells from plates can also be done at RT.
3. 8 M urea solidifies on ice, and therefore it is necessary to maintain the solution at RT (and therefore made fresh). The sample incubation is performed on ice, but we find that there is minimal freezing when this volume is incubated on ice for this amount of time. Occasional pipetting, but with care to ensure no bubbles, will also diminish freezing. Should the samples

solidify, this is not problematic. Once it returns to RT it will quickly thaw.

4. Upon pelleting, a pellet may or may not be visible, but it is essential to separate the supernatant. We suspect there are nucleotides in the pellet whose resolubilization causes the "goopiness" of the sample to return.
5. This total mass is a guideline which has been optimized for these stable cell lines and can be altered. This is also the case for the serial dilution in subheading 3.1 **step 18**, for which threefold is a suggestion. We have done fivefold dilutions which are also quite effective. Too much aggregated protein can fail to filter through the membrane causing it to clog. This is especially the case for transient overexpression.
6. Preventing the membrane from drying is extremely important to maintain the fidelity of the FT. We have also performed multiple washes as well as an uneven number of washes across samples and have found that more washes have less effect on the outcome than drying of the membrane. It is important to remember that the more the quantity of insoluble protein loaded, the slower the sample suction. The total mass of protein loaded in the first dilution might have to be adjusted to permit a clean FT.
7. Given that the samples are in urea, boiling can lead to carbamylation of the proteins. For the purposes outlined here, boiling has little impact, but can influence subsequent proteolytic digests, for example.
8. Although we use the tags to visualize the proteins, conformation or Htt epitope specific antibodies can also be used.

Acknowledgments

Harm H. Kampinga for helpful discussions and contributions toward the study leading up to this work. This work was supported by NIH grant NS077111 and NS063973, the Parkinson's Disease Foundation and the Hereditary Disease Foundation.

References

1. Balchin D, Hayer-Hartl M, Hartl FU (2016) In vivo aspects of protein folding and quality control. Science 353(6294):aac4354. https://doi.org/10.1126/science.aac4354
2. Eenjes E, Dragich JM, Kampinga HH, Yamamoto A (2016) Distinguishing aggregate formation and aggregate clearance using cell based assays. J Cell Sci. https://doi.org/10.1242/jcs.179978
3. Filimonenko M, Isakson P, Finley KD, Anderson M, Jeong H, Melia TJ, Bartlett BJ, Myers KM, Birkeland HC, Lamark T, Krainc D, Brech A, Stenmark H, Simonsen A, Yamamoto A (2010) The selective macroautophagic degradation of aggregated proteins requires the PI3P-binding protein Alfy. Mol Cell 38(2):265–279. https://doi.org/10.1016/j.molcel.2010.04.007

4. Yamamoto A, Cremona ML, Rothman JE (2006) Autophagy-mediated clearance of huntingtin aggregates triggered by the insulin-signaling pathway. J Cell Biol 172(5):719–731
5. Gossen M, Bujard H (1992) Tight control of gene expression in mammalian cells by tetracycline-responsive promoters. Proc Natl Acad Sci U S A 89(12):5547–5551
6. Los GV, Encell LP, McDougall MG, Hartzell DD, Karassina N, Zimprich C, Wood MG, Learish R, Ohana RF, Urh M, Simpson D, Mendez J, Zimmerman K, Otto P, Vidugiris G, Zhu J, Darzins A, Klaubert DH, Bulleit RF, Wood KV (2008) HaloTag: a novel protein labeling technology for cell imaging and protein analysis. ACS Chem Biol 3(6):373–382. https://doi.org/10.1021/cb800025k
7. Scherzinger E, Lurz R, Turmaine M, Mangiarini L, Hollenbach B, Hasenbank R, Bates GP, Davies SW, Lehrach H, Wanker EE (1997) Huntingtin-encoded polyglutamine expansions form amyloid-like protein aggregates in vitro and in vivo. Cell 90(3):549–558
8. Perutz MF, Johnson T, Suzuki M, Finch JT (1994) Glutamine repeats as polar zippers: their possible role in inherited neurodegenerative diseases. Proc Natl Acad Sci U S A 91(12): 5355–5358
9. Gusella JF, MacDonald ME (2000) Molecular genetics: unmasking polyglutamine triggers in neurodegenerative disease. Nat Rev 1(2): 109–115
10. Bujard H (1999) Controlling genes with tetracyclines [interview]. J Gene Med 1(5): 372–374
11. Huang J, Su W, Chen X, Cheng X, Dai Y, Han L, Liang D (2015) Doxycycline attenuates endotoxin-induced uveitis by prostaglandin E2-EP4 signaling. Invest Ophthalmol Vis Sci 56(11):6686–6693. https://doi.org/10.1167/iovs.15-17045
12. Iwasaki H, Inoue H, Mitsuke Y, Badran A, Ikegaya S, Ueda T (2002) Doxycycline induces apoptosis by way of caspase-3 activation with inhibition of matrix metalloproteinase in human T-lymphoblastic leukemia CCRF-CEM cells. J Lab Clin Med 140(6):382–386. https://doi.org/10.1067/mlc.2002.129308
13. Jantzie LL, Cheung PY, Todd KG (2005) Doxycycline reduces cleaved caspase-3 and microglial activation in an animal model of neonatal hypoxia-ischemia. J Cereb Blood Flow Metab 25(3):314–324. https://doi.org/10.1038/sj.jcbfm.9600025
14. Lai HC, Yeh YC, Ting CT, Lee WL, Lee HW, Wang LC, Wang KY, Lai HC, Wu A, Liu TJ (2010) Doxycycline suppresses doxorubicin-induced oxidative stress and cellular apoptosis in mouse hearts. Eur J Pharmacol 644(1–3):176–187. https://doi.org/10.1016/j.ejphar.2010.07.010
15. Mouratidis PX, Colston KW, Dalgleish AG (2007) Doxycycline induces caspase-dependent apoptosis in human pancreatic cancer cells. Int J Cancer 120(4):743–752. https://doi.org/10.1002/ijc.22303
16. Onoda T, Ono T, Dhar DK, Yamanoi A, Fujii T, Nagasue N (2004) Doxycycline inhibits cell proliferation and invasive potential: combination therapy with cyclooxygenase-2 inhibitor in human colorectal cancer cells. J Lab Clin Med 143(4):207–216. https://doi.org/10.1016/j.lab.2003.12.012
17. Onoda T, Ono T, Dhar DK, Yamanoi A, Nagasue N (2006) Tetracycline analogues (doxycycline and COL-3) induce caspase-dependent and -independent apoptosis in human colon cancer cells. Int J Cancer 118(5):1309–1315. https://doi.org/10.1002/ijc.21447
18. Sagar J, Sales K, Dijk S, Taanman J, Seifalian A, Winslet M (2009) Does doxycycline work in synergy with cisplatin and oxaliplatin in colorectal cancer? World J Surg Oncol 7:2. https://doi.org/10.1186/1477-7819-7-2
19. Sagar J, Sales K, Seifalian A, Winslet M (2010) Doxycycline in mitochondrial mediated pathway of apoptosis: a systematic review. Anti Cancer Agents Med Chem 10(7):556–563
20. Solomon A, Rosenblatt M, Li DQ, Liu Z, Monroy D, Ji Z, Lokeshwar BL, Pflugfelder SC (2000) Doxycycline inhibition of interleukin-1 in the corneal epithelium. Invest Ophthalmol Vis Sci 41(9):2544–2557
21. Song H, Fares M, Maguire KR, Siden A, Potacova Z (2014) Cytotoxic effects of tetracycline analogues (doxycycline, minocycline and COL-3) in acute myeloid leukemia HL-60 cells. PLoS One 9(12):e114457. https://doi.org/10.1371/journal.pone.0114457
22. Sourdeval M, Lemaire C, Brenner C, Boisvieux-Ulrich E, Marano F (2006) Mechanisms of doxycycline-induced cytotoxicity on human bronchial epithelial cells. Front Biosci 11: 3036–3048
23. Wang SQ, Zhao BX, Liu Y, Wang YT, Liang QY, Cai Y, Zhang YQ, Yang JH, Song ZH, Li GF (2016) New application of an old drug: antitumor activity and mechanisms of doxycycline in small cell lung cancer. Int J Oncol 48(4):1353–1360. https://doi.org/10.3892/ijo.2016.3375
24. Wu J, Liu T, Xie J, Xin F, Guo L (2006) Mitochondria and calpains mediate caspase-dependent apoptosis induced by doxycycline in

HeLa cells. Cell Mol Life Sci 63(7–8):949–957. https://doi.org/10.1007/s00018-005-5565-6

25. Wu W, Yu LH, Ma B, Xu MJ (2014) The inhibitory effect of doxycycline on cisplatin-sensitive and -resistant epithelial ovarian cancer. PLoS One 9(3):e89841. https://doi.org/10.1371/journal.pone.0089841
26. Yoon JM, Koppula S, Huh SJ, Hur SJ, Kim CG (2015) Low concentrations of doxycycline attenuates FasL-induced apoptosis in HeLa cells. Biol Res 48:38. https://doi.org/10.1186/s40659-015-0025-8
27. Zhao Y, Wang X, Li L, Li C (2016) Doxycycline inhibits proliferation and induces apoptosis of both human papillomavirus positive and negative cervical cancer cell lines. Can J Physiol Pharmacol 94(5):526–533. https://doi.org/10.1139/cjpp-2015-0481

Chapter 10

Monitoring Proteome Stress in Live Cells Using HaloTag-Based Fluorogenic Sensor

Yu Liu, Matthew Fares, and Xin Zhang

Abstract

Fluorescent folding sensor is a powerful tool to detect proteome stresses, including heat, osmotic, oxidative, and drug induced stresses. Monitoring proteome stress using these sensors allows us to dissect the mechanism of cellular stress and find therapeutics that ameliorate stress related diseases. Here we present a HaloTag-based fluorogenic proteome stress sensor (AgHalo) to robustly detect and quantify proteome stresses in live cells. We describe how proteome stresses are monitored in both bacterial and mammalian live cells using fluorescence confocal microscope and fluorescence plate reader.

Key words Proteome stress, Misfolding and aggregation, Fluorescent sensor, HaloTag, Confocal imaging, Fluorescence plate reader

1 Introduction

A proper protein homeostasis (proteostasis) is critical to cellular viability when subjected to diverse environmental stress factors, including environmental perturbations, chemical toxins, pathogen invasion, and aging [1–3]. These exogenous stress conditions impair the integrity of proteostasis and leads to global misfolding and aggregation of the endogenous proteome. As a result, cells accumulate a series of misfolded proteins in the forms of soluble oligomers, disordered or amorphous aggregates, and fibrils containing hydrogen-bonded β-sheet structures. These forms have been implied in a number of neurodegenerative disorders, metabolic disorders, and cardiovascular diseases [4–6]. To discern the mechanisms and therapeutic strategies toward these diseases, new methodologies are essential to visualize and quantify impairment of cellular proteostasis during stress.

To this end, it is highly desirable to develop proteome stress sensors that can visualize and quantify stress-induced proteostasis deficiency. One characteristic feature of stressed cells is the misfolding and aggregation of meta-stable fraction of cellular proteome

Cláudio M. Gomes (ed.), *Protein Misfolding Diseases: Methods and Protocols*, Methods in Molecular Biology, vol. 1873, https://doi.org/10.1007/978-1-4939-8820-4_10,

due to the loss of sufficient proteostasis capacity under stress conditions. Taking advantage of this feature, current proteome stress sensors utilize destabilized client proteins to represent the metastable proteome. Because of the compromised protein stability, the client protein sensors require sufficient proteostasis capacity and cellular surveillance to maintain their solubility, correct folding, and function. Upon proteome stress, cellular proteostasis deficiency can be reported via the misfolding and aggregation of the sensor client proteins.

This strategy has been used in multiple sensor systems [7], represented by the destabilized firefly luciferase sensor fused to GFP [8] and the destabilized retroaldolase sensor enabled by a fluorescent probe [9]. Based on the similar principle, we recently have developed a HaloTag-based fluorogenic proteome stress sensor system that can be used to visualize and quantify proteome stress in intact live cells [10, 11]. To achieve this goal, we evolved an established HaloTag mutant (K73T) that would aggregate in cells subjected to a variety of stress conditions. Previous work has demonstrated that the severity of proteome stress is positively proportional to the extent of Halo K73T aggregation. To enable a signal that can quantify the aggregation of Halo K73T in cells, we designed and synthesized a fluorogenic HaloTag ligand that contains an environment-sensitive fluorophore P1, sulfonylbenzoxadiazole (SBD). Upon forming the Halo K73T–P1 conjugation (hereinafter AgHalo sensor), the SBD moiety remains dark when AgHalo is folded. Aggregation of AgHalo, however, effectively turns on the fluorescence of the SBD moiety. The fluorescence intensity from the AgHalo sensor is quantitatively related to the extent of AgHalo aggregation. Thus, fluorescence intensity of the AgHalo sensor can be used to visualize and quantify the severity of proteome stress.

This protocol described how the AgHalo fluorogenic proteome stress sensor is used to monitor proteome stress in both bacterial *Escherichia coli* (*E. coli*) cells and human embryonic kidney (HEK293T) cells. The method provides details on visualization and quantification using confocal microscope and fluorescence microplate reader, respectively. Although experimental data are demonstrated using established stress factors, the general workflow can be followed for other stress conditions with appropriate modification.

2 Materials

Prepare all solutions using ultrapure water (prepared by the Barnstead GenPure xCAD Plus Ultrapure Water Purification System, to attain a sensitivity of 18 mΩ cm at 25 °C) and analytical grade reagents.

2.1 Plasmids

1. Plasmid for bacterial expression: pET29b(+) vector encoding Halo-K73T. Stock solution of plasmid is 50 ng/μL in ultrapure deionized water (*see* **Note 1**).
2. Plasmid for mammalian expression: plasmid encoding Halo-K73T is constructed based on the pHTN HaloTag® CMV-neo vector from Promega® (G7721) [10]. Stock solution of plasmid is 500 ng/μL in ultrapure deionized water (*see* **Note 1**).

2.2 Synthesis of Fluorogenic Probe P1

This section is adapted from ref. [10] (Scheme 1). All reagents and anhydrous solvents are of commercial grade and used as received unless stated otherwise. Reactions are monitored by thin layer chromatography (TLC) analysis using glass sheets precoated with silica gel 60 Å and detection by UV absorption at 254 or 365 nm. Flash column chromatography is performed using silica gel in the indicated solvent mixture. ^{1}H and ^{13}C NMR spectra are obtained on a Bruker NMR (400/100 MHz) spectrometer in the given solvent. Chemical shifts (δ) are reported in ppm relative to a $CDCl_3$ residual solvent peak or tetramethylsilane (TMS) internal standard. Coupling constants are provided in Hz. All ^{13}C NMR spectra are shown as proton-decoupled spectra. High-resolution mass spectra are recorded with a Waters Q-TOF Premier quadrupole/time-of-flight (TOF) mass spectrometer (*see* **Note 2**).

1. Compound 1: (Tert-butyl (2-(2-((6-chlorohexyl)oxy)ethoxy) ethyl)carbamate) followed previously reported protocol [12]. ^{1}H NMR (400 MHz, $CDCl_3$): δ 1.31–1.41 (m, 13H), 1.57–1.60 (m, 2H), 1.73–1.77 (m, 2H), 3.29–3.58 (m, 12H), 5.01 (s, 1H). ^{13}C NMR (100 MHz, $CDCl_3$): 156.08, 79.23, 71.36, 70.36, 70.30, 70.13, 45.11, 40.43, 32.62, 29.53, 28.51, 26.77, 25.51.
2. P1: The previously reported SBD-COOH precursor (1 eq, 0.5 mmol) is dissolved in 10 mL anhydrous DMF. The TFA-

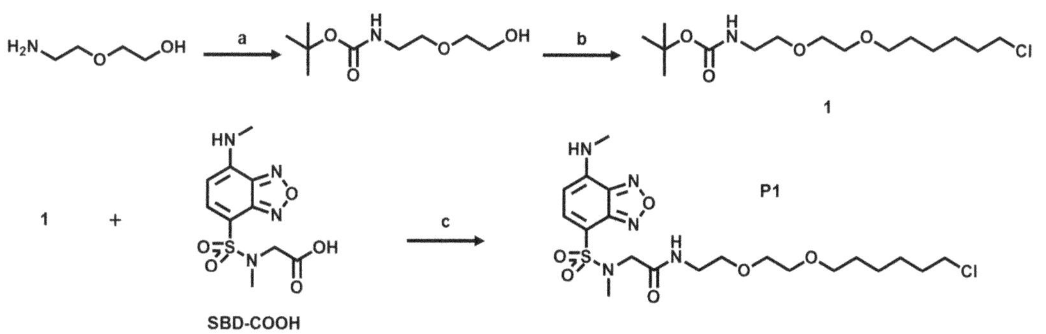

Scheme 1 Synthesis of P1. Conditions (a) Boc_2O (2 eq), Methanol, 2 h, RT, 90%. (b) 60% NaH (1 eq), 0 °C, 1 h; 1-Chloro-6-iodohexane (1.1 eq), dimethylformamide, RT, 24 h, 30%. (c) TFA deprotected **1** (1 eq), SBD-COOH (1 eq), EDC·HCl (2 eq), HOBt (2 eq), triethylamine (4 eq), dimethylformamide, RT, 24 h, 50–60%. (Adapted from ref. 10)

deprotected form of P1 after drying is used directly without further purification and added into the reaction solution (1 eq, 0.5 mmol, 126 mg for each reaction). 1-Hydroxybenzotriazole hydrate (HOBt) (2 eq, 1 mmol, 153 mg) and triethylamine (4 eq, 2 mmol, 202 mg) are introduced to the reaction mixture in one portion. *N*-(3-Dimethylaminopropyl)-*N′*-ethylcarbodiimide hydrochloride (EDC) (2 eq, 1 mmol, 192 mg) is added in one portion. The reaction is stirred for 24 h, protected from light. The reaction mixture is quenched by water and extracted with DCM. The organic fraction is dried in vacuo. Compounds are further purified by flash chromatography using 100% ethyl acetate as eluent. ^{1}H NMR (400 MHz, $CDCl_3$): δ 7.91 (d, J = 8.1 Hz, 1H), 7.13 (s, 1H), 6.38 (s, 1H), 6.07 (d, J = 8.1 Hz, 1H), 3.89 (s, 2H), 3.66–3.41 (m, 12H), 3.09 (d, J = 5.3 Hz, 3H), 2.83 (s, 3H), 1.72 (t, J = 7.1 Hz, 2H), 1.56 (t, J = 7.0 Hz, 2H), 1.43–1.31 (m, 4H). ^{13}C NMR (100 MHz, $CDCl_3$): 168.28, 145.87, 144.31, 142.42, 140.54, 136.11, 98.45, 71.30, 70.47, 70.09, 69.56, 53.88, 45.15, 39.28, 36.63, 32.55, 30.16, 29.69, 29.47, 26.72, 25.42. HRMS for [M+H+]: Calc. 506.1835, Obs, 506.1848.

2.3 Chemical Stressors

1. MG132: 10 mM DMSO stock solution, stored in −80 °C freezer (*see* **Note 3**).
2. Nilotinib: 20 mM DMSO stock solution, stored in −80 °C freezer (*see* **Note 3**).

2.4 Bacterial Culture

1. *E. coli* strain: One Shot BL21 Star (DE3) chemically competent *E. coli* cells (Thermo Fisher, C601003).
2. Growth medium: Luria broth (*see* **Note 4**).
3. Antibiotics: Kanamycin (*see* **Note 5**).
4. Isopropyl β-D-1-thiogalactopyranoside (IPTG): (*see* **Note 6**).
5. DPBS buffer: 2.67 mM KCl, 1.47 mM KH_2PO_4, 136.9 mM NaCl, 8.10 mM Na_2HPO_4, pH 7.2.
6. Cell density meter: WPA CO8000.

2.5 Mammalian Tissue Culture

1. Mammalian cell line: HEK293T (ATCC CRL-3216).
2. Culture dish: Poly-D-lysine coated 35 mm glass bottom dishes.
3. 12-well plates: 12-well cell culture microplate.
4. Medium: Dulbecco's Modified Eagle Medium (DMEM); Opti-MEM Reduced-Serum Medium; TrypLE Express, fetal bovine serum (FBS); penicillin–streptomycin–glutamine stock solution (PSQ) (*see* **Note 7**).
5. CO_2 incubator: HERAcell VIOS 160i CO_2 incubator.

2.6 Fluorescence Spectrophotometer and Microscope

1. Fluorescent microplate reader: Tecan Infinite M1000 PRO or any microplate readers.
2. Confocal microscope: Olympus FluoView™ FV1000 confocal microscope.
3. 96-well microplate for fluorescence intensity measurement: Black, opaque, round bottom 96-well microplate (Corning; Catalog No. 3356).

3 Methods

3.1 Detecting Proteome Stress in E. coli Cells Using Fluorescence Confocal Microscope

1. Figure 1a shows the experimental scheme. Plasmids encoding Halo K73T (pET29b+; kanamycin resistant) are transformed into *E. coli* BL21 DE3* cells. A single colony isis used to inoculate 5 mL of LB medium with kanamycin (50 μg/mL) as a starter culture that has been grown shaking at 37 °C for 14 h in 14 mL culture tubes (*see* **Note 8**).
2. Experiments are initiated by adding 0.5 mL of the starter culture into 20 mL of freshly prepared LB medium with kanamycin (50 μg/mL) in a 50 mL conical tube and allowed to grow at 37 °C with shaking (*see* **Note 9**). At an optical density at 600 nm (OD_{600}) of approximately 0.7, IPTG is added to a final concentration of 50 μM to express Halo K73T (*see* **Note 10**). These cultures continued to grow at 30 °C for 2 h (*see* **Note 11**). Cells are pelleted by centrifugation at 4000 × *g* for 5 min and supernatant medium is decanted.

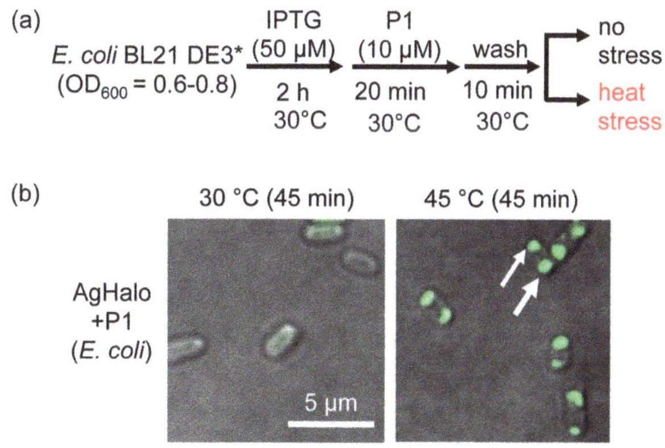

Fig. 1 Detecting heat-induced proteome stress in *E. coli* cells. (**a**) Experimental scheme. (**b**) The fluorescence of aggregated AgHalo sensor can be visualized by confocal microscopy. The AgHalo aggregates are visualized by the fluorescent punctate structures. P1 (10 μM) treated *E. coli* cells harboring AgHalo sensor are either maintained at 30 °C or stressed at 45 °C for 45 min prior to confocal imaging. Green: aggregates of AgHalo–P1 conjugate. White arrows highlight the formation of large fluorescent aggregated species at the polar locations in *E. coli* cells. (Figure adapted from ref. 10)

3. Labeling of Halo K73T is subsequently carried out by resuspending pellet with fresh LB medium containing kanamycin (50 μg/mL) and P1 (20 μM) till a final OD_{600} reading of 2.0. To wash away unbound ligands, cells are pelleted by centrifugation, supernatant medium is decanted, and the pellet is resuspended in fresh LB with kanamycin (50 μg/mL, warmed to 30 °C). Both labeling and washing should be carried out for 20 min at 30 °C with shaking. For best results, repeat the wash step once (*see* **Note 12**).
4. After the final wash incubation, cells are pelleted by centrifugation, supernatant media is decanted, and the pellet is resuspended in fresh LB with kanamycin (50 μg/mL).
5. 1 mL aliquots of each sample are incubated at desired temperatures (e.g., 30 or 45 °C) for 45 min in 1.5 mL eppendorf tubes in a static heat block. After heat stress the cells are pelleted by centrifugation, supernatant medium is decanted, and the pellet is resuspended in 300 μL of DPBS.
6. 5 μL of the cell suspension is added onto freshly prepared agarose pads for immediate acquisition of fluorescent images by the Olympus FluoView FV1000 confocal microscope with the UPlanS Apo 100×/1.4 oil objective. P1 is visualized with a blue argon laser of 488 nm (Fig. 1b).

3.2 Quantifying Proteome Stress in E. coli Cells Using Fluorescence Intensity

1. **Steps 1–5** of Subheading 3.1 are used to prepare the cells for experiments in this section.
2. 1 mL aliquots of each sample are then incubated at the desired temperatures (30, 42, and 45 °C) for either 0, 15, 30, 45, or 60 min. Cells are pelleted by centrifugation, supernatant medium is decanted, and the pellet is resuspended in DPBS (300 μL).
3. 100 μL of cell suspension is pipetted into a single well of a black, opaque, round bottom 96-well microplate. Fluorescence emission spectra from 470 to 750 nm (*Ex*: 440 nm) are measured using a microplate reader (Tecan Infinite M1000 PRO). Fluorescence intensity at 550 nm is used as a signal to evaluate proteome stress (*see* **Note 13**).
4. 100 μL of cell suspension is diluted to a final volume of 1 mL in DPBS to acquire an OD_{600} reading from the cell density meter (*see* **Note 14**).
5. The fluorescence signals are normalized to the cell density by dividing the fluorescence intensity of the sample by the matching OD_{600} readings. The normalized signal of each heated sample is then divided by the normalized signal of the 0 min nonstressed sample to determine the fold-of-change at each temperature and time point (Fig. 2) (*see* **Note 15**).

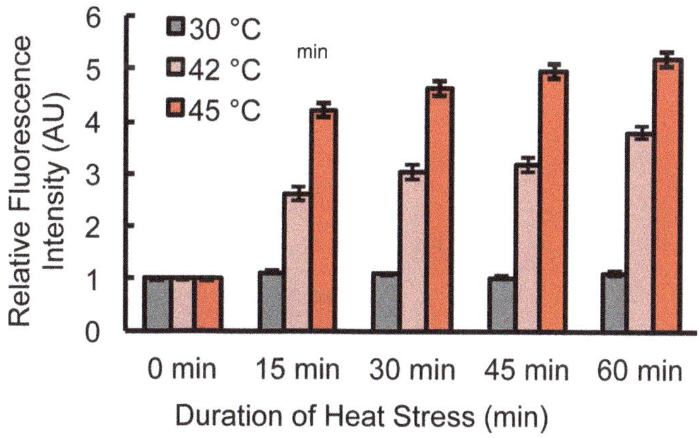

Fig. 2 Quantification of proteome stress in *E. coli* cells. Data is normalized to nonstressed cells. Error bars: standard error ($n = 3$). (Figure adapted from ref. 10)

3.3 Detecting Proteome Stresses in HEK293T Cells Using Confocal Microscope

1. Figure 3a shows the experimental scheme. HEK293T cells are seeded at 25% confluency 24 h prior to transfection in poly-D-lysine coated 35 mm glass bottom dishes (~8 × 10^5 cells in each dish). Cells are cultured in DMEM medium supplemented with 10% FBS and penicillin–streptomycin antibiotics until reached 50–60% confluency.
2. Transfection of the plasmid encoding Halo-K73T (AgHalo, 4 μg/35 mm dish) is carried out using X-tremeGene™ 9 DNA transfection reagent (Roche) according to the manufacturer's instructions in Opti-MEM® Reduced-Serum Medium. AgHalo protein is transiently expressed for 24 h prior to stress conditions (*see* **Note 16**).
3. For cotranslational probe labeling, P1 in DMSO stock (2 mM) is diluted into fresh DMEM as 10× stock solution (10 μM) and introduced to the medium at the time of transfection (**step 3**). The final labeling concentration of P1 is 1 μM. For posttranslational probe labeling, cells are incubated with 10 μM of P1 in fresh DMEM medium 24 h after transfection. Labeling reaction should be carried out for 30 min at 37 °C. Excess P1 is removed by incubating cells in fresh DMEM medium for 30 min at 37 °C. For best results, repeat the wash step one more time (*see* **Note 17**).
4. To test proteome deficiency with stress, cells are incubated with fresh DMEM medium containing desired chemical stressors at indicated concentrations. In our experiments, we used 10 μM of MG132 and 50 μM of Nilotinib as a set of established chemical stressors. Cells are subjected to stress conditions for desired duration at 37 °C in the incubator. For heat stress, cells are incubated for 2 h in an incubator that is prewarmed to 42 °C (*see* **Note 18**).

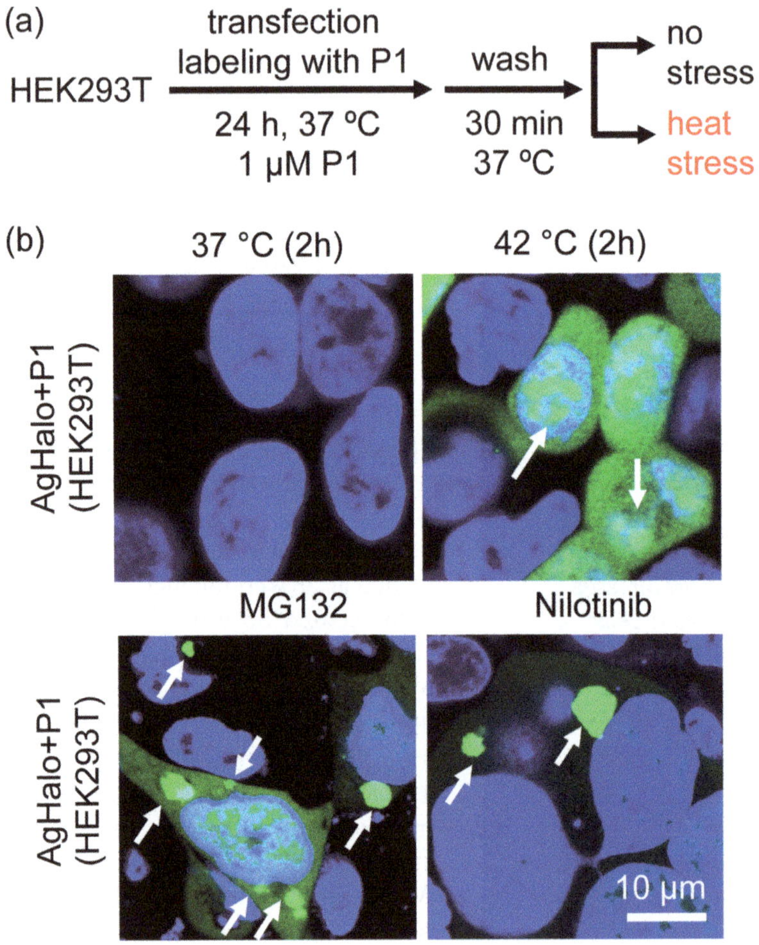

Fig. 3 Detecting heat-induced proteome stress in HEK293T cells. (**a**) Experimental scheme. (**b**) Live cell imaging of AgHalo sensor in stressed HEK293T cells. Both diffuse and punctate (highlighted by white arrows) fluorescent regions are observed, presumably arising from soluble misfolded oligomers and insoluble aggregates of AgHalo, respectively. Green: the AgHalo sensor, Blue: nucleus using Hoechst 33342. (Figure adapted from ref. 10)

5. Immediately after stress, live cell samples are visualized by confocal fluorescence microscope. Nucleus staining is achieved by incubating cells with Hoechst 33342 (0.1 μg/mL) for 30 min prior to imaging. Confocal images are acquired using Olympus FluoView™ FV1000 confocal microscope. P1 fluorescence is visualized with blue argon (458 or 488 nm) laser and nuclear staining is visualized using violet laser (405 nm). Fluorescence images are shown in Fig. 3b (*see* **Note 19**).

3.4 Quantifying Proteome Stresses in HEK293T Cells Using Fluorescence Intensity

1. **Steps 1–4** of Subheading 3.3 are used to prepare the cells for experiments in this section.
2. Cells are harvest to measure the fluorescence intensity for quantification of proteome stress. To harvest, aspirate DMEM medium using vacuum and add 0.2 mL of trypLE directly in

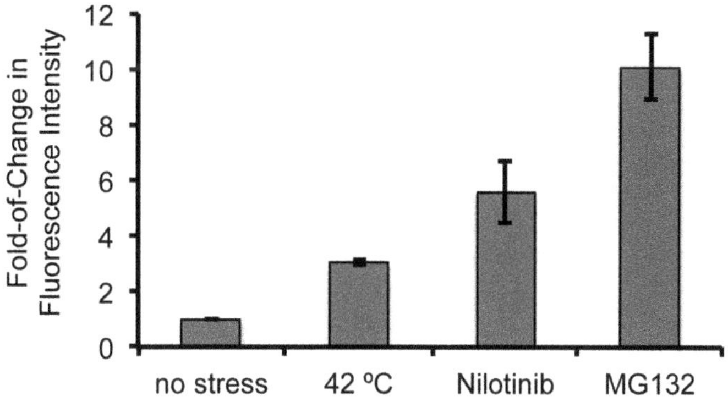

Fig. 4 Quantification of proteome stress in HEK293T cells. Data is normalized to nonstressed cells. Error bars: standard error ($n = 3$). (Figure adapted from ref. 10)

the dish (spray on top of cells). After 2 min incubation at 37 °C, neutralize cell suspension with 0.8 mL of DMEM medium. Harvested cells are pelleted by centrifugation to remove DMEM medium, and resuspended in 500 μL DPBS buffer.

3. 100 μL of cell suspension is pipetted into a single well of a black, opaque, round bottom 96-well microplate. Fluorescence emission spectra from 470 to 750 nm (*Ex*: 440 nm) are measured using a microplate reader (Tecan Infinite M1000 PRO). Fluorescence intensity at 550 nm is used as a signal to evaluate proteome stress (*see* **Note 13**).
4. 10 μL of cell suspension is used to acquire cell number count using the BioRad TC20 automated cell counter (*see* **Note 20**).
5. The fluorescence signals are normalized to the cell number by dividing the fluorescence intensity of the sample by cell number. The normalized signal of each stressed sample is then divided by the normalized signal of the nonstressed sample to determine the fold-of-change (Fig. 4) (*see* **Note 15**).

4 Notes

1. Both plasmids could be made available upon request addressed to XZ.
2. P1 could be made available upon request addressed to XZ. P1 is less stable in DMSO stock solution than in solid form. DMSO solution needs to be stored in aliquots of small volumes to prevent multiple freeze–thaw cycles. Store the DMSO solution in −20 °C freezer. Avoid long-term light exposure during experiments.
3. MG132 is chemically unstable in DMSO solution. Freshly prepared MG132 is required to achieve best performance in

concentration dependence experiments. DMSO solution needs to be stored in aliquots of small volumes to prevent multiple freeze–thaw cycles. We store the DMSO aliquot in −80 °C freezer. We generally do not use MG132 DMSO stock older than 6 months.

4. All LB should be autoclaved before use and kept sealed while being allowed to cool. Antibiotics should only be added to LB once it has cooled to a reasonable temperature (i.e., a standard glass bottle containing it can be comfortably held by a person indefinitely without any heat protective gloves).
5. Kanamycin powder is stable at room temperature. For ease of use, 1000× stocks (50 mg/mL) are made in H_2O and stored at −20 °C. These stocks are still fairly stable, and can be used through several freeze–thaw cycles.
6. IPGT powder should be stored in at −20 °C and protected from light. IPTG is less stable in solution. A 1 M stock is prepared in H_2O and either used fresh or stored at −20 °C for no more than 6 months. Frozen IPTG stocks should be considered "single use" and never refrozen after a thaw, thus appropriate aliquots should be made.
7. FBS and PSQ need to be stored in aliquots of small volumes to avoid multiple freeze–thaw cycles.
8. Growth periods at this stage should be kept between 12 and 16 h. If the growth of the starter culture is too short, insufficient growth may occur, resulting in slow growth of the experimental culture. If the starter culture growth is too long, cells may behave somewhat abnormally during experimentation.
9. Should a different experimental growth volume be required, the starter culture should be diluted 1/40. Less dilution (i.e., 1/20) will give faster growth, be less reliable results.
10. If IPTG concentrations are significantly lower, very little protein will be produced. If IPTG concentrations are significantly higher, the overproduction of the Halo K73T will result in premature aggregation and potentially high background signal.
11. 30 °C provides a good balance between cellular growth and protein folding. If cells are grown at higher temperatures, much of the protein produced may be in misfolded state, leading to higher background signal. At least 2 h of protein production is recommended to allow for high signal to noise ratio.
12. Labeling can also be carried out in situ by adding the ligand at the same time that IPTG is added. For this procedure, at least 20 μM of ligand should be used to ensure efficient labeling.

The volume of LB used for the wash should be approximately equal to the starting culture volume. For maximally efficient wash steps, any LB should be prewarmed to 30 °C. Decanting is achieved with a vacuum line, as any remaining LB will result in background fluorescence. Simultaneously, care must be taken not to disturb the cellular pellet with the vacuum.

13. We found that the top-reading mode produces more consistent fluorescence intensities than the bottom-reading mode. When loading samples into a microplate, be extremely careful to avoid the formation or addition of air bubbles.
14. Sample should be diluted with LB to attain an accurately measurable OD (i.e., a 1/6 dilution that has a measured OD of 0.8 means the original sample has an OD of ~4.8).
15. The absolute fluorescent intensity may vary greatly based on experimental conditions, such as amount of Halo in cells, concentration of P1, and labeling efficiency. The relative fluorescence increase from unstressed cells (defined as fold-of-change), however, provides a generally robust measure that is independent on the cellular level of the Halo–P1 conjugate. Thus, a nonstressed control sample should be included in every experiment.
16. Any commercially available transfection reagents could be readily applied to this experiment. Transfection efficiency using varying reagents and mammalian cell lines should be optimized. However, transfection efficiency has only moderate effects on the fluorescence fold-of-change as previously demonstrated [10].
17. Concentration of P1 should be optimized for different mammalian cell lines. However, this concentration has only moderate effects on the fluorescence fold-of-change as previously demonstrated [10]. During the wash step, it is critical not to disturb adherent cells. Mechanically disturbing adherence cells introduce undesired stress that could induce fluorescence increase in the nonstressed control.
18. Under severe stress conditions, adherent cells tend to detach from the culture dish. It is critical not to adherent cells during the wash step. Cells that have already detached should be removed because cell debris tends to cluster P1 and emit fluorescence.
19. Cell debris tends to cluster P1 and emit fluorescence. Thus, they should be differentiated from stressed live cells that contain AgHalo aggregates.
20. Other cell counters should also be applicable to this experiment, as long as the user's manual is strictly followed.

Acknowledgment

This work has been supported by the Burroughs Wellcome Fund Career Award at the Scientific Interface (X.Z.), Paul Berg Early Career Professorship (X.Z.), Lloyd and Dottie Huck Early Career Award (X.Z.).

References

1. Balch WE, Morimoto RI, Dillin A, Kelly JW (2008) Adapting proteostasis for disease intervention. Science 319(5865):916–919. https://doi.org/10.1126/science.1141448
2. Powers ET, Morimoto RI, Dillin A, Kelly JW, Balch WE (2009) Biological and chemical approaches to diseases of proteostasis deficiency. Annu Rev Biochem 78:959–991. https://doi.org/10.1146/annurev.biochem.052308.114844
3. Kaushik S, Cuervo AM (2015) Proteostasis and aging. Nat Med 21(12):1406–1415. https://doi.org/10.1038/nm.4001
4. Chiti F, Dobson CM (2006) Protein misfolding, functional amyloid, and human disease. Annu Rev Biochem 75:333–366. https://doi.org/10.1146/annurev.biochem.75.101304.123901
5. Aguzzi A, O'Connor T (2010) Protein aggregation diseases: pathogenicity and therapeutic perspectives. Nat Rev Drug Discov 9(3):237–248. https://doi.org/10.1038/nrd3050
6. Jensen MB, Jasper H (2014) Mitochondrial proteostasis in the control of aging and longevity. Cell Metab 20(2):214–225. https://doi.org/10.1016/j.cmet.2014.05.006
7. Radwan M, Wood RJ, Sui X, Hatters DM (2017) When proteostasis goes bad: protein aggregation in the cell. IUBMB Life 69(2):49–54. https://doi.org/10.1002/iub.1597
8. Gupta R, Kasturi P, Bracher A, Loew C, Zheng M, Villella A, Garza D, Hartl FU, Raychaudhuri S (2011) Firefly luciferase mutants as sensors of proteome stress. Nat Methods 8(10):879–884. https://doi.org/10.1038/nmeth.1697
9. Liu Y, Zhang X, Chen W, Tan YL, Kelly JW (2015) Fluorescence Turn-on folding sensor to monitor proteome stress in live cells. J Am Chem Soc 137(35):11303–11311. https://doi.org/10.1021/jacs.5b04366
10. Liu Y, Fares M, Dunham NP, Gao Z, Miao K, Jiang X, Bollinger SS, Boal AK, Zhang X (2017) AgHalo: a facile fluorogenic sensor to detect drug-induced proteome stress. Angew Chem Int Ed 56(30):8672–8676. https://doi.org/10.1002/anie.201702417
11. Fares M, Li Y, Liu Y, Miao K, Gao Z, Zhai Y, Zhang X (2018) A molecular rotor-based Halo-Tag ligand enables a fluorogenic proteome stress sensor to detect protein misfolding in mildly stressed proteome. Bioconjug Chem 29(1):215–224. https://doi.org/10.1021/acs.bioconjchem.7b00763
12. Pulsipher A, Griffin ME, Stone SE, Hsieh-Wilson LC (2015) Long-lived engineering of glycans to direct stem cell fate. Angew Chem Int Ed 54(5):1466–1470. https://doi.org/10.1002/anie.201409258

Chapter 11

Quantification of Protein Aggregates Using Bimolecular Fluorescence Complementation

Vibha Prasad and Aaron Voigt

Abstract

Neuronal accumulations of the presynaptic protein α-synuclein represent a hallmark of Parkinson's disease (PD). Multiple system atrophy and dementia with Lewy bodies are other neurodegenerative diseases in which α-synuclein aggregates are the main pathological characteristic. This is why these diseases are summarized as synucleinopathies. The abnormal α-Synuclein accumulation eventually results in the formation of Lewy bodies and Lewy neurites in surviving neurons of the brain. α-Synuclein aggregation has been widely researched in vitro but little is known about α-synuclein aggregation in vivo. This is why it is still unclear whether α-synuclein accumulation and/or which of the resulting aggregates (soluble oligomers or insoluble fibrils) actually trigger neurotoxic events. In our study, we employed the robust genetic makeup of *Drosophila melanogaster* in combination with bimolecular fluorescence complementation and developed an in vivo assay allowing to determine abundance of soluble α-synuclein aggregates (most likely oligomers). Insoluble α-synuclein aggregates (most likely fibrils) are detected by filter retardation assay. We provide an experimental outline to verify the reported assay system. The described method is easy, inexpensive, fast, and sensitive. Altogether, the described assay system is ideal for large-scale screening approaches aiming to test/identify compounds/conditions that affect α-synuclein aggregation in vivo.

Moreover, the experimental outline reported here could serve as a template to analyze aggregation of proteins other than α-synuclein. As accumulation of specific proteins is a hallmark of almost every neurodegenerative disease, only minor changes to our protocol should be necessary for a similar analysis. We have put special focus on critical evaluation steps to consider when adapting our protocol to other aggregate-forming proteins.

Key words BiFC, Neurodegeneration, Protein aggregation, Filter Retardation assay

1 Introduction

Neurodegenerative diseases comprise a whole range of disorders. In disease situation, neuronal dysfunction and severe neuron loss can be observed in affected patients. Both neuron loss and dysfunction are usually associated with severe phenotypic changes (e.g., dementia). Interestingly, in most cases neuronal dysfunction/decline in these diseases is accompanied by the presence of

Cláudio M. Gomes (ed.), *Protein Misfolding Diseases: Methods and Protocols*, Methods in Molecular Biology, vol. 1873, https://doi.org/10.1007/978-1-4939-8820-4_11,

misfolded proteinaceous deposits [1–3]. Accordingly, these neurodegenerative diseases in which misfolded and aggregated proteins are the main pathological hallmark are summarized under the umbrella term aggregopathies [4]. The most well-known and most common neurodegenerative diseases, which display protein aggregates, are Alzheimer's disease and Parkinson's disease [5, 6]. Additional neurodegenerative diseases presenting proteinaceous deposits are for example, Lewy body dementia, frontotemporal dementia, amyotrophic lateral sclerosis, Huntington's disease, spinocerebellar ataxias, and prion diseases [7–14].

Here we will focus on α-synuclein, the aggregating protein in Parkinson's disease. Specific mutations in the *SNCA* gene that codes α-synuclein seem to enhance aggregation and will cause Parkinson's disease in mutant allele carriers [15, 16]. This underlines the importance of α-synuclein and its aggregation in Parkinson's disease. Under physiological conditions α-synuclein is an intrinsically disordered protein, as it retains an unfolded conformation in solution in cell-free systems, the aggregation of α-synuclein can be triggered by a number of factors such as pH change, metal ions, temperature change and many more [17–19]. In contrast to the highly disordered and unfolded state in solution, α-synuclein forms highly ordered amyloid structures in the aggregated state [20, 21]. Although ordered antiparallel fashion, overall α-synuclein aggregates are heterogeneous and dynamic. A clear differentiation of specific aggregation states in living cells is not possible. Only detailed analysis, for example by electron microscopy, allows for analyzing differences in the composition and structure of α-synuclein aggregates [22–24]. Apart from such detailed analysis, the following broad groups of α-synuclein aggregation states are used: oligomers, fibrils and Lewy bodies. Oligomers (dimer to x-mer) are soluble. Soluble oligomers aggregate further into insoluble, amyloidogenic fibrils (of high molecular weight). Fibrils eventually cluster to form microscopic visible Lewy bodies [25, 26].

In an aggregated form, α-synuclein assembles in an antiparallel fashion. We took advantage of the organized assembly of α-synuclein aggregates. By applying bimolecular fluorescence complementation (BiFC), we were able to measure and quantify (in relative amounts) aggregated α-synuclein. Under favorable conditions, α-synuclein monomers orient themselves in an antiparallel manner while forming a dimer (*see* Fig. 1). In this case one strand of α-synuclein is tagged with a nonfluorescent part of the yellow fluorescent protein (Venus) at the C-terminus. The nonfluorescent counterpart of Venus is fused to another α-synuclein at its N-terminus. Dimerization of α-synuclein will bring the two Venus parts in close proximity, resulting in the formation of a fluorescent Venus protein. Thus, an increase in the Venus fluorescence can be interpreted as a measure of α-synuclein aggregation. We show that

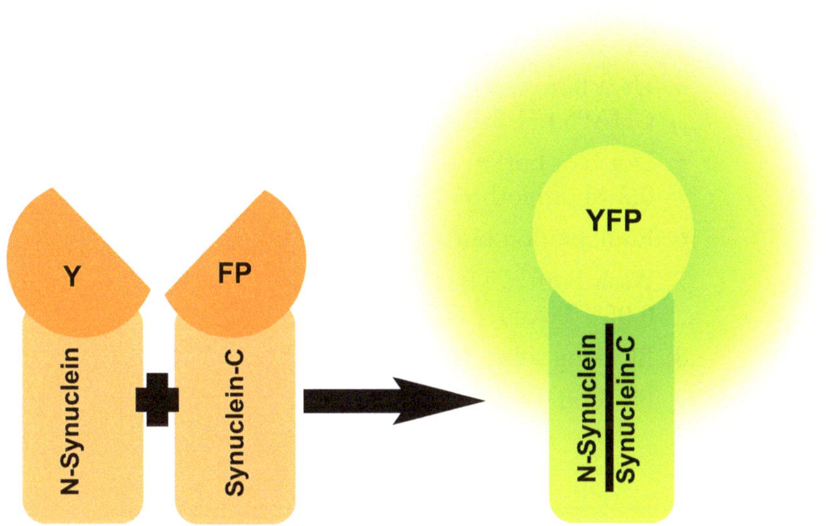

Fig. 1 Cartoon illustrating bimolecular fluorescence complementation driven by dimerization of α-synuclein. The α-synuclein variants are fused to the nonfluorescing parts of a fluorescent protein. Upon α-synuclein aggregation in an antiparallel fashion, the nonfluorescent parts of the Venus protein are in close proximity forming a fluorescent Venus protein

α-synuclein-BiFC yields a quick and sensitive assay to detect α-synuclein aggregation in a quantifiable fashion [27–29].

Here we present a simple, fast, inexpensive, and sensitive assay that allows for quick separation of α-synuclein aggregates into soluble lower molecular weight aggregates and the insoluble higher molecular weight proteins. The abundance of α-synuclein aggregates in the soluble protein fraction is quantified via detection of Venus fluorescent signal intensities while the insoluble higher molecular weight aggregates of α-synuclein were analyzed via the filter retardation assay [30–33]. This simple outline of the overall experimental approach allows for large-scale analysis and screening approaches (e.g., to test potential α-synuclein aggregation inhibitors). Moreover, the technique reported here can be used with minor adaptations to analyze aggregation of other proteins as well.

2 Materials

Prepare all buffers freshly and store them at room temperature (unless otherwise specified) using double distilled water.

1. Radioimmunoprecipitation assay buffer (RIPA Buffer): 0.05 M Tris-Cl pH-8.0, 0.5% sodium deoxycholate, 0.15 M NaCl, 0.1% SDS, 1% Triton, 1 tablet complete without EDTA (*see* **Note 2**).

2. Polypropylene rounded screw-capped tubes (2.0 mL, Sarstedt, Germany).
3. Ceramic microbeads (1.4 mm in diameter, Cayman Chemical, Germany).
4. Urea buffer: 0.03 M Tris buffer, 7 M urea, 2 M thiourea, 4% CHAPS pH-8.5.
5. Dot blot buffer: 0.5 M Tris pH 6.8, 0.4% SDS, 20% glycerol, 0.2 M dithiothreitol (DTT).
6. Equilibration buffer: 1× TBS in 0.1% SDS.
7. Wash buffer: TBS (Tween 20): 0.025 M Tris, 0.14 M NaCl, 0.05% Tween, pH 7.5.
8. Blocking solution: 5% skim milk in 1× TBS (Tween 20).
9. Fly food for assays: 2% agar and 4% sucrose in double distilled water.

3 Methods

3.1 Preparation of Fly Head Lysates

1. Place up to 20 flies labeled 1.5 mL Eppendorf tube. Snap-freeze in liquid nitrogen.
2. To separate the fly heads from its body, strike the Eppendorfs onto a hard surface in a quick motion (will separate heads from the remaining parts of the body).
3. Empty the contents of Eppendorf tube on a white surface (e.g., facial tissue).
4. Use forceps to carefully transfer the separated fly heads into the prelabeled screw-capped tubes containing ceramic microbeads (0.3–0.4 g/vial). Add RIPA buffer with a volume of 10 μL/fly head. Do not use more than 30 heads per vial (*see* **Notes 1** and **2**).
5. Lyse heads in a speed mill programmed: 2 min grinding with 2 min of resting interval followed by an additional 2 min of grinding. Spin for 20 s in a tabletop centrifuge to get rid of foam.
6. In the meantime, prepare labeled Eppendorf tubes. Carefully transfer as much as possible of the lysates to the fresh tubes (*see* **Note 3**).

3.2 Separation of Soluble and Insoluble Fraction from Fly Head Lysates

1. Spin the fly head lysates in tabletop centrifuge at 13,000 × *g* for 2 min at a temperature of 4 °C.
2. Collect the supernatant (soluble protein fraction) into a fresh-labeled Eppendorf tube. Keep pellet as it represents the RIPA-insoluble fraction. Determine protein concentration of the soluble fraction using the DC protein assay (Bio-Rad) (*see* **Notes 4–6**).
3. Wash pellet thrice by adding RIPA buffer (10 μL per fly head) to the pellet followed by centrifugation at 4 °C, 13,000 × *g* for

20 min each. The supernatant is discarded after the wash and replaced with fresh RIPA buffer (*see* **Note 6**).

4. Lastly, measure the total volume of the RIPA-soluble fraction and add the same volume of urea buffer to the pellet (*see* **Note 7**). Briefly vortex and spin at 13,000 × *g* for 20 min at 4 °C. Take supernatant and transfer to fresh Eppendorf tube. This fraction is considered the RIPA-insoluble fraction (for continuation *see* Method 3.5).

3.3 Relative Measure of Soluble Protein Fraction: Fluorescence

1. Use a Tecan M200/Fluorescence reader and Nanoquant plate to quantify the signal. Adjust the lysate concentration to 1.5 μg/μL protein by dilution with RIPA buffer. Set the following parameters for excitation and emission in case of measuring Venus signal intensities: 485 nm/530 nm. Use a total volume of 2 μL in RIPA buffer to blank all slots of the Nanoquant plate. Afterward load 2 μL of your sample to determine Venus signal intensities. Measure each sample at least in triplicate (*see* **Notes 8–10**).
2. Normalize the lysate fluorescence against a given control sample and plot the data as arbitrary units in any software (e.g., GraphPad Prism).

3.4 Preliminary Evaluation and Validation of the BiFC System: A Quality Control Check

The method described below can be adapted to other aggregation-prone proteins. This is an elementary step prior to carrying out an aggregation assessment protocol.

1. The BiFC system aids in relative quantification of the amount of aggregates. To assess the fluorescent signal impediment by disturbing agents in lysates (e.g., autofluorescent agents), conduct a dilution series (e.g., 100%, 80%, 60%, 40%, 20%, and 0) using the lysis buffer/non-Venus containing lysates. Measure the Venus fluorescence signal of all dilutions at 530 nm. The lysis buffer measures the background noise and helps in normalizing the samples to yield aggregate measure in arbitrary units. There should be a more or less linear regression (Fig. 2).
2. The linear increase of Venus signal intensity is proportional with time. In flies, we measured Venus fluorescent signals on 1, 5, and 10 days post eclosion. Expect a more or less linear increase in Venus signal over time. This can be also used to estimate a time point for future analyses. A similar approach can be performed using confocal microscopy on tissue expressing BiFC proteins. Also here, expect an increase of Venus signal intensities over time.

3.5 Relative Quantification of the Insoluble Protein Fraction

1. To determine the relative amount of aggregated α-synuclein in the insoluble protein fraction, work with a protein concentration varying from 5 to 150 μg. Based on the soluble protein fraction concentration (DC protein assay), pipet the same volume of insoluble protein fraction into labeled Eppendorf tubes.

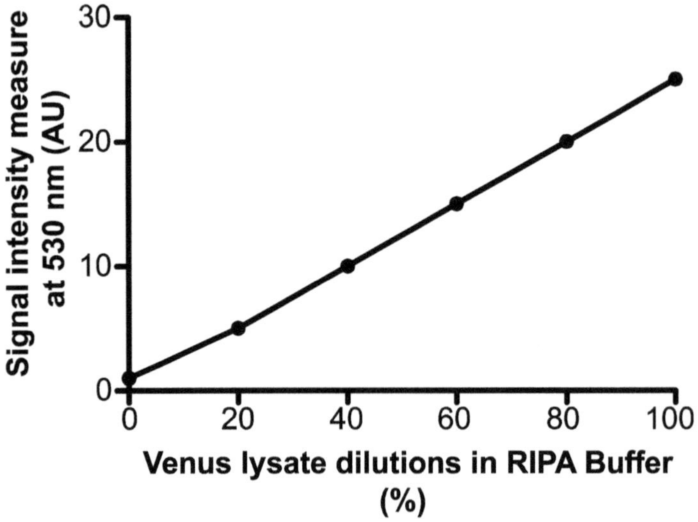

Fig. 2 Evaluation and validation of the BiFC system. Above is the representation of Venus fly head lysate dilutions in RIPA buffer. Similar results should be obtained when analyzing aggregating proteins other than α-synuclein

2. Adjust the volume to 25 μL using RIPA buffer, if necessary. Sonicate (we used the Ultrasonic Cleaner VWR for sonication) the tubes for 10 min.
3. Centrifuge the tubes with samples using a tabletop centrifuge for 30 min at 13,000 × *g* at 4 °C.
4. Add 7 μL of dot blot buffer. Place the tubes at 4 °C (*see* **Note 11**).
5. Set up the dot-blot apparatus. Cut the nitrocellulose paper to the required dimensions and incubate in 1× TBS, 0.1% SDS solution. Membrane porosity plays a vital role in aggregate retention and therefore a preliminary membrane test is a requirement to check whether the candidate aggregates are retained and can be detected on the membrane. It is a prerequisite to additionally load samples in increasing concentrations. This serves as a quality control, as FRA results may vary. Only quantify FRA results when an increase in aggregate load correlates with an increase in volume loaded.
6. Keep a tray ready with 1× TBST buffer for column equilibration and another trough for blocking with 5% skimmed milk.
7. After membrane incubation for about 2–3 min, assemble the apparatus and close the screws tightly. Connect the outlet pipe of the apparatus to a vacuum pump. Keep the pump running on low speed (*see* **Notes 12** and **13**).

8. Incubate the blotting columns with 200 μL of equilibration buffer and turn up the pump speed to the maximum. Watch the buffer run through the column under the influence of gravitational forces and vacuum. Thereafter, repeat this process two more times. Lower the vacuum speed to a minimum.
9. Load the entire sample volume into the corresponding slots of the precast dot-blot apparatus. Immediately after the sample is sucked through the membrane, wash slot with 200 μL of the 1× TBST buffer and turn up the vacuum speed to a maximum (*see* **Notes 14–17**).
10. Once the lysates run through the column, disassemble the apparatus. Carefully take out the nitrocellulose paper. Label a corner of the nitrocellulose paper with the sample name and then cut the membrane if it has unused paper (*see* **Note 18**).
11. Block the membrane in 5% skimmed milk for 1 h at room temperature.
12. Dilute the corresponding antibody in 5% skim milk to achieve the designated concentration and incubate the membrane overnight under agitation at 4 °C.
13. The following day, wash the membrane twice for 10 min in 1×-TBST buffer at room temperature.
14. Incubate the membrane with appropriate concentration of an HRP-coupled secondary antibody.
15. Subsequently, carry out membrane washes for 10 min with 1×-TBST wash buffer (*see* **Note 19**).
16. Detect signals using the electrochemiluminescence (ECL).

3.6 Application of the BiFC System on Established Cellular Degradation Pathways

1. In our Parkinson fly model system; we investigated the effect of established drugs that affect the cellular degradation pathways.
2. We tested MG132, a known inhibitor of the ubiquitin proteasomal system. After administration of the drug with the fly food over a period of 5 days, the corresponding fly head lysates were subjected to the protocol described above. Drug-treated flies were compared to solvent only-treated flies (control).
3. The BiFC method provided us with information of the soluble protein fraction measure, while the filter retardation assay gave us information about the insoluble protein aggregate content in the tested sample (Fig. 3).
4. As expected, the treated sample showed a slight increase in the soluble protein fraction and a very high increase in the insoluble protein fraction versus the control sample.

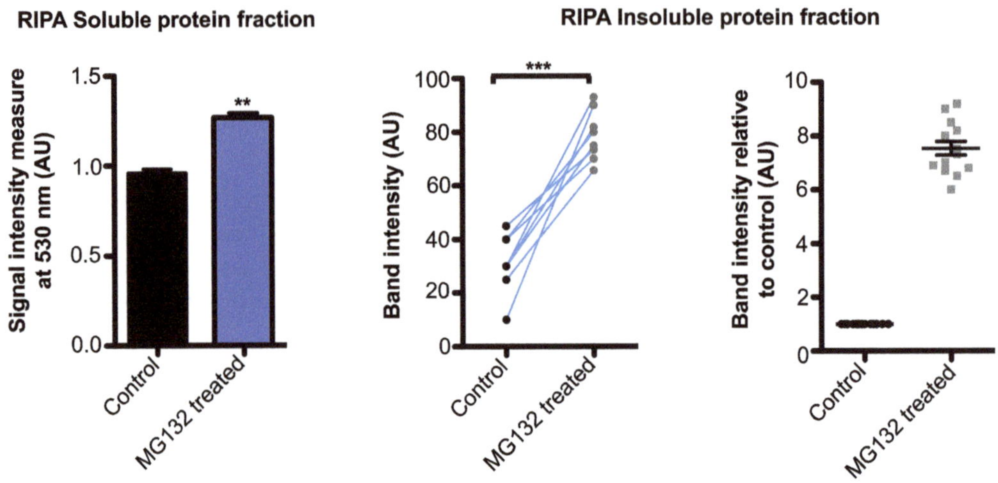

Fig. 3 Representation of MG132 treatment on BiFC tagged α-synuclein protein. The above results demonstrate the effect of the drug MG132 (ubiquitin proteasomal system inhibitor) on soluble α-synuclein aggregate fraction (measured by determination of Venus signal intensities) and insoluble α-synuclein aggregate abundance determined by filter retardation assay. In the above figure, the insoluble fraction (unnormalized data) is represented in the form of a line plot while the normalized data is represented by the scatter plot. Significance in *t*-test, $^{**}P < 0.01$, $^{***}P < 0.001$

4 Notes

1. Carefully transfer all the fly heads (after a count) to the bottom of the screw-capped tube in RIPA buffer. Ensure that all the fly heads are introduced into RIPA buffer before lysis to make sure that the lysis is complete.
2. The number/weight of microbeads is estimated to fill the base of each screw-capped tube. After some experience, fill beads by rough measure without weighing.
3. During and after the lysate formation, the tubes are warm and it is a prerequisite to transfer the lysate immediately to a new tube prior to the protein fraction separation. This will ensure sample integrity for further analysis.
4. Ensure the temperature is maintained at 4 °C at all times.
5. Carefully pipette the supernatant to separate (from pellet) and collect the soluble protein fraction from the lysate, leaving the debris behind. The debris consists of the fly head cuticle.
6. Conduct a minimum of three washes before adding the proportionate volume of urea buffer at the final step. In case you establish this method for a protein other than α-synuclein, collect washing fractions and analyze in Western blot for the abundance of your protein. It should be present in the soluble fraction, absent in the washes (especially two and three), and reappear in the insoluble fraction.

7. The prepared urea buffer is stored at room temperature and can be used for 3–4 months.
8. It is necessary to take a minimum of three measurements for each lysate sample when using the Nanoquant.
9. Ensure that the drop of lysate does not collapse on the Nanoquant plate during measurement, to avoid inconsistency in fluorescent signal reading.
10. Take the fluorescence measurements the same day as the lysate preparation, as fluorescence is a sensitive parameter and temperature change affects the reading. Rethawed lysate gives a reduced fluorescence measure possibly due to altered sample properties.
11. Keep the samples maintained at 4 °C at all times. Use a freshly prepared dot-blot solution for good results.
12. On beginning with the filter retardation assay, ensure to connect the apparatus to a vacuum pump at minimal speed so as to clear off residual air from the blotting columns.
13. Tighten the screws to a maximum to make sure that the lysate/wash buffer flow rate is smooth and uniform.
14. The nitrocellulose membrane should cover all the slots on the dot blot apparatus (irrespective of the number of slots in use), as this will ensure that the pressure is uniform throughout. Failure to cover the entire surface of the apparatus will alter the flow rate of the lysate. Use optimal membrane for your proteins of interest. For example, 0.2 μm cellulose acetate paper which has been successfully tested for tau aggregates (cos-7 cells) and washed thrice with 1% SDS prior to antibody incubation [34]. With regard to tau aggregates (in vitro), it has been proven that polyvinylidenefluoride membrane with a pore size of 0.45 μm works better than the cellulose acetate membranes while the Nitrocellulose membrane of 0.20 μm porosity comes highly recommended [35].
15. Use slots preferably at the top row and bottom row of the dot blot apparatus.
16. Load the samples in increasing concentrations to directly have a quality test on each individual membrane. Use only membranes for quantification in case the increase in loaded protein amount is reflected by an increase in signal intensities.
17. Do not overload the slots as the overflow will contaminate the neighboring chambers and result in artifacts or false positive results.
18. Ensure to clean inlet pipe of the vacuum pump after each use.
19. Wash the blots well, after and in-between the antibody incubations so as to obtain clean and clear results.

5 Data Processing

The filter retardation assay blots are relatively quantified using Image J software. The raw data is gathered and the treated sample should be presented in relation to the normalized control sample as scatter plot. The raw data should be represented as a line plot followed by statistical analysis (control sample versus the treated sample; two-tailed Student's *t*-test, paired analysis; $***P < 0.001$, $**P < 0.01$, $*P < 0.05$) to determine significance between the control sample versus the treated sample.

Acknowledgment

This work is supported by the Interdisciplinary Center for Clinical Research (IZKF) Aachen within the Faculty of Medicine at RWTH Aachen University (Clinical Neuroscience, Project N7-3).

References

1. Rangel-Barajas C, Coronel I, Floran B (2015) Dopamine receptors and neurodegeneration. Aging Dis 6(5):349–368
2. Irvine GB, El-Agnaf OM, Shankar GM, Walsh DM (2008) Protein aggregation in the brain: the molecular basis for Alzheimer's and Parkinson's diseases. Mol Med 14(7–8):451–464
3. Chouhan AK, Guo C, Hsieh Y-C, Ye H, Senturk M, Zuo Z, Li Y, Chatterjee S, Botas J, Jackson GR, Bellen HJ, Shulman JM (2016) Uncoupling neuronal death and dysfunction in Drosophila models of neurodegenerative disease. Acta Neuropathol Commun 4:62
4. Dohm CP, Kermer P, Bähr M (2008) Aggregopathy in neurodegenerative diseases: mechanisms and therapeutic implication. Neurodegener Dis 5:321–338
5. Gibrat C, Saint-Pierre M, Bousquet M, Levesque D, Rouillard C, Cicchetti F (2009) Differences betwwen subacute and chronic MPTP mice models: investigation of dopaminergic neuronal degeneration and alpha-synuclein inclusions. J Neurochem 109:1469–1482
6. Mhyre TR, Boyd JT, Hamill OW, Maguire-Zeiss KA (2012) Parkinson's disease. Subcell Biochem 65:389–455
7. Kovacs G (2015) Molecular pathological classification of neurodegenerative diseases: turning towards precision medicine. Int J Mol Sci 17:189
8. Bertram L, Tanzi RE (2005) The genetic epidemiology of neurodegenerative disease. J Clin Investig 115(6):1449–1457
9. Rademakers R, Cruts M, van Broeckhoven C (2004) The role of Tau (MAPT) in frontotemporal dementia and related tauopathies. Hum Mutat 24:277–295
10. McKeith I (2004) Dementia with Lewy bodies. Dialogues Clin Neurosci 6(3):333–341
11. Grad LI, Rouleau GA, Ravits J, Cashman NR (2017) Clinical spectrum of amyotrophic lateral sclerosis (ALS). Cold Spring Harb Perspect Med 7:a024117
12. Treusch S, Cyr DM, Lindquist S (2009) Amyloid deposits: protection against toxic protein species? Cell Cycle 8(11):1668–1974
13. Mark MD, Krause M, Boele H-J, Kruse W, Pollok S, Kuner T, Dalkara D, Koekkoek S, De Zeeuw CI, Herlitze S (2015) Spinocerebellar ataxia type 6 protein aggregates cause deficits in motor learning and cerebellar plasticity. J Neurosci 35(23):8882–8895
14. Cobb NJ, Surewicz WK (2009) Prion diseases and their biochemical mechanisms. Biochemistry 48(12):2574–2585
15. Klein C, Westenberger A (2012) Genetics of Parkinson's disease. Cold Spring Harb Perspect Med 2(1):a008888
16. Nuytemans K, Theuns J, Cruts M, Van Broeckhoven C (2010) Genetic etiology of Parkinson disease associated with mutations in the SNCA, PARK2, PINK1, PARK7, and LRRK2 genes: a mutation update. Hum Mutat 31(7):763–780
17. Narkiewicz J, Giachin G, Legname G (2014) In vitro aggregation assays for the characterization

of α-synuclein prion-like properties. Prion 8(1):19–32
18. Uversky VN, Eliezer D (2009) Biophysics of Parkinson's disease: structure and aggregation of α-synuclein. Curr Protein Pept Sci 10(5):483–499
19. Breydo L, Wu JW, Uversky VN (2012) α-Synuclein misfolding and Parkinson's disease. Biochim Biophys Acta 1822(2):261–285
20. Moriarty GM, Olson MP, Atieh TB, Janowska MK, Khare SD, Baum J (2017) A pH dependent switch promotes β-synuclein fibril formation via glutamate residues. J Biol Chem 292(39):16368–16379
21. Kuen-Phon W, Weinstock DS, Narayanan C, Levy RM, Baum J (2009) Structural reorganization of α-synuclein at low pH observed by NMR and REMD simulations. J Mol Biol 391(4):784–796
22. Rodriguez JA, Ivanova MI, Sawaya MR, Cascio D, Francis Reyes DS, Sangwan S, Guenther EL, Johnson LM, Meng Zhang LJ, Arbing MA, Nannega B, Hattne J, Whitelegge J, Brewster AS, Messerschmidt M, Boutet S, Sauter NK, Gonen T, Eisenberg D (2015) Structure of the toxic core of α-synuclein from invisible crystals. Nature 525(7570):486–490
23. Boassa D, Berlanga ML, Yang M-L, Terada M, Hu J, Bushong EA, Hwang M, Masliah E, George JM, Ellisman MH (2013) Mapping the subcellular distribution of alpha-synuclein in neurons using genetically encoded probes for correlated light and electron microscopy: implications for Parkinson's disease pathogenesis. J Neurosci 33(6):2605–2615
24. Bousset L, Pieri L, Ruiz-Arlandis G, Gath J, Jensen PH, Habenstein B, Madiona K, Olieric V, Anja B, Meier BH, Melki R (2013) Structural and functional characterization of two alpha-synuclein strains. Nat Commun 4:2575
25. Scott Kim W, Kågedal K, Halliday GM (2014) Alpha-synuclein biology in Lewy body diseases. Alzheimers Res Ther 6(5):73
26. Luk KC, Song C, O'Brien P, Stieber A, Branch JR, Brunden KR, Trojanowski JQ, Lee VM (2009) Exogenous α-synuclein fibrils seed the formation of Lewy body-like intracellular inclusions in cultured cells. Proc Natl Acad Sci U S A 106(47):20051–20056
27. Miller KE, Kim Y, Huh W-K, Park H-O (2015) Bimolecular fluorescence complementation (BiFC) analysis: advances and recent applications for genome-wide interaction studies. J Mol Biol 427(11):2039–2055
28. Lai HT, Chiang CM (2013) Bimolecular fluorescence complementation (BiFC) assay for direct visualization of protein-protein interaction in vivo. Bio Protoc 3(20):e935
29. Kerppola TK (2006) Design and implementation of bimolecular fluorescence complementation (BiFC) assays for the visualization of protein interactions in living cells. Nat Protoc 1(13):1278–1286
30. Wanker EE, Scherzinger E, Heiser V, Sittler A, Eickhoff H (1999) Membrane filter assay for detection of amyloid-like polyglutamine-containing protein aggregates. Methods Enzymol 309:375–386
31. Waelter S, Boeddrich A, Lurz R, Scherzinger E, Lueder G, Lehrach H, Wanker EE (2001) Accumulation of mutant Huntingtin fragments in aggresome-like inclusion bodies as a result of insufficient protein degradation. Mol Biol Cell 12:1393–1407
32. Dou F, Netzer WJ, Tanemura K, Li F, Hartl FU, Takashima A, Gouras GK, Greengard P, Xu H (2002) Chaperones increase association of tau protein with microtubules. Proc Natl Acad Sci U S A 100(2):721–726
33. Brady OA, Meng P, Zheng Y, Mao Y, Fenghua H (2011) Regulation of TDP-43 aggregation by phosphorylation and p62/SQSTM1. J Neurochem 116:248–259
34. Zhang X, Zhang YW, Liu S, Bulloj A, Tong GG, Zhang Z, Liao F-F, Huaxi X (2006) Tumor suppressor PTEN affects tau phosphorylation: deficiency in the phosphatase activity of PTEN increases aggregation of an FTDP-17 mutant Tau. Mol Neurodegener 1:7
35. Chang E, Kuret J (2008) Detection and quantification of tau aggregation using a membrane filter assay. Anal Biochem 373(2):330–336

Chapter 12

Screening Protein Aggregation in Cells Using Fluorescent Labels Coupled to Flow Cytometry

Salvador Ventura and Susanna Navarro

Abstract

Protein misfolding and aggregation into amyloid conformations have been described to underlie the onset of several human neurodegenerative diseases. Although a large number of biophysical approaches are available to study amyloids in vitro, we still need robust methods to address their self-assembly in living cells. In this context, simple cellular models, like bacteria and yeast, expressing recombinant amyloidogenic proteins are emerging as convenient systems for studying the formation of protein inclusions, their toxicity, propagation, and interactions. We describe here a simple and fast flow cytometry method able to detect intracellular inclusions, as well as to analyze the distribution of the amyloidogenic protein of interest in intact cells. Using specific fluorescent amyloid-dyes, such as thioflavin-S and ProteoStat, or the fusion of fluorescent molecules, such as GFP, the technique can be applied in the quantification of intracellular amyloid content, for the screening of antiamyloidogenic compounds, and to test epigenetic or environmental conditions able to modulate amyloid deposition in vivo.

Key words Protein aggregation, Amyloid, Protein inclusions, Bacteria, Yeast, Flow cytometry

1 Introduction

Protein misfolding and amyloid deposition have been linked to a wide number of human disorders ranging from neurodegenerative diseases such as Alzheimer's (AD) and Parkinson's (PD) diseases, to nondegenerative amyloidoses such as type II diabetes, systemic amyloidoses, and cataracts [1–4]. Despite the polypeptides behind these pathologies are different and unrelated, in all cases they experience an aberrant protein transition to form cross-β fibrillar structures within the affected cell population [5]. The growth of amyloid structures compromises the proteostasis machinery, and results in intracellular cytotoxicity. Cell-to-cell transmission of these aggregates can propagate this toxic effect in the tissue vicinity.

There is an unmet medical need to advance our understanding of protein aggregation mechanisms in order to facilitate the

Cláudio M. Gomes (ed.), *Protein Misfolding Diseases: Methods and Protocols*, Methods in Molecular Biology, vol. 1873,
https://doi.org/10.1007/978-1-4939-8820-4_12,

development of new therapeutic approaches against these pathologies. Although we have in our hands a large number of techniques to characterize amyloid aggregation in vitro [6–8], the study of amyloids in a physiological context has been far more challenging, mainly due to the compositional complexity of the crowded cellular milieu. In this context, unicellular organisms, such as bacteria or yeast, have emerged as cheap and powerful model systems to get insights into the mechanisms behind protein aggregates formation in simple, but still relevant, physiological contexts [9, 10].

On the one hand, bacteria provide an affordable model system for the analysis of protein aggregation since they can produce recombinantly insoluble protein deposits, known as inclusion bodies (IBs), displaying structural properties characteristic of amyloids [11, 12]. They offer a unique opportunity to characterize the features of intracellular protein aggregates [13], to assess the impact of protein aggregation on cell fitness [14], and for the screening of aggregation modulators for therapeutic purposes [15, 16]. On the other hand, baker's yeast has become an ideal eukaryotic model organism to dissect the specific molecular pathways that become unbalanced in different human diseases. As in bacteria, the recombinant expression of amyloidogenic proteins in yeast usually leads to the formation of β-sheet-rich structures similar to those described in neurodegenerative diseases. The remarkable conservation of the proteostasis machinery among eukaryotes [17–21] makes yeast a privileged model to elucidate the molecular alterations in such devastating disorders.

Fluorescence-based techniques, such as fluorescence microscopy and flow cytometry (FC), allow us to develop new approaches to characterize cell protein aggregation using fluorescently tagged molecules or fluorescent amyloid dyes in unicellular model systems. In particular, the capability of FC to process thousands of cells in a high-throughput manner using simultaneously multiple fluorescence channels makes this method ideal to target intracellular protein aggregation. Here we provide a detailed protocol for the characterization of the aggregates formed by recombinant amyloidogenic proteins in living and fixed bacterial and yeast cells using FC.

2 Materials

Filter all buffers before resuspending cells through a 0.2 μm filter to avoid small contaminant particles that could be detected by FC.

2.1 Culture and Preparation of Bacteria and Yeast Cells

1. Yeast rich media (YPD): 20 g/L bacteriological peptone, 10 g/L yeast extract, 20 g/L glucose.
2. Yeast synthetic drop-out media (SD −ura).
3. LB media: 10 g/L tryptone, 5 g/L yeast extract, 5 g/L NaCl.
4. Galactose.

5. Antiobiotics depending on the resistance of the selected plasmid.
6. Isopropyl β-D-1-thiogalactopyranoside (IPTG).
7. Phosphate-buffered saline (PBS):137 mM NaCl, 2.7 mM KCl, 10 mM Na_2HPO_4, 2 mM KH_2PO_4.
8. 0.5 M ethylenediaminetetraacetic acid (EDTA).
9. 37% paraformaldehyde (PFA) (*see* **Note 1**).

2.2 Staining of Cells

1. Thioflavin S (Th-S).
2. ProteoStat (Enzo Life Sciences).
3. SYTO 9 stain (Life Technologies).
4. Propidium iodide.
5. CellROX Deep Red Reagent (Life Technologies).
6. PBS buffer.

2.3 Fusion to Fluorescent Reporters

Application of the most commonly used fluorescent protein tags, spectral properties, recommended excitation lasers and detector filters on flow cytometry (Table 1) [22].

2.4 Equipment

1. Flow cytometer, equipped with the appropriate lasers for excitation and bandpass filters for detection; BD Canto or an equivalent.
2. Thermostatic incubator shaker.
3. FlowJo version xV0.7, FACSDiva Sofware v5.0 o v7.0.

Table 1
Fluorescent protein tags commonly used in flow cytometry

Fluorescent protein	Color	Brightness	Ex_{MAX} (nm)	Em_{MAX} (nm)	Recommended laser (nm)	Recommended bandpass filters (nm)
EGFP	Green	33.6	484	507	488	510/21, 530/30 (FL-1)
EYFP	Yellow	50.9	514	527	488, 505, 514, 517, 532	530/30, 550/30, 556/21, 575/26, 585/42 (FL-1, FL-2)
ECFP	Cyan	13.0	439	476	405, 440, 457	450/50, 460/50, 485/22 with 488 nm block (SSC)
DsRed mCherry	Short-red	3.5 15.8	553	583	532, 550, 561	585/42, 590/20, 610/20 (FL-3)

4. GraphPad Prism 6.05 (GraphPad Software, San Diego, CA, USA).
5. Centrifuge.
6. Tubes appropriate for holding cells for the flow cytometer.

3 Methods

3.1 Preparation of Cells for Staining

1. Grow bacteria or yeast cells in the appropriate media (LB or YPD) and temperature 37 °C or 30 °C, respectively, in the presence of the suitable antibiotic, under agitation conditions. When cells reach the log phase induce the expression of recombinant protein according to the plasmid promoter. The induction time depends on the expression rate of respective protein variants.
2. Collect cells by centrifugation at 3000 × *g*, and wash twice with PBS.
3. Detection of intracellular aggregates can be performed with living or fixed cells. However, fixation can affect the integrity of the aggregates.

   (a) Living cells.

   - Take 1 mL of induced cell cultures and determine the OD_{600}.
   - Dilute cells to the desired density in PBS.
   - Transfer the cell suspension to an appropriately labeled tube and store at room temperature, protected from the light (*see* **Note 2**).

   (b) Fixed cells.

   When choosing a fixation method, consider the fact that ethanol-based fixation can compromise the integrity of the protein aggregates leading to their complete loss. On the contrary, PFA does not significantly affect large aggregates.

   - Take 1 mL of washed induced cells and centrifuge at 3500 × *g* for 6 min at room temperature.
   - Aspirate carefully the supernatant and resuspend the cell pellet in 3.7% PFA (diluted in PBS) for at least 20 min. In case of alcohol-based fixation, resuspend cells in cold 75% ethanol for 1 h at 4 °C (*see* **Note 3**).
   - Centrifuge at 3500 × *g* for 5 min at room temperature.
   - Aspirate the supernatant and wash the cell pellet twice with PBS.
   - Transfer the cell suspension to an appropriately labeled tube and store at room temperature, protected from the light until needed.

3.2 Monitoring the Presence of Intracellular Protein Aggregates

FC is able to measure both the optical and fluorescence features of a cell in suspension. In this sense, it is similar to a fluorescence microscope but rather than providing an image it generates a quantitative evaluation of a set of parameters, including size (forward scatter FSC), complexity (side-scatter SSC), and relative fluorescence intensity of a fluorochrome (Fig. 1). Within a short time, the flow cytometer allows the analysis of thousands of cells, thus allowing the statistical data of the desired parameters.

Traditionally, flow cytometry has been used for the detection of cell antigens using fluorochromes coupled to antibodies. However, new applications of FC for the analysis of different states and distribution of protein aggregation in cell model systems have been developed exploiting fluorescent amyloid dyes and fluorescently tagged proteins.

3.2.1 Optimizing the Strategy to Measure Protein Aggregation by Flow Cytometry. Rapid Characterization of Aggregates in Living Cells by Th-S Staining

Intracellular inclusions can be detected with the thioflavin-T dye thanks to its ability to stain β-sheet rich protein aggregates. One of its derivatives is thioflavin-S (Th-S), a fluorescence benzothiazol salt dye also reported to bind to amyloid structures. The binding of the dye induces a shift in excitation (385–450 nm) and emission (445–482 nm) spectra. Its cell permeability makes it suitable to stain living cells. Once the dye penetrates the cell membranes, it accumulates into amyloid protein deposits [23].

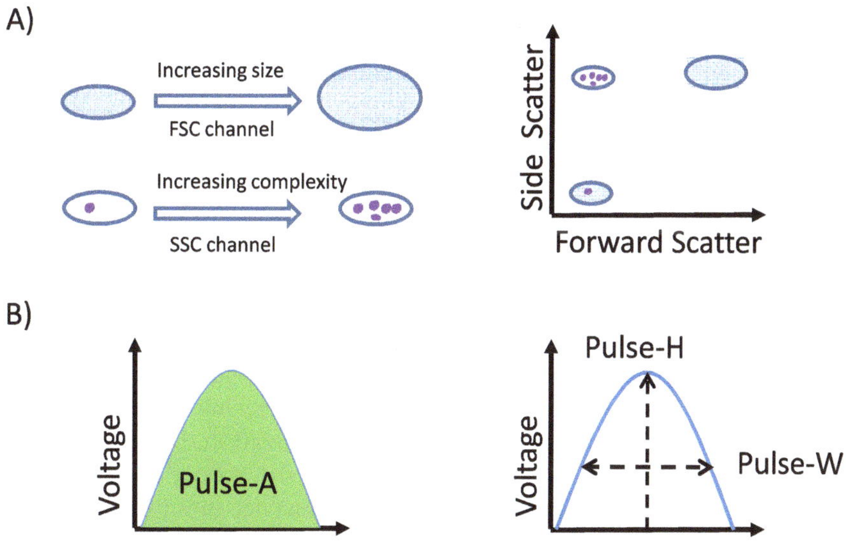

Fig. 1 Cells forward and side scattered light, as well as emitted fluorescence by stained cells or by cells expressing fluorescently tagged proteins are the main parameters detected by flow cytometry. (**a**) As cells go through the laser beam a detector in front of the light beam measures forward scatter, and several detectors at the sides measure side scatter, providing information about the cells' size and granularity (left panel). Dot plot of SSC vs FSC displaying the usual distribution for small, small-complex, and big cells (right panel). (**b**) The photomultipliers (PMTs) detect the fluorescence from positively stained cells. PMTs measure the pulse area of the voltage generated by emitted photons from fluorescent cells (left panel). The pulse area of the emitted fluorescence can be dissected into two parameters: pulse height (pulse-H), and pulse width (pulse-W) (right panel)

Procedure for staining cells with Th-S:

1. Cells prepared as described in the aforementioned Subheading 3.1 are diluted to an $OD_{600\ nm}$ of 0.1 and incubated for 1 h in the presence of a final concentration of 125 μM Th-S.
2. Cells are centrifuged at 1500 × *g* for 5 min, and resuspended in the same volume of PBS.
3. Acquisition is performed following the instructions given in Subheading 3.3 in a flow cytometer equipped with a 355 nm laser source and a long-pass filter of 530/30 nm in a logarithmic amplification scale.

3.2.2 ProteoStat Staining Coupled to Flow Cytometry

Although the traditionally dyes to detect aggregates in vivo are thioflavin-based (Th-T, Th-S), the ProteoStat dye is becoming increasingly used, as it is more specific and sensitive than Th-S [24]. ProteoStat is an excitable 488 nm red fluorescent molecular rotor dye able to detect protein inclusions both in fixed and living cells. It provides a larger dynamic range and improves the discrimination between amyloid-like and nonordered protein aggregates.

Procedure for staining cells with ProteoStat:

1. Cells expressing protein aggregates are prepared as it is described in Subheading 3.1, and are diluted to an $OD_{600\ nm}$ of 0.1 (*see* **Note 4**).
2. Resuspend cells in PBS containing 1/5000 dilution of ProteoStat dye, and incubate them for 30 min at room temperature. When dealing with living cells, incubation should be performed at 37 °C.
3. Acquisition is performed following the instructions given Subheading 3.3 in a flow cytometer equipped with a 488 nm laser source, and a band pass filter of 585/42 nm (PE-A or FL-2 channel) in a logarithmic amplification (Fig. 2).

3.2.3 Fusion to Fluorescent Reporters

Since the discovery of the green fluorescent protein (GFP) in *Aequorea victoria* many other derivatives have been evolved as a powerful toolbox for cell biology studies. Fluorescent protein tags have notably contributed to the determination of gene expression levels and subcellular protein localization. Because the fluorescence of GFP-tagged proteins require the proper folding of the fluorescent reporter, the relative fluorescence intensity of GFP-fused proteins, relative to a GFP control, can be used as a proxy of their solubility [25]. Accordingly, it has been reported that the in vivo fluorescence of GFP-tagged amyloidogenic polypeptides is inversely correlated with their aggregation propensity [26]. All in all, fluorescence reporters come out as excellent reagents to track the in vivo formation of protein aggregates, either by microscopy or FC. Again, FC allows for faster and simpler analysis of a large number of cells, providing accurate statistical data.

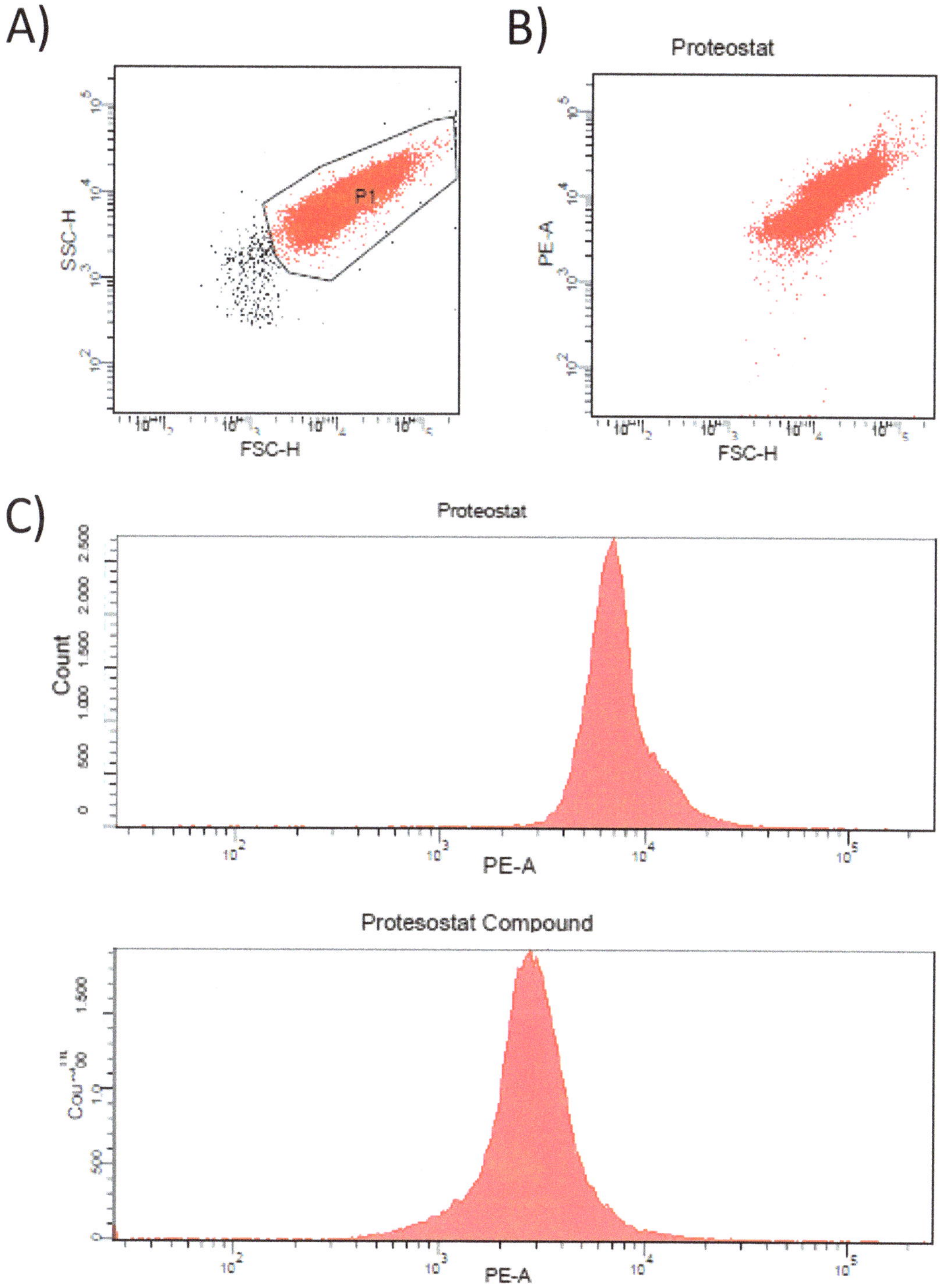

Fig. 2 Flow-cytometric analysis of *E. coli* cells expressing intracellular protein inclusions. In this experiment induced cells expressing a recombinant aggregation-prone protein were grown in the presence of an aggregation inhibitor, and later stained with the ProteoStat dye, in vivo. (**a**) Dot plot of SSC vs FSC showing nontreated cell population. In red are represented selected intact cells (gated in P1), and in black small particles and cell debris. (**b**) Dot plot of untreated cells (previously gated in P1), showing the size (FSC-H) and the emitted fluorescence signal of the ProteoStat dye (PE-A channel). (**c**) Histograms of the ProteoStat fluorescence signal detected in PE-A channel in the absence (upper panel) or presence of an aggregation inhibitor compound (lower panel). The difference between the median fluorescence intensity of untreated and treated cells correlates with the intracellular solubility of the protein. Cells were analysed in a FACSCanto flow cytometer (BD Biosciences)

1. Bacterial or yeast cells should be transformed with the appropriate plasmid to express the desired protein fused to the selected fluorescence reporter (*see* Table 1). There are reported many variants of GFP and other fluorescent proteins (FP) that differ in their relative foldability and intracellular solubility. For that reason, it is advisable to choose the most suitable FP form for your specific protein system (*see* **Note 5**).
2. Grow and induce cell cultures as indicated in the Subheading 3.1.
3. Wash cells twice with PBS, and dilute them at an $OD_{600\ nm}$ of 0.1.
4. Acquire the relevant parameters with settings fitted to the FP tag spectral properties (*see* Subheading 3.3.1) (*see* **Note 6**).

3.2.4 Combining Aggregation Reporters with Other Markers

It is possible to combine protein aggregation detection with other reporter markers on the presence of reactive oxygen species, proliferation, cell cycle or apoptosis.

The most popular combination might be the cell viability kit composed of SYTO Green and propidium iodide (PI). This method permits the discrimination of cells with damaged membranes (stained with PI) from the total population (stained with SYTO Green). The application of the cell viability staining allows to correlate the presence of intracellular aggregates with its cytotoxicity (Fig. 3). Formation of protein aggregates can be also correlated with the cellular oxidative stress level using CellROX Deep Red Reagent (Thermo Scientific).

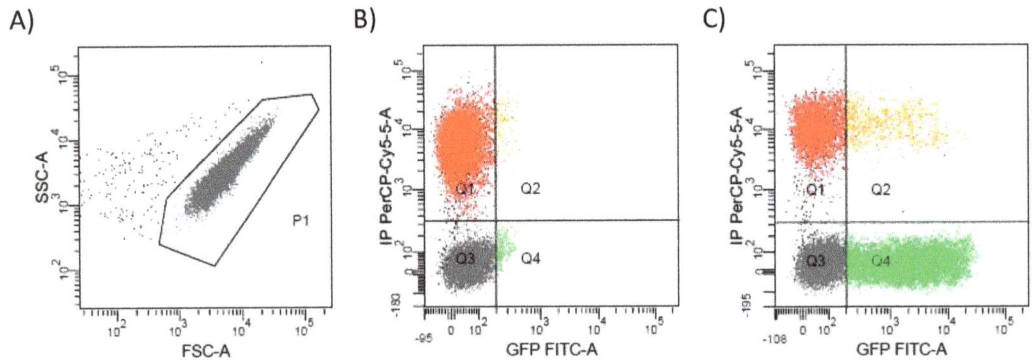

Fig. 3 Monitoring the relationship between the expression of a recombinant protein fused to GFP and the cellular integrity. (**a**) Dot plot of FSC vs SSC displaying the yeast population where intact cells are gated in P1 (grey). (**b** and **c**) Dot plots (FITC vs PerCP-CY5-5-A) of noninduced and induced cells, respectively. Protein expression levels, from P1 cells, are detected by the fluorescence emission of the fused GFP in the FITC-A channel, and cells with damaged membranes are detected by the staining with propidium iodide in the PerCP-Cy5-5-A channel. The quadrant analysis of the dot plot shows intact viable cells that do not express protein, designed as Q3 (lower left), viable cells expressing protein, Q4 (lower right), nonviable cells with disrupted plasma membranes that do not express protein, Q1 (upper left), and expressing protein Q2 (upper right). Cells were analysed in a FACSCanto flow cytometer (BD Biosciences). Each plot represents 50,000 bacterial cells

1. SYTO Green and PI are added to cells expressing the amyloidogenic protein at a final concentration of 10 μM, just prior to FC analysis.
2. Acquire cells using a 488 nm laser source and a long-pass filter of 530/30 nm for SYTO Green signal, and with a 488 nm laser source and a band pass filter of 530/30 nm for PI fluorescence. Previously, acquire data for unstained cells and single-color positive controls.
3. Optionally, for determining oxidative stress, add CellROX reagent at a final concentration of 5 μM to the cells and incubate for 30 min at 30 °C or 37 °C. Acquisition is performed with a 635 nm red laser and a deep red emission 660/20 BP filter.

3.2.5 Screening of Aggregation Inhibitors and Modulator Compounds Using Flow Cytometry

Yeast and prokaryotic cells are described to be suitable models for the screening of aggregation modulators since they share physiological similarities with higher eukaryotic cells.

The ability of amyloid fluorescent dyes, such as Th-S and ProteoStat, to bind intracellular aggregates provides an effective and affordable means to identify aggregation inhibitor compounds by monitoring the impact of these molecules on the overall cellular fluorescence by FC [16, 27].

1. Incubate cells in the presence of the desired concentration of putative inhibitor 30 min before induction of protein expression to allow proper penetration of the compound in the cells.
2. Proceed as recommended in Subheadings 3.1, and 3.2.1 or 3.2.2 (*see* Fig. 2).

3.3 Instrument Configuration

Perhaps the most important factor to ensure that your protein aggregation related protocol is successful is the correct setup of the instrument. First, it is important to know the specific capabilities of the flow cytometer you use, as the sensitivity might vary from model to model. Sensitivity is an intrinsic property of the instrument and it is related with the quality and potency of the excitation light source. A fluorescence sensitivity measurement must be accompanied with a complete definition of the instrument configuration. Included should be a description of the laser (type, wavelength, and output power), the associated beam conditioning and steering optics, excitation optics, collection optics, optical filters and their configuration, flow cell, and sheath velocity, to name some important aspects of a flow cytometer which influence the instrument's sensitivity.

Basic flow cytometers, such as BD FACScan (BD Biosciences), are equipped with only one blue argon laser, but most of the modern cytometers are equipped with three or more lasers, with colors ranging from ultraviolet (355 nm), violet (405 nm), blue (488 nm), yellow (561 nm) to red (640 nm). Besides lasers, other crucial components are the optical filters, since they define which colors can be simultaneously detected.

3.3.1 Setup of the Flow Cytometer Parameters and Sample Acquisition

1. Ensure your flow cytometer is equipped with the appropriate lasers and filters setups: For Th-S use a 355 nm laser source, and a 530/30 nm bandpass filter; for ProteoStat use a 488 nm laser source, and a 585/42 nm bandpass. In case of fluorescent protein tags refer to Subheading 2.3 to check the recommended lasers and bandpass filters. Detect fluorescent signal with logarithmic amplification.
2. Make sure that the flow cytometer is able to collect the height, area, and weight parameters for the selected channel.
3. Select a total number of 10,000 and 50,000 events to be recorded for yeast and bacterial cells, respectively.
4. Select slow flow rate for cells intake (*see* **Note 7**).
5. Create and effective work sheet panel where acquired cells will be displayed: a properly work sheet panel can save you significant time on analyzing data and ensures that the experiment is correctly performed (Fig. 2).
   (a) Create a scatter plot in the work sheet with the parameters side scatter-area (SSC-A) vs forward side scatter-area (FSC-A) in linear scale (*see* **Note 8**). This plot allows to define the cell population (gated as P1) and remove cell debris (*see* **Note 9**).
   (b) Make a dot plot with the parameters: FITC-A (area) vs FSC-A (area), and FITC-H (height) vs FITC-W (width). Select to represent only the previous gated population (P1).
   On these plots, a threshold of fluorescence can be established and the rest of the stained population is gated as P2.
   (c) Make a histogram plot to represent the fluorescence intensity of P2 cells vs counts. The fluorescence intensities data of the selected fluorochromes would allow to compare the aggregated state of the proteins of interest in different cell samples.
   (d) Create a statistical table displaying the mean fluorescent intensity (MFI) and median fluorescence with its correspondent standard error and/or coefficient of variance to quantify differences between samples.
   (e) Set all compensation levels to zero.
6. Run, first, unstained cells to adjust the gain of SSC-A and FSC-A channels in such a way, that the smallest cells fit into the lower left corner of the scatter plot. Then, gate your population (P1) in order to remove dead cells, detritus and/or aggregated cells.

7. Run negative control cells and adjust the photomultipliers (PMTs) voltages using the FSC-A vs fluorochrome-A dot plot so that negative control cells fall within the lower-left corner of the plot. This fluorescence background of unstained cells determines the threshold of the fluorochrome signal. Create a new gate (P2) above this region to define fluorescent positive cells.

   Negative controls can be either nonstained cells and noninduced cells or cells transformed with empty plasmid in the case of fused-fluorescent molecules.

8. Check the gain of the PMTs of the selected fluorochrome with positive control samples exhibiting the maximum fluorescent signal in such a way that cells fall into P2. Use the most fluorescent cells which will correspond to the most aggregation-prone variants, since they bind more Th-S and ProteoStat, or to untreated cells during aggregation inhibitors screening experiments (*see* **Note 10**).

   In the case multicolor assays is performed, *see* Subheading 3.3.2 for compensation.

9. Once all the parameters have been defined, samples can be run and statistics data obtained for further analysis (*see* Subheading 3.4). Save the experiment as a FCS file (*see* **Note 11**).

10. As bacteria and yeast are small cells, caution should be taken since they might contaminate the flow cytometer fluidics. It is critical to thoroughly clean the system after running our samples. Flush the system with 5% sodium hypochlorite solution for 5 min. Change tube and flush distilled water for additional 5 min. This is especially important when working with living bacteria or yeast cells (*see* **Note 12**).

3.3.2 Compensation on Multicolor Assays

When performing a multicolor assay, for example checking the cell viability using propidium iodide dye, sample preparation and setup of the cytometer parameters are not the only factors to consider. Compensation constitutes an indispensable adjustment that can notably affect the accuracy and quality of the data. Whenever there is more than one labeling, the fluorescent emission signal of one fluorochrome can be artifactually detected by a detector designed to measure a signal from another, spectrally unrelated, fluorochrome. This phenomenon is called spillover. Provided that the level of spillover follows a linear function, the compensation process is achieved equalizing the mean signal for all fluorochromes and in all detectors, via a single-color control.

In order to correctly compensate the spillover of the affected cells populations:

1. Prepare 300 μL of the appropriate controls consisting of a negative and positive sample for each fluorochrome to be analyzed.

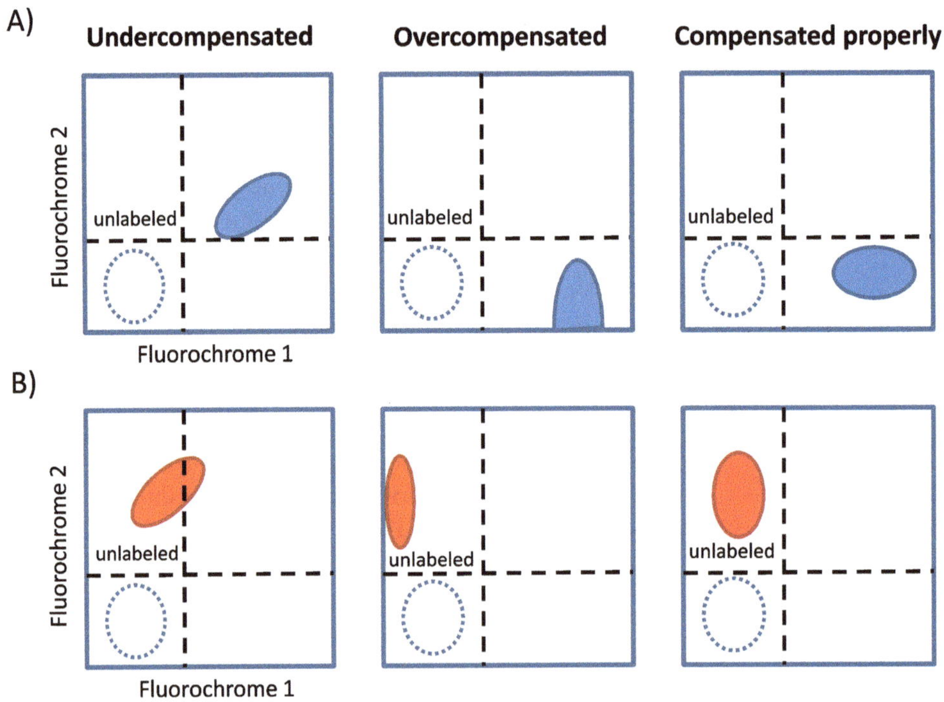

Fig. 4 Diagram showing correct vs incorrect compensation for cells stained with two unrelated fluorochromes with different emission colors. The fluorescence overlap can be corrected by adjusting compensation settings. (**a**) Dot plots showing the correction of the spillover of positively stained cells with fluorochrome 1 (GFP detected in the FITC channel), and (**b**) positively stained cells with fluorochrome 2 (ProteoStat detected in the PE-A channel). (left) Not enough fluorescence subtraction. F1 positive population is greater than negative F1 population in F2 channel; (middle) too much subtraction. F2 fluorescence of negative F1 population is greater than that of the positive F1 population; (right) median fluorescence of the F1 negative and positive populations are equal in F2 channel. Correct compensation is only achieved once the negative and positive populations show the same fluorescent median values

2. Ensure that the background fluorescence (autofluorescence) of the negative and positive sample is comparable.
3. Adjust compensation using exactly the same fluorochrome, i.e., do not use GFP to compensate Alexa 488 or FITC, even if they are all green.
4. Analyze cells as soon as possible after staining, minimizing the exposure to light.
5. Make a scatter plot with the paired fluorochromes (FL1 vs FL2) and create a statistics table displaying the media fluorescence intensity (MFIs) values.

   To compensate samples, adjust the compensation settings until both positive and negative controls are aligned in the neighboring channels, meaning that the media fluorescence intensities (MFIs) values are equal in both channels (Fig. 4).
6. When performing compensation it is important to use biexponential display rather than the traditional log scale, in order to

visualize better the compensated data and to detect more easily possible errors (*see* **Note 13**).

Do not compensate manually using visual cues without considering the statistical data.

3.4 Flow Cytometry Data Analysis

As mentioned above, one the most relevant advantages of flow cytometry is its high-throughput ability to analyze large cellular populations, providing robust statistical data, which is essential to attain an objective interpretation of fluorescence based results. Once the fluorescent cells have been run according to the above described steps it is time to initiate the data analysis procedure. For this aim, there are two main representation graphics to present flow cytometry data: dot plots (density plot, contour plots, bivariate plots, scatter graphs) and histograms.

3.4.1 Analysis of the Fluorescence Intensity of the Signal

1. Load the experiment in FCS format. Create an analysis sheet with the dot plot (FSC-A vs SSC-A) corresponding only to intact cells, without any contribution of cell debris, (P1).
2. Create a dot plot (FSC-A vs Fluorochrome-A). Set the threshold for the background signal, above which the fluorescence signal will be considered significant.
3. Change the selected sample to stained cells and create a second gate P2 which would only include positive fluorescent cells.
4. Additionally, the gated P2 population can be represented on a histogram plot (counts vs Fluorochrome-A) in a log scale. Histograms are useful when comparing mean fluorescence intensities between samples, rather than between fluorochromes. Moreover, histogram overlays are an exceptional way to illustrate differences between the fluorescence profiles of different cell samples.
5. The most recommendable way to compare fluorescence signals between samples, either corresponding to different expressed aggregation-prone proteins or variants or to the analysis of the antiaggregational potential of different molecules, is to obtain the median fluorescent intensity (MFI) value, together with the associated standard deviation (SD). Because fluorescence is expressed in terms of arbitrary units, it is highly recommended to convert the data to a fold-over background, or resolution metric (RD) value before you move on to hypothesis testing.

$$R_{\mathrm{D}} = \frac{\mathrm{Median}_{\mathrm{positive}} - \mathrm{Median}_{\mathrm{control}}}{\mathrm{SD}_{\mathrm{positive}} + \mathrm{SD}_{\mathrm{control}}}$$

The higher precision of the RD factor comes from the fact that it not only measures the the difference between samples but also accounts for the spreading of the data. These kind of normalizations are specially necessary whenever multiple experiments are carried out over time and should be compared.

6. Once the RDs have been calculated, a T-test is advisable to confirm differences between protein variants or between compound-treated and untreated cells.

3.4.2 Analysis of the Aggregate Load and Distribution

Differences in the localization of GFP-tagged proteins can be monitored in yeast cells using FC. This analysis allows to distinguish whether intracellular recombinantly expressed proteins are located at the soluble/diffuse or aggregated fractions of the cell by applying Pulse Shape Analysis [28] (Fig. 5).

1. Aggregation-prone proteins fused to GFP can present different forms when expressed in yeast cells, ranging from a single large insoluble protein deposit (IPOD), to multiple foci distributed randomly in the cytoplasm as well as soluble species. It is possible to infer the coexistence of these conformations in the cell population with a scatter plot of FSC-A vs GFP-A. Although cells usually display, approximately, the same size (FSC-A), the cell fluorescence signal (FITC-A) allows to affirm the existence of different protein aggregation states in cells. In

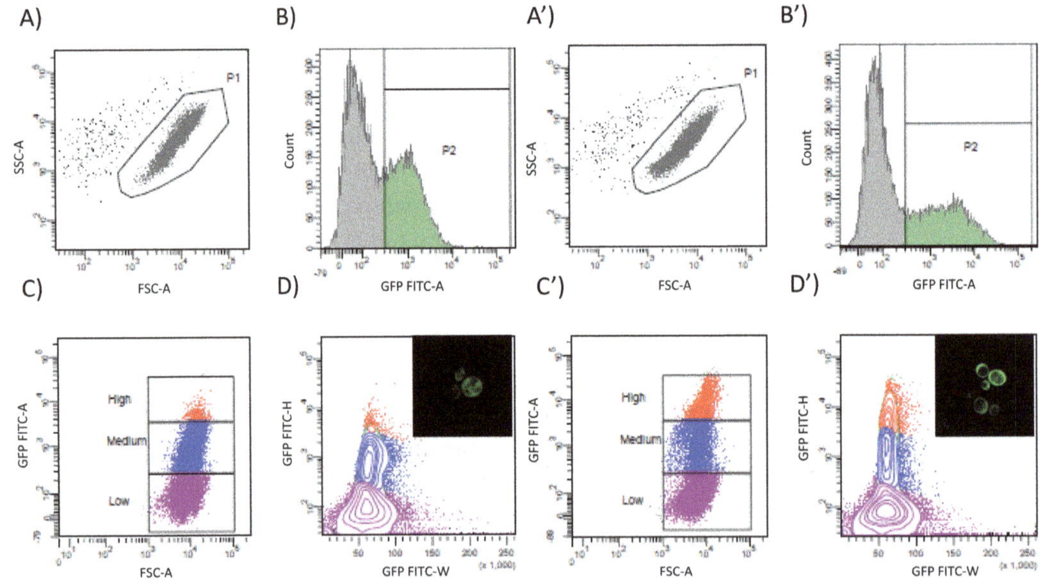

Fig. 5 Flow cytometry analysis of cultured yeast cells expressing two protein variants with different aggregation propensities fused to GFP. Left and right panels correspond to the soluble and the aggregation-prone variants, respectively. (**a**, **a′**) Dot plot (FSC-A vs SSC-A). Detritus are removed from the cell population gating intact cells in P1. (**b**, **b′**) Histogram of the GFP fluorescence signal displays two different types of cells; nonfluorescent cells (not expressing the recombinant protein), and fluorescent cells (expressing the recombinant protein) gated in P2. (**c**, **c′**) Pseudocolor dot plot of FSC-A vs FITC-A showing the size of cells relative to fluorescent protein signal. In this distribution, it is possible to distinguish three different population containing low, medium, or high protein load. (**d**) Pseudocolor contour scatter plot (FITC-W vs FITC-H) showing the distribution of the fluorescence signal inside cells according to their pulse height and width. Confocal microscope images are embedded in the dot plot to visualize the distribution of the fluorescent protein. Cell acquisition was performed in a FACSCanto flow cytometer (BD Biosciences). Each plot represents 20,000 events

this manner, in a heterogeneous population, cells can be gated into different groups, usually three, according to the fluorescence levels of both the soluble and aggregated protein they contain (FITC-A). A subsequent dot plot (FITC-W vs FITC-H) provides complementary information on the cellular distribution of the fluorescent protein content. Soluble proteins distributed all over the cytoplasm display a high fluorescence signal but higher FITC-W values than foci. Dot plot of cells containing foci exhibits a sharper FITC-W signal together with high fluorescence intensity (FITC-A).

2. To represent the cell protein aggregation load as a frequency of its distribution, export data of the Fluorochrome-W read-out to an excel file (*see* **Note 14**).
3. For pulse-shape analysis, open the file in GraphPad Prism and tabulate the fluorescent data as relative frequency in percentages (Column analysis > frequency distribution). Fit to a Gaussian curve the frequency distribution plot and determine amplitude, SD and mean of the fluorochrome-W signal. When comparing cells in which the target protein remains soluble with those containing insoluble foci, the diffusely distributed protein presents a significantly larger amplitude of the GFP-W signal, higher SD values, and a wider pulse signal distribution.

   To compare these parameters between different samples, Student's t-test and ANOVA analysis are used, depending on whether we compare two or more samples simultaneously.

4 Notes

1. Special care must be taken with paraformaldehyde as it is a known carcinogen. Avoid inhalation, ingestion, or contact with skin, eyes, or mucous membranes.
2. Living cells should be analyzed within 1 h after their collection.
3. Cells can be used immediately after fixation or stored at 4 °C for ≤3 days.
4. Noninduced cells should be also prepared as a background fluorescence control.
5. Solubility of protein variants can be potentially predicted by aggregation predictors. Among them, we advise the use of AGGRESCAN [29] (http://bioinf.uab.es/aggrescan.), when only the sequence is available and AGGRESCAN3D [30] (http://biocomp.chem.uw.edu.pl/A3D/), if we know the structure of the protein or have a reliable structural model.
6. It is highly recommended that prior to any attempt of a more detailed analysis, you run on the flow cytometer a sample of

the culture with and without the fluorescently tagged protein of interest to determine whether the approach is suitable to your system.

7. Pay attention to the flow rate. For small cells as bacteria and yeast it is better to set up a low flow rate. High flow rates usually promote the intake of aggregated cells.
8. When establishing a gating strategy, consider adding a pulse gate (height vs area) to discard the contribution of cell aggregates.
9. It might be difficult to discern and remove dead cells from the population by a FSC vs SSC gating alone; therefore it is possible to combine staining with a live dead exclusion marker, such as propidium iodide or a cell viability kit, to help gating only healthy cells.
10. Despite the appropriate lasers and filters being available, if detection is not possible, then the optical alignment of the system may be suboptimal. This can be assessed and corrected by using quality control beads. These microsphere beads with known sizes and/or fluorescence levels allow for the system to automatically characterize, track, and report measurements of the laser(s). A need to increase the amount of voltage for a particularly fluorophore, to achieve previously attained intensities, would indicate a loss of laser power, particularly if it occurs across all detectors for a given laser. If it is concerning to a single fluorophore, it would indicate that the specific detector is damaged. Moreover, an increase of the coefficient of variation of the bead population, for example with respect to fluorescence intensity, it would indicate that the sensitivity of the system has decreased and would suggest that there is either a misalignment of the laser or an issue with detection optics. If this is the case, then contact the manufacturer of the system. If detection is still not possible, then it would be an issue due to a scarce expression level of the construct and/or fluorescent tag chosen.
11. Since the instrument settings for an experiment can be stored an uploaded again, determination of gain, compensation and fluorescent threshold must be always performed at the beginning of any particular project.
12. If your flow cytometer is equipped with a cell sorting system, it is possible to isolate cellular populations to ensure the purity of a phenotypically defined feature for its further analysis.
13. In the case that the detection of the recombinantly or chemically labeled protein is too low, it might be an issue of the protein expression level. Perform a SDS-PAGE and or Western blot analysis to ensure the presence of your protein of interest.

14. Exporting flow cytometry raw data can be done with FlowJo software or FCS extract 1.02 software (http://research.stowers-institute.org/efg/ScientificSoftware/Utility/FCSExtract/).

Acknowledgment

S.V. has been granted an ICREA ACADEMIA award. SV is supported by the Spanish Ministry of Economy and Competitiveness (MINECO), through project BIO2016-78310-R. We thank the flow cytometry facility (SCAC) at the Institut de Biotecnologia i Biomedicina (IBB) for their technical advice.

References

1. Invernizzi G, Papaleo E, Sabate R, Ventura S (2012) Protein aggregation: mechanisms and functional consequences. Int J Biochem Cell Biol 44:1541–1554
2. Chiti F, Dobson CM (2006) Protein misfolding, functional amyloid, and human disease. Annu Rev Biochem 75:333–366
3. Rapezzi C, Quarta CC, Riva L et al (2010) Transthyretin-related amyloidoses and the heart: a clinical overview. Nat Rev Cardiol 7:398–408
4. Westermark P, Bergström J, Solomon A et al (2003) Transthyretin-derived senile systemic amyloidosis: clinicopathologic and structural considerations. Amyloid 10(Suppl 1):48–54
5. Nelson R, Eisenberg D (2006) Structural models of amyloid like fibrils. Adv Protein Chem 73:235–282
6. Mishra R, Sjölander D, Hammarström P (2011) Spectroscopic characterization of diverse amyloid fibrils in vitro by the fluorescent dye Nile red. Mol BioSyst 7:1232–1240
7. Sabate R, Rodriguez-Santiago L, Sodupe M et al (2013) Thioflavin-T excimer formation upon interaction with amyloid fibers. Chem Commun (Camb) 49:5745–5747
8. Klunk WE, Jacob RF, Mason RP (1999) Quantifying amyloid by Congo red spectral shift assay. Methods Enzymol 309:285–305
9. Kayed R, Head E, Thompson JL et al (2003) Common structure of soluble amyloid oligomers implies common mechanism of pathogenesis. Science 300:486–489
10. Lee H-J, Lee S-J (2002) Characterization of cytoplasmic α-Synuclein aggregates. J Biol Chem 277:48976–48983
11. García-Fruitós E, Sabate R, de Groot NS et al (2011) Biological role of bacterial inclusion bodies: a model for amyloid aggregation. FEBS J 278:2419–2427
12. de Groot NS, Sabate R, Ventura S (2009) Amyloids in bacterial inclusion bodies. Trends Biochem Sci 34:408–416
13. Villaverde A, Corchero JL, Seras-Franzoso J, Garcia-Fruitós E (2015) Functional protein aggregates: just the tip of the iceberg. Nanomedicine 10:2881–2891
14. Villar-Pique A, de Groot NS, Sabaté R et al (2012) The effect of Amyloidogenic peptides on bacterial aging correlates with their intrinsic aggregation propensity. J Mol Biol 421:270–281
15. Villar-Piqué A, Espargaró A, Sabaté R et al (2012) Using bacterial inclusion bodies to screen for amyloid aggregation inhibitors. Microb Cell Factories 11:55
16. Navarro S, Carija A, Muñoz-Torrero D, Ventura S (2016) A fast and specific method to screen for intracellular amyloid inhibitors using bacterial model systems. Eur J Med Chem 121:785–792
17. Moosavi B, Mousavi B, Macreadie IG (2015) Yeast model of amyloid-β and tau aggregation in Alzheimer's disease. J Alzheimers Dis 47:9–16
18. Tenreiro S, Munder MC, Alberti S, Outeiro TF (2013) Harnessing the power of yeast to unravel the molecular basis of neurodegeneration. J Neurochem 127:438–452
19. Yang J, Hao X, Cao X et al (2016) Spatial sequestration and detoxification of Huntingtin by the ribosome quality control complex. elife 5:e11792
20. Braun RJ, Büttner S, Ring J et al (2010) Nervous yeast: modeling neurotoxic cell death. Trends Biochem Sci 35:135–144

21. Figley MD, Gitler AD (2013) Yeast genetic screen reveals novel therapeutic strategy for ALS. Rare Dis 1:e24420
22. Telford WG, Hawley T, Subach F et al (2012) Flow cytometry of fluorescent proteins. Methods 57:318–330
23. Espargaró A, Sabate R, Ventura S (2012) Thioflavin-S staining coupled to flow cytometry. A screening tool to detect in vivo protein aggregation. Mol BioSyst 8:2839
24. Navarro S, Ventura S (2014) Fluorescent dye ProteoStat to detect and discriminate intracellular amyloid-like aggregates in *Escherichia coli*. Biotechnol J 9:1259–1266
25. Waldo GS, Standish BM, Berendzen J, Terwilliger TC (1999) Rapid protein-folding assay using green fluorescent protein. Nat Biotechnol 17:691–695
26. de Groot NS, Ventura S (2006) Protein activity in bacterial inclusion bodies correlates with predicted aggregation rates. J Biotechnol 125:110–113
27. Pouplana S, Espargaro A, Galdeano C et al (2014) Thioflavin-S staining of bacterial inclusion bodies for the fast, simple, and inexpensive screening of amyloid aggregation inhibitors. Curr Med Chem 21:1152–1159
28. Hidalgo IH, Fleming T, Eckstein V et al (2016) Characterization of aggregate load and pattern in living yeast cells by flow cytometry. BioTechniques 61:137–148
29. de Groot NS, Castillo V, Graña-Montes R, Ventura S (2012) AGGRESCAN: method, application, and perspectives for drug design. Methods Mol Biol 819:199–220
30. Zambrano R, Jamroz M, Szczasiuk A et al (2015) AGGRESCAN3D (A3D): server for prediction of aggregation properties of protein structures. Nucleic Acids Res 43(W1): W306–W313

Chapter 13

Induction of Cu/Zn Superoxide Dismutase (SOD1) Aggregation in Living Cells

Edward Pokrishevsky, Jeremy Nan, and Neil R. Cashman

Abstract

Amyotrophic lateral sclerosis (ALS) is a fatal neurodegenerative disease characterized by progressive degeneration and loss of motor neurons that appears to spread through the neuroaxis in a spatiotemporally restricted manner. In the familial form of ALS, the presence of any one of over 180 inherited mutations in the gene that encodes Cu/Zn superoxide dismutase (SOD1) leads to its eventual misfolding and aggregation. Once the pathological SOD1 seed is formed, it can continue growing into a larger aggregate through nucleation of other SOD1 substrate molecules. To date, there is no effective and rapid method to study the nucleation of SOD1 or to test therapeutics against this mechanism in living cells. In this chapter, we describe a series of protocols used to study induced aggregation of SOD1 in a simple but robust cell culture model. This assay can also be used to evaluate the potential therapeutic efficacy of small molecules targeting the induced aggregation mechanisms of SOD1.

Key words Protein aggregation, Cu/Zn superoxide dismutase, SOD1, Reporter protein, Live-cell microscopy, Aggregate quantification, Cell culture, Transfection

1 Introduction

Amyotrophic lateral sclerosis (ALS), also known as Lou Gehrig's disease, is a terminal motor neuron disease characterized by systematic paralysis of the muscles of the limbs, speech, swallowing, and respiration, due to the progressive degeneration of innervating motor neurons [1, 2]. Although the etiology of disease is not well understood, approximately 20% of genetically linked ALS is caused by mutations in Cu/Zn superoxide dismutase (SOD1), an intracellular protein whose function is to convert highly reactive superoxide radicals into hydrogen peroxide and oxygen [3, 4]. Previous studies have demonstrated that ubiquitinated SOD1 inclusions are a common pathological feature in all ALS patients with SOD1 mutations [4]. Interestingly, coexpression of human wild-type SOD1 with mutant SOD1 accelerates onset of disease in mouse models of ALS, implying that wild-type SOD1 is a substrate for

Cláudio M. Gomes (ed.), *Protein Misfolding Diseases: Methods and Protocols*, Methods in Molecular Biology, vol. 1873,
https://doi.org/10.1007/978-1-4939-8820-4_13,

aggregate seeding by mutant SOD1, a mechanism that can ultimately cause neurotoxicity in disease [5–7]. Mounting evidence also indicates that ALS propagates through prion-like mechanisms, with pathological SOD1 playing a key role in this process: (1) aggregates composed of ALS-causing mutant SOD1 can traverse between neuronal cells in culture and nucleate aggregation of cytoplasmic and soluble mutant SOD1 protein [8]; (2) mutant SOD1- or TDP-43-induced misfolded human wild-type SOD1 can propagate intercellularly [9, 10]; (3) spinal cord injection of homogenates prepared from paralyzed $SOD1^{G93A}$ mice can induce widespread spinal inclusion pathology [11, 12]; and (4) spinal cord homogenates prepared from SOD1-familal ALS induce SOD1 aggregation in cultured cells [13].

While SOD1 normally forms a homodimer, it can be kindled to break down into oxidized monomers. This, in turn, exposes their hydrophobic residues and triggers self-aggregation, a process potentially mediated through cysteines [14]. Previous studies, including those from our laboratory, have used cumbersome immunoprecipitation and immunofluorescence studies to demonstrate that SOD1 can be triggered to misfold by mutant SOD1, or by pathological TDP-43 or FUS [15, 16]. However, it was recently shown that intraspinal injection of $SOD1^{G85R}$-YFP mice with spinal cord homogenates prepared from paralyzed $SOD1^{G93A}$ transgenic mice can trigger disease in the recipient mice, and trigger aggregation of $SOD1^{G85R}$-YFP [11]. Based on this finding, we sought to simplify our methodology of studying the nucleation features of SOD1 by generating a fluorescence-based assay using SOD1-based reporter proteins to screen for induced aggregation of SOD1. In this chapter, we provide details on the usage of $SOD1^{G85R}$-GFP (referred to as SOD1-GFP onwards) as a reporter protein for the study of induced SOD1 aggregation, and methodological details how this assay can be used for screening small molecules (Fig. 1).

2 Materials

All commercial buffers and reagents were stored in accordance with the manufacturer's instructions. In-house buffers and solutions were prepared using distilled water purified using the Milli-Q Synthesis A10 water purification system (Millipore, USA).

2.1 Tissue Culture

- 100 mm and 24-well cell culture dishes, treated polystyrene.
- Black 96-well plates (*see* **Note 1**).
- #1.0 tissue culture coated 8-well borosilicate chamber coverglasses.
- Disposable serological pipets, polystyrene (5, 10, 25 mL).
- Dulbecco's Modified Eagle's Medium (DMEM) supplemented with 10% (vol/vol) heat-inactivated fetal bovine serum,

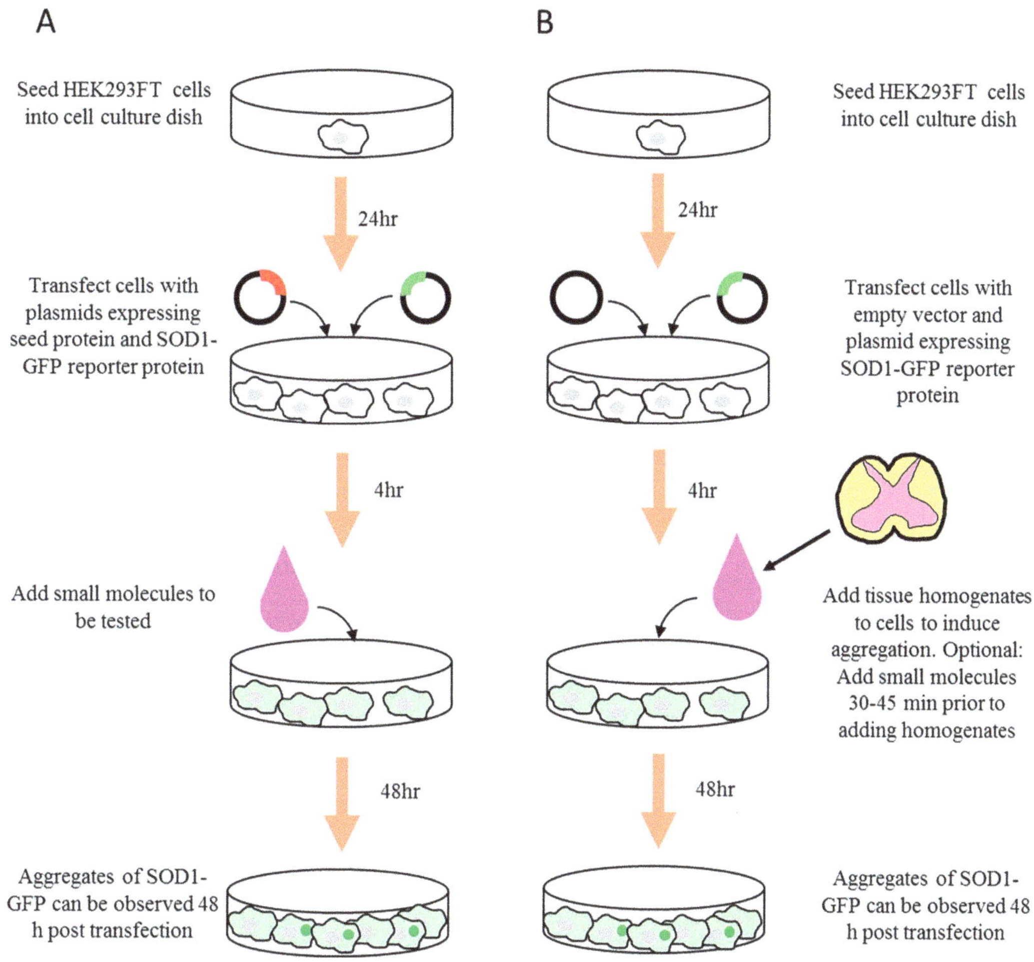

Fig. 1 Flowchart of SOD-GFP induced aggregation. (**a**) Induction of SOD1-GFP reporter aggregation through the expression of a seeding protein. Small molecules, if tested, should be added dropwise to the cells 4–6 h post transfection. (**b**) Induction of SOD1-GFP reporter aggregation using tissue homogenate. Tissue homogenates should be added to the cells as part of a transfection mixture approximately 4–6 h post reporter protein transfection. Small molecules, if tested, should be added 30–45 min prior to the addition of homogenates

100 U/mL penicillin, 100 μg/mL streptomycin, and 2 mM L-glutamine (Gibco by ThermoFisher Scientific, USA). Warm medium to 37 °C before use.

- Human embryonic kidney cells (HEK293FT; ATCC, USA).
- DH Autoflow CO_2 air-jacketed incubator. Set to 37 °C, 5% CO_2.

2.2 Mammalian Cell Transfection

- Disposable serological pipettes, polystyrene (5, 10, 25 mL).
- 1.5 mL snap-cap centrifugation tubes.
- Opti-MEM® I Reduced serum medium (Gibco by ThermoFisher Scientific, USA).

- Lipofectamine™ LTX Reagent (ThermoFisher Scientific, USA).
- Plasmid DNA stored at 4 °C for short term, or −20 °C for long term.

2.3 Tissue Homogenization

- Flash-frozen spinal cord tissue.
- Analytical balance.
- Phosphate buffered saline (PBS), pH 7.4, without calcium chloride or magnesium chloride supplemented with EDTA-free protease inhibitor cocktail (*see* **Note 2**).
- Tissue homogenizer with disposable probes (Omni International, USA).
- Probe-type sonicator.
- Centrifuge.
- 1.5 mL snap-cap centrifugation tubes.
- 15 mL polysterene conical centrifugation tubes.
- Bicinchoninic acid (BCA) assay.

2.4 Drug Preparation

- Analytical balance.
- 1.5 mL snap-cap centrifugation tubes.
- Opti-MEM® I Reduced serum medium (Gibco by ThermoFisher Scientific, USA).
- Tissue-culture grade sterile dimethyl sulfoxide (DMSO).
- 0.22 μm PES filter.

2.5 Immuno fluorescence

- Permeabilization buffer: 0.3% Triton X-100 in PBS.
- Blocking buffer: 10% normal goat serum in PBS, filtered.
- Staining buffer: 2% normal goat serum in PBS, filtered.
- 0.22 μm PES filter.

2.6 Cell Lysis and Differential Centrifugation

- Gentle lysis buffer: 0.5% sodium deoxycholate, 0.5% Triton X-100 in PBS, supplemented with 1× protease inhibitors (*see* **Note 2**).
- Centrifuge.
- Ultracentrifuge capable of spinning microcentrifuge tubes at 100,000 × *g*.
- Ultracentrifuge tubes.
- SDS-sample buffer with β-mercaptoethanol.
- SOD1 (Enzo Life Sciences, USA) and GFP (Santa Cruz Biotechnology, USA) specific antibodies.

3 Methods

3.1 Cotransfection of Cells with Mutant SOD1 DNA and Reporter Protein Constructs

1. Seed each well of a 24- or 96-well plate with 500 or 100 μL of HEK293FT cells, respectively, to a 30% confluency.
2. On the following day, set up transfection reactions by combining the following in a microcentrifuge tube: 100 μL Opti-MEM® media, 500 ng seed protein DNA, and 100 ng reporter protein DNA per well for 24-well plates or 20 μL Opti-MEM® media, 100 ng seed protein DNA, and 20 ng reporter protein DNA per well for 96-well plates.
3. Vortex transfection mixture before adding 1.75 or 0.35 μL of Lipofectamine™ LTX per well reaction for 24-well or 96-well plates, respectively.
4. Vortex and let the reaction incubate for 30 min at room temperature before adding dropwise to the cells.
5. Incubate for 24 h in a cell-culture incubator and monitor for growth and expression.
6. Incubate for an additional 24 h in cell-culture incubator.
7. Process the cells for the desired assay.

3.2 Tissue Extraction and Induction of SOD1-GFP Aggregation Using Human Homogenates

1. Cut ~0.1 g of flash frozen human spinal cord tissue (C- or T-spine) and place in a microcentrifuge tube.
2. Add 900 μL of cold PBS supplemented with protease inhibitors.
3. Homogenize 3× for 20 s with 40 s breaks (on ice).
4. Sonicate once for 15 s.
5. Spin down homogenized and sonicated tissue at 1000 × *g* for 5 min.
6. Remove supernatant into a fresh tube.
7. Measure total protein concentration in each homogenate using a standard BCA assay, and adjust for total protein concentration between the samples.
8. Aliquot into fresh tubes and freeze at −80 °C.
9. In order to induce SOD1-GFP aggregation, plate HEK293FT cells in 96- or 24-well plate.
10. On the following day transfect the cells with one part reporter protein, five parts empty vector (*see* Subheading 3.1).
11. 4–6 h following transfection, add the homogenates dropwise to the cells (*see* **Note 3**).
12. Incubate for 48 h in a cell-culture incubator.
13. Visualize using a microscope. We utilized an inverted microscope (Carl Zeiss AG, Germany) with motorized stage and autofocus adjustment:

(a) Place the 96-well plate in a plate holder and cover with live-cell chamber lid supplying humidified 5% CO_2 and maintained at 37 °C.

(b) Use a template with global focus points (*see* **Note 4**).

(c) Set up tile-based acquisition to cover full wells.

3.3 Immuno fluorescence

1. Place cells on sterilized tissue-culture glass coverslips (#1.5) in a 24-well plate prior to transfection (*see* Subheading 3.1 for protocol).
2. On the day of analysis, wash 2× with cold PBS.
3. Fix in 4% paraformaldehyde (in PBS, pH 7.4) for 15 min at room temperature (*see* **Note 5**).
4. Wash once with PBS.
5. Permeabilize for 5 min at room temperature using permeabilization buffer.
6. Incubate with blocking buffer for 1 h at room temperature.
7. Incubate for 1 h at room temperature with primary antibodies diluted in staining buffer.
8. Wash 2× in PBS.
9. Incubate for 1 h at room temperature with secondary antibodies diluted in staining buffer. Protect from light.
10. Wash 1× with PBS.
11. Counterstain DNA using 2 μg/mL bis-benzimide H33342 trihydrochloride (Hoechst 33342) for 5 min at room temperature. Protect from light.
12. Wash 2× with PBS.
13. Mount coverslips on a glass slide in a drop of Fluoromount-G (SouthernBiotech, USA).
14. Visualize cells using fluorescence microscopes.

3.4 Time-Course Microscopy of Induced SOD1-GFP Aggregation

1. Plate cells in chambered coverglasses in a total volume of 0.5 mL per well and let cells attach for 24 h.
2. On the day of transfection, follow transfection protocol outlined in Subheading 3.1.
3. 16–24 h after transfection place the chamber in a preheated live-cell imaging chamber (*see* **Note 6**), and monitor for expression of SOD1-GFP.
4. Once fluorescence of the reporter protein is visible, adjust focus and set acquisition points (*see* **Note 7**).
5. Adjust exposure time in GFP channel to allow for more reporter protein expression to be captured without visible saturation (*see* **Note 8**).
6. Image for up to 72 h.

3.5 Flow Cytometry for SOD1-GFP Inclusion Counting

1. Plate cells in 12-well plates.
2. On the day of analysis, wash 2× with cold PBS.
3. Incubate for 5 min with permeabilization solution (*see* **Note 9**).
4. Collect the cells by gentle pipetting and transfer into microcentrifuge tubes.
5. Centrifuge for 5 min at 1000 × *g*.
6. Transfer supernatants into flow cytometry tubes and store on ice. Protect from light.
7. Proceed for flow cytometry collection.
8. Cells should be detected in the 488 channel.

3.6 Blocking of SOD1 Aggregation Using Small Molecules

1. Weigh out appropriate amount of chemical powder needed to make a 100 mM stock solution (*see* **Note 10**).
2. Filter-sterilize stock solution into a fresh sterile tube.
3. Further dilute stock solution if needed using Opti-MEM® media.
4. Preplate and transfect as previously described in Subheading 3.1. These studies can be performed in a 24- or 96-well format, or in live cell chambers, and volumes should be adjusted accordingly.
5. 4–6 h following transfection, add the small molecule stock solution dropwise to the desired final concentration.
6. Incubate for 48 h in a cell-culture incubator.
7. Visualize using microscopy or flow cytometry.

3.7 Inclusion Counting Algorithm Using Microscopy

1. Acquire images using a fluorescence microscope.
2. Export as a high resolution JPG or uncompressed TIFF files.
3. Set up an ImageJ-based algorithm to count inclusions (*see* Fig. 2):
   (a) Convert the image to 8-bit grey scale format.
   (b) Adjust contrast and brightness.
   (c) Perform local background subtraction (*see* **Notes 11** and **12**).
   (d) Adjust threshold based on negative and positive controls (*see* **Note 13**).
   (e) Perform inclusion count based on predefined sizes (*see* **Note 14**).
4. Set up a secondary Image-J based algorithm to count total GFP fluorescence:
   (a) Convert the image to 8-bit grey scale format.
   (b) Adjust contrast and brightness.
   (c) Perform local background subtraction. Adjust radius (*see* **Note 15**).

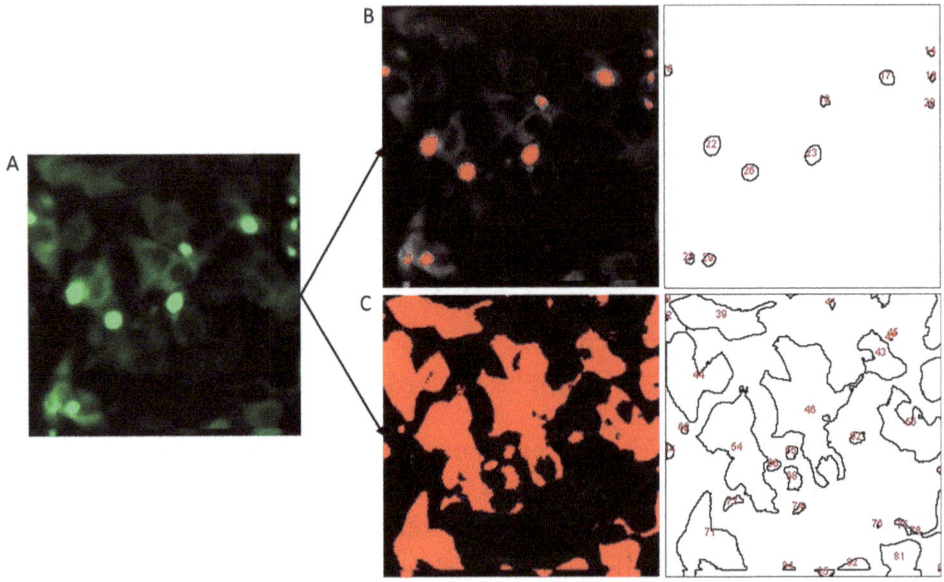

Fig. 2 Quantification of GFP inclusions using a computerized algorithm. (**a**) A fluorescence microscopy micrograph of HEK293FT cells 48 h post cotransfection with nontagged SOD1^{G85R} seed and SOD1^{G85R}-GFP reporter protein. (**b**) Micrographs are first converted to 8-bit grey-scale images and local background is subtracted to emphasize inclusions and remove background fluorescence. Threshold is then set to identify inclusions (red in micrograph), followed by particle counting that identifies only thresholded inclusions of predetermined size. (**c**) Total reporter protein fluorescence is quantified using a similar process, but with lower threshold and more permissive size to capture entire cell fluorescence. Once all the parameters are adjusted based on representative images for each experiment, the rest of the micrographs are analyzed in a batch-form using the same set of optimized parameters

(d) Adjust threshold to include all transfected cells.

(e) Quantify the total area of fluorescent cells based on predefined parameters (*see* **Note 16**).

5. Normalize the area occupied by GFP in inclusion form to total GFP.

3.8 Cell Lysis, Differential Centrifugation, and Immunoblotting

1. Collect the cells inside the tissue culture hood (*see* **Note 17**).
2. Lyse the cells for 10 min on ice using 100–500 μL of lysis buffer (exact volume depends on number of cells).
3. Centrifuge the tubes for 5 min at 1000 × *g* in a prechilled centrifuge.
4. Transfer the supernatant containing soluble proteins into fresh tubes.
5. Perform differential centrifugation (*see* **Note 18**).

6. Perform gel electrophoresis and probe SOD1 and GFP (*see* **Note 19**).

4 Notes

1. To avoid skewing our results due to the edge effect, we plate cells in the entire plate, but do not use the wells in the perimeter rows and columns for our experiments. This allows us to test up to 60 conditions per plate with minimal distortion.
2. 50× protease inhibitor cocktail is prepared by dissolving one tablet of cOmplete, EDTA-free Protease Inhibitor Cocktail (Roche, Switzerland) in 1.0 mL of PBS. The microcentrifuge tube is then vortexed until powder is dissolved. Store at 4 °C for immediate use, or freeze for longer storage. Avoid repeated freeze-thaw cycles.
3. Each homogenate aliquot should only be used for one experiment. In order to prepare the homogenates, add 1.5 μL Lipofectamine 2000 (ThermoFisher Scientific, USA) and 1 μL of each homogenates to 20 μL serum reduced media. Incubate the complex for 20 min and add dropwise to cells.
4. We generated a 96-well plate template based on the exact dimensions of our optical plate. The template contains four global focus points generating a 2-dimensional focal plane. The focus is self-adjusted during image acquisition for each well based on this plane. The template is calibrated for every new plate, and each focal point is adjusted in order to determine the focal plane for the given plate. The software, together with the Definite Focus technology (Carl Zeiss AG, Germany), maintains each well in focus throughout the acquisition.
5. The reporter protein is based on a fusion between SOD1 and GFP, a light sensitive molecule. Therefore, if reporter protein is present, the plate should be placed under aluminum foil during all incubations.
6. The live cell imaging culture chamber should be humidified, kept at 37 °C, and maintain 5% CO_2 for healthy cell growth. Cells can be maintained and imaged for up to 72 h under these conditions.
7. For live cell microscopy, we select one point per well, around which we form a 3 × 3 tile region. The focus is maintained throughout the imaging session by compensating for potential drift using the Definite Focus technology (Carl Zeiss AG, Germany). Image acquisition and tile stitching is performed using Zen Blue (Carl Zeiss AG, Germany).
8. In order to accurately visualize and quantify induced aggregation of the reporter protein, saturation should be observed at

the very most in the center of the aggregates alone. Ideally, it is best to avoid saturation altogether. During the initial setup of imaging settings, increase the exposure time and adjust the dynamic range in order to clearly identify the cells and determine the best focus. Once the settings are set, lower the exposure time in such way that the right shoulder of the exposure peak reaches up to approximately one fifth of the total available dynamic range. Rapid protein expression will continue to occur from 24 to 72 h post transfection, leading to higher expression levels of the reporter protein.

9. 0.3% weight/volume saponin in PBS. Commercially available saponin is typically available in powder form (8–25% sapogenin; Sigma-Aldrich, USA). Before preparing the 0.3% solution of saponin in PBS, confirm the purity of the saponin used. Cells may not be properly lysed if insufficient amount of saponin is added, resulting in lower saponin concentration and in some of the diffused GFP being retained by the cells.
10. Due to the nature of various chemical compounds, while some dissolve well in aqueous solutions, others may require organic solvents such as DMSO. We dissolve the water-soluble compounds in Opti-MEM® media (Gibco by ThermoFisher Scientific, USA) in order minimize the dilution of the cell culture media.
11. Despite the high degree of experimental reproducibility, certain parameters within the algorithm have to be verified and adjusted for every biological repeat. Once these parameters are set based on positive and negative controls, they should not be changed during the sample analysis.
12. For background subtraction, we use the rolling ball method. Radius may vary between biological repeats. We typically keep it within the range of 15–30 pixels. Once the most appropriate radius is selected based on controls, do not change it for sample analysis.
13. Threshold depends on the brightness of the samples which can vary due to biology (e.g., slight variation in transfection efficiencies or cell density) or acquisition parameters. The lower value of the threshold function is typically set to 60–120. For dimmer samples, threshold should be set close to 60. For brighter samples it should be set to the higher end. Once the most appropriate threshold is selected based on controls, do not change it during sample analysis.
14. For size exclusion, we analyze images in their pixel dimensions. We set the lower limit to 4–6 pixels in order to avoid capturing noise as false positive signal. The upper limit is set to 100–150 pixels, depending on the size of the larger inclusions in the control samples. Higher limit should be smaller than the size of intact cells (including dying cells). Once the most appropri-

ate inclusion size is selected based on controls, do not change it during sample analysis.

15. We perform background subtraction for the quantification of total GFP signal using the rolling ball method with a radius of 50 pixels. This allows for the removal of out-of-focus fluorescence light captured during image acquisition.
16. In order to quantify the total amount of GFP fluorescence, we set the minimum and maximum sizes to 5 and infinite number of pixels, respectively. This guarantees that we capture the signal even if cells are in close proximity to each other and appear to clump.
17. In order to collect the cells, we discard the media, perform two washes with cold PBS and scrape the cells off the plate using a cell culture scraper. Cells are then collected in 1.0 mL of cold PBS and transferred into fresh tubes. Alternatively, cells can be collected by first discarding the media, followed by gently pipetting of the cells in 1.0 mL of cold PBS using a micropipette. Cells are then centrifuged at 1000 × *g*, and the supernatant is discarded.
18. We centrifuge the samples in prechilled centrifuges. Following the brief spin at 1000 × *g*, transfer a small volume of the original lysates into fresh tubes and mix with SDS-sample buffer (prespin control). Centrifuge the remaining sample for 30 min at 20,000 × *g*, and transfer the supernatants into fresh ultracentrifuge 1.5 mL tubes. Resuspend the pellets in SDS-sample buffer. Centrifuge the supernatant for 1 h at 100,000 × *g*. Remove the supernatant into fresh tubes, and add the appropriate amount of SDS-sample buffer to the pellet. If the final pellet does not easily resuspend in the sample buffer, a brief sonication step may be required (2 × 10 s).
19. Although any electrophoresis gels can be employed, we utilize the Novex® 4–20% Tris-Glycine Gels (ThermoFisher Scientific, USA). In order to monitor for the presence of multiple proteins on one membrane, we cut the PVDF membrane along the 25 kDa weight band and probe the separate membranes with antibodies against SOD1 or GFP. Alternatively, two duplicate gels can be run and transferred simultaneously, and stained with SOD1 or GFP specific antibodies.

Acknowledgments

We thank Ran Ha Hong and Destiny Lu-Cleary for their contribution. This work was supported by grants from ALS Canada, Canadian Institutes of Health Research (CIHR), Brain Canada, and Webster Foundation.

References

1. Cleveland DW (1999) From Charcot to SOD1: mechanisms of selective motor neuron death in ALS. Neuron 24(3):515–520
2. Grad LI, Rouleau GA, Ravits J, Cashman NR (2017) Clinical spectrum of amyotrophic lateral sclerosis (ALS). Cold Spring Harb Perspect Med 7(8). https://doi.org/10.1101/cshperspect.a024117
3. Beckman JS, Koppenol WH (1996) Nitric oxide, superoxide, and peroxynitrite: the good, the bad, and ugly. Am J Phys 271(5 Pt 1):C1424–C1437
4. Rosen DR, Siddique T, Patterson D, Figlewicz DA, Sapp P, Hentati A, Donaldson D, Goto J, O'Regan JP, Deng HX et al (1993) Mutations in Cu/Zn superoxide dismutase gene are associated with familial amyotrophic lateral sclerosis. Nature 362(6415):59–62. https://doi.org/10.1038/362059a0
5. Wang L, Deng HX, Grisotti G, Zhai H, Siddique T, Roos RP (2009) Wild-type SOD1 overexpression accelerates disease onset of a G85R SOD1 mouse. Hum Mol Genet 18(9): 1642–1651. https://doi.org/10.1093/hmg/ddp085
6. Deng HX, Shi Y, Furukawa Y, Zhai H, Fu R, Liu E, Gorrie GH, Khan MS, Hung WY, Bigio EH, Lukas T, Dal Canto MC, O'Halloran TV, Siddique T (2006) Conversion to the amyotrophic lateral sclerosis phenotype is associated with intermolecular linked insoluble aggregates of SOD1 in mitochondria. Proc Natl Acad Sci U S A 103(18):7142–7147. https://doi.org/10.1073/pnas.0602046103
7. Prudencio M, Durazo A, Whitelegge JP, Borchelt DR (2009) Modulation of mutant superoxide dismutase 1 aggregation by co-expression of wild-type enzyme. J Neurochem 108(4):1009–1018. https://doi.org/10.1111/j.1471-4159.2008.05839.x
8. Munch C, O'Brien J, Bertolotti A (2011) Prion-like propagation of mutant superoxide dismutase-1 misfolding in neuronal cells. Proc Natl Acad Sci U S A 108(9):3548–3553. https://doi.org/10.1073/pnas.1017275108
9. Grad LI, Yerbury JJ, Turner BJ, Guest WC, Pokrishevsky E, O'Neill MA, Yanai A, Silverman JM, Zeineddine R, Corcoran L, Kumita JR, Luheshi LM, Yousefi M, Coleman BM, Hill AF, Plotkin SS, Mackenzie IR, Cashman NR (2014) Intercellular propagated misfolding of wild-type Cu/Zn superoxide dismutase occurs via exosome-dependent and -independent mechanisms. Proc Natl Acad Sci U S A 111(9):3620–3625. https://doi.org/10.1073/pnas.1312245111
10. Pokrishevsky E, Grad LI, Cashman NR (2016) TDP-43 or FUS-induced misfolded human wild-type SOD1 can propagate intercellularly in a prion-like fashion. Sci Rep 6:22155. https://doi.org/10.1038/srep22155
11. Ayers JI, Fromholt S, Koch M, DeBosier A, McMahon B, Xu G, Borchelt DR (2014) Experimental transmissibility of mutant SOD1 motor neuron disease. Acta Neuropathol 128(6): 791–803. https://doi.org/10.1007/s00401-014-1342-7
12. Bidhendi EE, Bergh J, Zetterstrom P, Andersen PM, Marklund SL, Brannstrom T (2016) Two superoxide dismutase prion strains transmit amyotrophic lateral sclerosis-like disease. J Clin Invest 126(6):2249–2253. https://doi.org/10.1172/JCI84360
13. Pokrishevsky E, Hong RH, Mackenzie IR, Cashman NR (2017) Spinal cord homogenates from SOD1 familial amyotrophic lateral sclerosis induce SOD1 aggregation in living cells. PLoS One 12(9):e0184384. https://doi.org/10.1371/journal.pone.0184384
14. Cozzolino M, Amori I, Pesaresi MG, Ferri A, Nencini M, Carri MT (2008) Cysteine 111 affects aggregation and cytotoxicity of mutant Cu,Zn-superoxide dismutase associated with familial amyotrophic lateral sclerosis. J Biol Chem 283(2):866–874. https://doi.org/10.1074/jbc.M705657200
15. Grad LI, Guest WC, Yanai A, Pokrishevsky E, O'Neill MA, Gibbs E, Semenchenko V, Yousefi M, Wishart DS, Plotkin SS, Cashman NR (2011) Intermolecular transmission of superoxide dismutase 1 misfolding in living cells. Proc Natl Acad Sci U S A 108(39):16398–16403. https://doi.org/10.1073/pnas.1102645108
16. Pokrishevsky E, Grad LI, Yousefi M, Wang J, Mackenzie IR, Cashman NR (2012) Aberrant localization of FUS and TDP43 is associated with misfolding of SOD1 in amyotrophic lateral sclerosis. PLoS One 7(4):e35050. https://doi.org/10.1371/journal.pone.0035050

Chapter 14

A Cell Model for HSP60 Deficiencies: Modeling Different Levels of Chaperonopathies Leading to Oxidative Stress and Mitochondrial Dysfunction

Cagla Cömert, Paula Fernandez-Guerra, and Peter Bross

Abstract

Besides providing the majority of ATP production in cells, mitochondria are also involved in many other cellular functions and are central for cellular stress signaling. Mitochondrial dysfunction induces not only inherited mitochondrial disorders but also contributes to neurodegenerative diseases, cancer, diabetes, and metabolic syndrome. The HSP60/HSP10 molecular chaperone complex facilitates folding of mitochondrial proteins and is thus an important factor for many mitochondrial functions. To model different degrees of oxidative stress and mitochondrial dysfunction we here describe a HEK293 derived Flp-In cell system with stable insertion and tunable expression of HSP60 cDNA carrying a dominant negative mutation. When expressed the dominant negative HSP60 mutant is incorporated into endogenously encoded HSP60/HSP10 complexes and impairs chaperone activity of the HSP60/HSP10 complex in a dose dependent manner. Using this system, different levels of oxidative stress and mitochondrial dysfunction challenges can be generated depending on the induction level of the mutant HSP60 cDNA insert. Here we describe our system and pertinent analysis methodology for use in studies of mitochondrial chaperone deficiency and resulting effects of increased production of reactive oxygen species and mitochondrial dysfunction.

Key words Heat shock protein 60 deficiency, Cell models, Image cytometry, Mass spectrometry

1 Introduction

Mitochondria are the cellular components that carry out oxidative phosphorylation in the cell resulting in ATP production for the variety of cellular functions [1]. Mitochondria are also involved in many diverse functions beyond ATP production, and their dysfunction not only induces inherited mitochondrial disorders but also contributes to neurodegenerative diseases, cancer, obesity, and metabolic syndrome [2, 3].

Oxidative phosphorylation is the major contributor to the production of reactive oxygen species (ROS) [4]. Superoxide dismutase enzymes (SODs) catalyze the conversion of superoxide

Cláudio M. Gomes (ed.), *Protein Misfolding Diseases: Methods and Protocols*, Methods in Molecular Biology, vol. 1873, https://doi.org/10.1007/978-1-4939-8820-4_14,

into hydrogen peroxide, a less reactive oxidant. We have previously demonstrated that folding of superoxide dismutase SOD2 is highly dependent on the HSP60/HSP10 chaperone complex [5, 6].

As a molecular chaperone in the mitochondrial matrix space HSP60 forms a barrel shaped isolation chamber together with HSP10 and facilitates the efficient folding of proteins [7]. Impairment of the HSP60/HSP10 system by mutations causes mitochondrial dysfunctions that lead to different neurodegenerative diseases [8].

Stable or transient cell transfection methods are highly utilized in protein research. However, it is difficult to precisely regulate the expression levels of proteins in these methods. Flp-In method provides a feasible way to regulate the expression levels of cDNA inserts [9]. Therefore, to investigate the relationship between HSP60/HSP10 protein folding system, ROS, and mitochondrial dysfunction, we have generated Flp-In-293 cells with stable insertion and tunable expression of HSP60 cDNA with a dominant negative mutation [10]. When expressed the dominant negative ATPase deficient HSP60 mutant is incorporated into complexes with endogenous HSP60 and abolishes the chaperone activity of HSP60/HSP10 system. With this system different levels of oxidative stress and mitochondrial dysfunction challenges can be generated depending on the induction level of the cDNA insert. The effects of the induction can be monitored with different methods (Fig. 1a). Here we describe our methods and comments for this system and its possible uses in analyses of the effects of oxidative stress, and mitochondrial dysfunction.

2 Materials

2.1 Induction of Expression

1. Flp-In-HEK293 cell line with cDNA insertion encoding HSP60 WT and Asp423Ala, respectively [10] (*see* **Note 1**).
2. 1 mg/mL tetracycline in ethanol. Store the working solution at −20 °C. Tetracycline is light sensitive.
3. Dulbecco's Modified Eagle's Medium (DMEM) supplemented with; 0.29 mg/mL glutamine, 0.1 streptomycin/mL, 100 units penicillin/mL, 10% FBS, 15 μg/mL blasticidin (Invitrogen, USA), and 100 μg/mL hygromycin (Invitrogen, USA) (*see* **Note 2**).
4. Trypsin–EDTA (0.5×).
5. CO_2 incubator.
6. T25 and 6-well cell culture plates.

2.2 Image Cytometry Measurements

1. NucleoCounter NC-3000 full-spectrum image cytometer (Chemometec, Denmark).

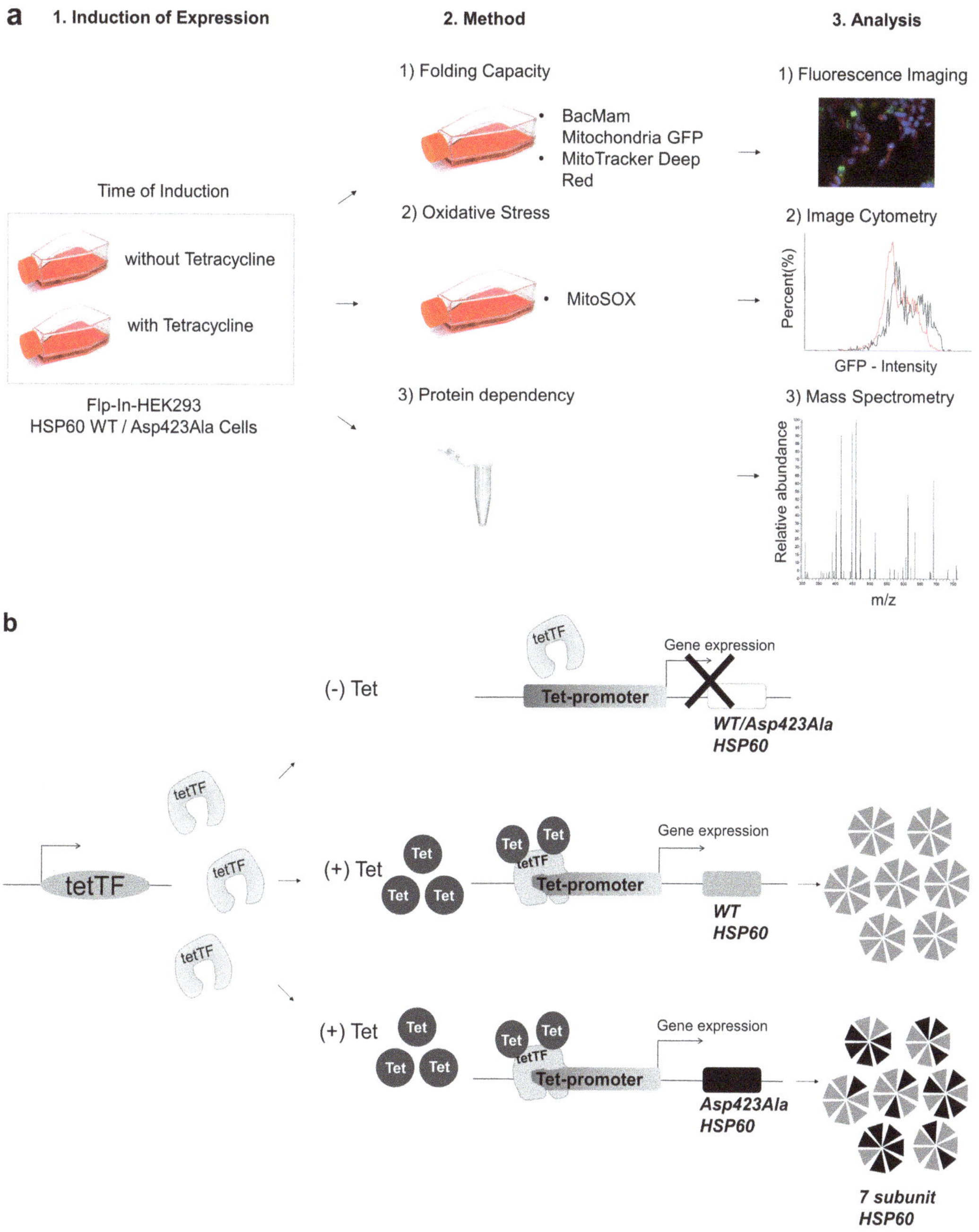

Fig. 1 Overview of the methodology. (**a**) General overview of analysis of HSP60 deficiencies with HEK293 Flp-In HSP60 WT/Asp423Ala cells. Tetracycline induced cells are analyzed with fluorescence imaging, image cytometry and mass spectrometry. (**b**) Mechanisim of tetracycline-on system in HEK293 Flp-In HSP60 cells. Without tetracycline induction HSP60 is not expressed. Upon tetracycline induction, tetracycline-controlled transcription factor (tetTF) is bounded by tetracycline and tetTF can initiate the gene expression of WT or dominant negative Asp423Ala HSP60. Gene expression, also the severity of folding deficiency and mitochondrial dysfunction, can be adjusted by treatment of different concentrations of tetracycline. Tetracycline treatment at higher concentrations results in higher expression of the insert. In the cells containing dysfunctional dominant negative Asp423Ala HSP60 insertion, tetracycline concentration determines the folding efficiency by replacing the WT subunits of the HSP60 protein with dysfunctional subunits

2. Two chamber NC-Slide A2™ (Chemometec, Denmark).
3. NucleoView NC-3000™ software (Chemometec, Denmark).

2.2.1 BacMam

1. CellLight® Mitochondria-GFP, BacMam 2.0 (Molecular Probes, Thermofisher Scientific, USA; C10600).
2. 10 μg/mL Hoechst-33342 (Chemometec, Denmark) in PBS.
3. 1 mM MitoTracker™ Deep Red FM (Thermofisher Scientific, USA; M22426) stock solution in HBSS.

2.2.2 MitoSOX

1. 47 mM antimycin stock solution in 70% ethanol.
2. 5 mM MitoSOX Red mitochondrial superoxide indicator (Molecular Probes, Thermofisher Scientific, USA) stock solution: 50 μg MitoSOX dissolved in 13 μL dimethylsulfoxide (DMSO)
3. HBSS.
4. PBS.
5. 10 μg/mL Hoechst-33342 (Chemometec, Denmark) in HBSS.
6. RedDot2 (Biotium, USA) diluted 1000 times in the sample (concentration not specified).
7. Thermoblock.

2.2.3 Mitochondrial Membrane Potential

1. JC-1 (Chemometec, Denmark). Mitochondrial membrane potential probe.
2. 1 μg/mL DAPI (Chemometec, Denmark) in PBS.
3. Carbonyl cyanide 3-chlorophenylhydrazone (CCCP; Molecular Probes).
4. PBS.

2.3 Mass Spectrometry

2.3.1 Cell Disruption and Protein Measurement

1. Urea Lysis buffer: 1 M Urea, 100 nM NH_4 carbonate.
2. Centrifuge.
3. Sonifier (Branson 250, Branson Ultrasonics, USA).
4. Bradford reagent.
5. Microplate Reader.
6. Sample buffer 6×: 1 M Tris–HCl, pH 6.8, 600 μM DTT, SDS, bromophenol blue 4 mg/mL, 30% glycerol.

2.3.2 SDS-PAGE

1. Criterion™ TGX Stain-Free™ Precast Gels, Any kD™ (Bio-Rad Laboratories, USA).
2. Coomassie Brilliant Blue R250 staining buffer: 0.5 g Brilliant Blue R250 in 50% ethanol and 10% acetic acid.
3. Destaininig buffer: 5% ethanol, 7% AcN in milliQ water.

2.3.3 In-Gel Digestion of Proteins

1. Protein Low Bind 1.5 mL Eppendorf tube (Eppendorf, Germany).
2. Destaining solution: 2 mg/mL ammonium bicarbonate in 50% acetonitrile (can be stored at 4 °C for 2 months).
3. Reduction buffer: 50 mM TCEP in digestion buffer (prepare fresh).
4. 5× Alkylation buffer: 500 mM Iodoacetamide (IAA) in water. Dilute 5× stock buffer to 1× with digestion buffer. Do not store the 5× stock or 1× working alkylation buffers.
5. Acetonitrile.
6. Activated trypsin solution: Prepare stock trypsin solution from 100 μg lyophilized trypsin powder (Promega; V5280) in 100 μL 50 mM acetic acid. Store the aliquots at 20 °C. Limit the freeze-thaw cycles. Prepare working trypsin solution by diluting the stock 1:10 in Baker water. Prepare fresh.
7. Digestion buffer: 2 mg/mL ammonium bicarbonate in Baker water.
8. 80% AcN, 1% TFA in Baker water.
9. Baker water.
10. 100% AcN.
11. Speedvac vacuum concentrator system.

2.3.4 Peptide Purification

1. Pierce™ C18 Spin Columns (Thermo Scientific; 89873).
2. Wash buffer; Equilibration buffer: 5% AcN, 0.5% TFA in Baker water.
3. Activation buffer: 50% methanol in Baker water.
4. Elution buffer: 70% AcN in Baker water.
5. Protein Low Bind 1.5 mL Eppendorf tube (Eppendorf, Germany).
6. Speedvac vacuum concentrator system.

3 Methods

A general overview of the methods is provided in Fig. 1a.

3.1 Induction of Expression

Flp-In T-Rex System is used to stably express functional (wild-type) and a mutated (p.Asp423Ala) HSP60 in HEK293 cell line. The HSP60-p.Asp423Ala variant is an artificially produced mutant that abolishes the ATPase function of HSP60 and has a dominant negative effect on the complex. When treated with tetracycline Flp-In HEK293 cells expresses wild-type or Asp423Ala HSP60 (*see* Fig. 1b). The level of expression of the respective HSP60 cDNA insert can be adjusted by titrating the tetracycline concentration. High concentration of tetracycline treatment (1 μg/mL) for long time (72–96 h) results in a halt in proliferation of the HSP60-Asp423Ala dominant negative mutant insertion containing cells whereas it does not affect the cells containing the HSP60-wild-type insertion [10].

First steps include the thawing protocol. If you already have the cells ready to seed for experiments, move to **step 8**.

1. Add 1 mL medium at 37 °C to a cryotube containing frozen cells and pipette up and down to start thawing and suspending the cells (*see* **Note 3**).
2. Add the cells to 10 mL 37 °C medium containing 15 mL Falcon tube.
3. Centrifuge for 3 min at 400 × *g*.
4. Remove the medium and resuspend in 5 mL complete DMEM without hygromycin and blasicidin; seed the cells in T25 flask and incubate in CO_2 incubator at 37 °C (*see* **Note 4**).
5. Next day, change the medium to remove dead cells.
6. When the cells reach to confluency, wash the cells with PBS once, trypsinize and transfer the cells to T75. In this step also start to add the additional antibiotics, hygromycin and blasticidin, which are required for selection for presence of the cells' insertions.
7. Passage cells prior to using them for the experiments to remove the dead cells and to select the hygromycin and blasticidin resistant cells.
8. Seed appropriate numbers of cells according to the size of the plate and duration of the treatment. For example; seed 8×10^4 to 10^5 in 1 mL DMEM for each well of P24 plate, 2×10^5 to 3×10^5 cells in 2 mL DMEM for each well of a P6 plate and 3×10^5 to 7×10^5 cells in 5 mL DMEM for each T25 flask.
9. Add 50 nM/mL or 1 μM/mL tetracycline to the medium prior to seeding to induce the expression of the Flp-In insert.
10. Induce the cells between 48 h to 7 days. Treat the cells with fresh tetracycline every other day to ensure that the concentration of tetracycline in the culture medium is maintained (*see* **Note 5**).

3.2 Image Cytometry Measurements

3.2.1 BacMam and MitoTracker® Deep Red

CellLight® Mitochondria-GFP BacMam is a mitochondria targeted GFP construct and it provides a mitochondrial localization to GFP. BacMam does not infect human therefore it is safe to use in biosafety level I laboratories. GFP requires the assistance of HSP60 for its folding and will fluorescent only when it is folded. Thus, the folding of mito-GFP can reflect the folding efficiency of HSP60. Therefore, it is expected to observe less fluorescence when HSP60 is not working, as the Asp423Ala HSP60 dominant negative mutant. MitoTracker® Deep Red is used as a control to determine if the mitochondrial membrane potential is affected by the other treatments. Interpretation of the results is explained in Fig. 2a.

1. Seed cells to P24 well plate and induce expression of the respective HSP60 variant protein with 50 ng/mL or 1 μg/mL tetracycline as described in Subheading 3.1.
2. After induction with tetracycline for 48 h, remove the medium and replace it with the medium without tetracycline. Cells are easy to detach at this step. Do not wash with PBS.
3. Treat the cells at 70% confluence with 30 μL BacMam solution per 400 μL of medium for 16 h.

$$\text{Volume of BacMam Reagent}(\text{mL}) = \frac{\text{number of cells} \times \text{desired MOI}}{1 \times 10^{8}\ \text{particles / mL}}.$$

4. After 16 h, prepare 1 mM stock solution by dissolving 50 μg MitoTracker® Deep Red in 9.2 μL DMSO. Prepare 100 nM working solution in HBSS (*see* **Note 6**).
5. Carefully remove BacMam containing solution.
6. Carefully add MitoTracker® Deep Red working solution to each well and incubate for 30 min at 37 °C.
7. Remove the MitoTracker® Deep Red solution.
8. Treat the cells with 10 μg/mL Hoechst in 400 μL PBS for 15 min at 37 °C. Hoechst stain is used to detect total number of the cells.
9. Wash with 500 μL PBS carefully.
10. Add 100 μL trypsin-PBS-EDTA to each well.
11. Harvest the cells by adding 100 μL culture medium and collect the cells from each well into an Eppendorf tube.
12. Load 30 μL of each sample to NC-Slide A2™.
13. Run the protocol to measure green fluorescence levels on Nucleocounter NC-3000. The fluorescence from the mitochondrial GFP is detected with excitation at 475 nm and emission at 560/35 nm. The Hoechst is detected at excitation at 365 nm and emission at 470/55 nm. Deep Red is detected at Excitation 630 nm and Emission 740/60 nm. For an example result *see* Fig. 2b.

3.2.2 MitoSOX™ Red Mitochondrial Superoxide Indicator (MitoSOX)

1. HSP60 expression is induced with 5 nM/mL or 50 nM/mL tetracycline for different time points (48 h, 96 h, 7 days) as described in Subheading 3.1. Tetracycline was not added to the medium used for MitoSOX positive controls.

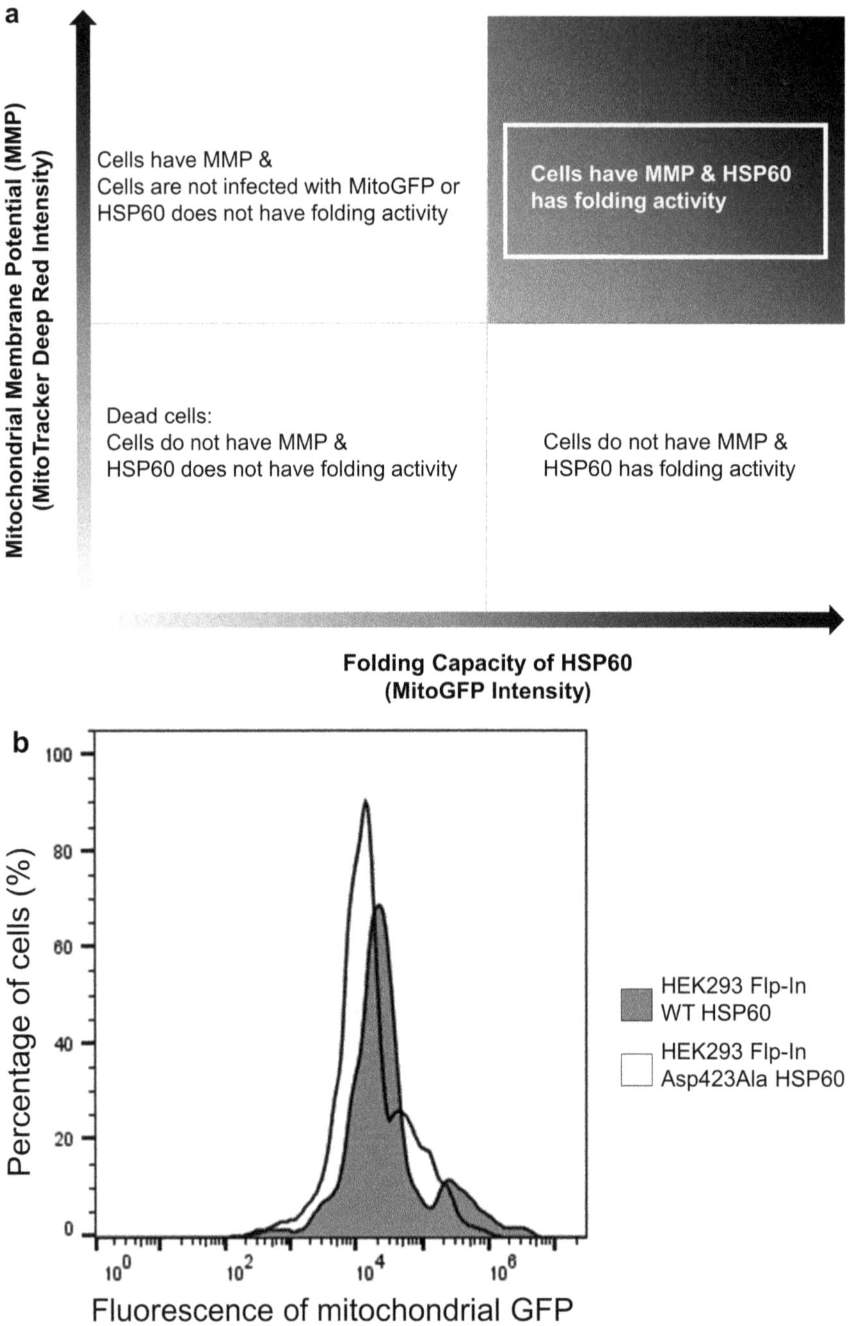

Fig. 2 Interpretation guide and example result from an experiment measuring HSP60/HSP10 folding capacity. (**a**) Interpretation of folding capacity of HSP60 by BacMam® and MitoTracker®. Shaded area represents the cells with positive signal from both BacMam and MitoTracker. These cells have mitochondrial membrane potential (MMP) associated with MitoTracker and have folding activity in accordance with BacMam intensity. (**b**) An example of image cytometry analysis comparing the folding capacity of HEK293 Flp-In HSP60 WT (gray) and Asp423Ala (white) cells induced with 1 μg/mL tetracycline for 48 h, respectively. In the dysfunctional Asp423Ala-HSP60 expressing cells fluorescence signal is lower than the WT-HSP60 expressing cells due to the disruption of the protein folding

2. To confirm MitoSOX sensivity to increased levels of superoxide, prepare positive control with 150 μM antimycin solution in serum-free medium and treat the noninduced Flp-In wild type HSP60 HEK 293 cells for 10 min in the incubator (*see* **Note 7**).
3. Remove the medium from wells and wash with 500 μL or 1 mL PBS for P6 or T25 flasks once (*see* **Note 8**).
4. Incubate all cells with 5 μM MitoSOX Red in at least 1 mL HBSS for 30 min in the incubator.
5. Remove the MitoSOX solution and carefully wash cells with PBS once.
6. Harvest cells with trypsin-EDTA (500 μL/well). At this stage cells are easy to detach so incubate for about a minute at room temperature and stop the tyrpsin with DMEM (1 mL/well).
7. Centrifuge the samples for 5 min at 400 × *g*.
8. Remove the liquid and dissolve cells with 100 μL of 10 μg/mL Hoechst in HBSS. Incubate for 15 min at 37 °C in the dark and in the thermoblock by mildly shaking at 300 rpm.
9. Prepare 1/100 dilution of RedDot™ 2 in PBS and add 1.5 μL to each sample just before each measurement to a final dilution of 1/1000.
10. Load ~30 μL of the sample to each of the two chambers in a NC-Slide A2™.
11. Run the protocol for MitoSOX on Nucleocounter NC-3000. Fluorescence from MitoSOX is detected at 475 nm and emission at 675/75 nm, for Hoechst at 365 nm and emission at 470/55 nm and for RedDot2 at 630 nm and emission at 740/60 nm (*see* **Note 9**).

3.2.3 Mitochondrial Membrane Potential (MMP)

1. Seed the cells in T25 flasks and induce HSP60 expression with the desired concentrations of tetracycline for different time intervals (e.g., 48 h, 96 h, 7 days) as described in Subheading 3.1.
2. One flask with noninduced Flp-In wild type HSP60 HEK 293 cells will be used as positive control. Treat the cells with CCCP (Molecular Probes) for 10 min at 37 °C with the final concentration 50 μM. CCCP disrupts mitochondrial membrane potential and therefore is used as a positive control for mitochondrial membrane potential disruption.
3. Count the cells of each flask and take 1 million cells from each into an Eppendorf tube. Centrifuge the cells for 5 min at 400 × *g* and remove the medium (*see* **Note 10**).
4. Resuspend the cells in 1 mL complete DMEM to each sample and pipet carefully until the cell suspension is homogenous.

5. Add 12.5 μL JC-1 to the sample with final concentration 2.5 μg/mL and incubate in thermoblock at 37 °C for 15 min.
6. Wash the cells with 1 mL PBS.
7. Centrifuge samples for 5 min at 400 × *g* to remove PBS.
8. Discard supernatant.
9. Repeat **steps 6–8**.
10. Resuspend the cell pellet with 250 μL DAPI in PBS (1 μg/ mL) solution. DAPI is used to detect cells.
11. Measure MMP by using NucleoCounter NC-3000 system (*see* **Note 11**).

3.3 Mass Spectrometry (See Note 12)

3.3.1 Cell Disruption and Protein Measurement

1. Grow cells and induce the expression of the HSP60 insert with tetracycline in T25 flasks as described in Subheading 3.1.
2. Wash the cells with 1 mL PBS, remove the PBS.
3. Repeat **step 2**.
4. Add 3 mL cold PBS and remove the cells from surface with a scrapper and collect it to a 15 mL Falcon tube (*see* **Note 13**).
5. Repeat **step 4** twice and centrifuge the total of 9 mL sample at 400 × *g* and 4 °C for 3 min to pellet the cells.
6. Discard the supernatant and resuspend the pellet in 60–100 μL urea lysis buffer.
7. Treat sample using a sonifier with a microtip with 3 cycles with 5 pulses of sonication. Cool samples on ice for a minute between each cycle (*see* **Note 14**).
8. Centrifuge cell lysates at 15,000 × *g* for 30 min at 4 °C.
9. Carefully remove the supernatant without disturbing the pellet and transfer to a new eppendorf tube.
10. Measure protein concentration with Bradford assay.

3.3.2 SDS-PAGE

1. Aliquot 30 μg of protein in an Eppendorf tube and adjust to 1× sample buffer by adding 6× sample buffer stock.
2. Load 30 μg protein from each sample on an Any kD™ sodium dodecyl sulfate (SDS) polyacrylamide gel.
3. Run the gel at low voltage (~100 V) until the protein pass the stacking part of the gel and just after they enter to the running gel, before the separation of the proteins begins.
4. Remove the gel from the tank.
5. Stain the SDS-polyacrylamide gel with Coomassie Brilliant Blue for 15 min.
6. Destain the gel with destaining solution twice for 10 min in a rotatory shaker to remove the excess dye. Continue destaining overnight in a rotary shaker (*see* **Note 15**).

7. Cut the region containing stained proteins out with the help of a scalpel.
8. Cut the gel pieces into smaller pieces and collect them into protein low binding Eppendorf tubes.
9. To destain the coomassie dye completely, incubate the gel pieces with 500 μL destaining solution and incubate with shaking at 400 rpm at 37 °C for twice 30 min. Remove the destaining solution and repeat the destaining once. Presence of Coomassie in peptides may interfere with mass spectrometric analysis in later steps.
10. Remove the destaining solution.
11. Prepare fresh reducing buffer and incubate the gel pieces with 90 μL for 10 min at 60 °C.
12. Remove excess reducing buffer.
13. Prepare fresh alkylation buffer, add 90 μL to gel pieces, and incubate samples in the dark for 50 min (*see* **Note 16**).
14. Wash samples with destaining buffer for 15 min at 37 °C, to remove residual reduction and alkylation buffers.

3.3.3 In-Gel Digestion of Proteins

1. Shrink the gel pieces with 150 μL acetonitrile (AcN) for 15 min. If the gel pieces are not completely white after this step, add more AcN and incubate again for 15 min (*see* **Note 17**).
2. Remove AcN and air dry the sample for 5–10 min.
3. Prepare activated trypsin solution (1:10 diluted in digestion buffer) and add 60 μL to gel pieces.
4. Incubate the gel pieces with activated trypsin solution for 15 min at room temperature with shaking (*see* **Note 18**).
5. Add an additional 75 μL digestion buffer to the sample and incubate overnight (10–17 h) at 34 °C or 4 h at 37 °C with shaking. Liquid should cover the gel pieces.
6. Transfer the digestion buffer that contains the peptides into a new protein low binding Eppendorf tube.
7. Remove AcN and other chemicals from the peptides by drying the samples with SpeedVac lyophilization (*see* **Note 19**).
8. Store the samples at −20 °C.

3.3.4 Peptide Purification

1. Dissolve the samples in 100 μL wash buffer (*see* **Note 20**).
2. Activate the column with 200 μL activation buffer. Centrifuge at 1500 × *g* for 1 min.
3. Add 200 μL equilibration buffer. Centrifuge at 1500 × *g* for 1 min. Discard flow-through.
4. Repeat **step 3**.
5. Load the sample from **step 1** to the column.

6. Place column into a new protein low bind Eppendorf tube and centrifuge at 1500 × *g* for 1 min.
7. Load the flow-through to the column again and centrifuge at 1500 × *g* for 1 min. Discard the flow-through.
8. Add 200 μL wash buffer and centrifuge at 1500 × *g* for 1 min. Discard flow-through.
9. Repeat **step 8**.
10. Place column into a new protein low bind Eppendorf tube. Add 20 μL elution buffer and centrifuge at 1500 × *g* for 1 min. Do not discard the flow-through.
11. Repeat **step 10** with the same collection tube and obtain 40 μL elute in total.
12. Lyophilize the samples.

    After this step samples are ready for mass spectrometric analysis.

4 Notes

1. For detailed description of the design and production of Flp-In cells *see* ref. 9.
2. The cells with the HSP60 cDNA Flp-In insertions are blasticidin and hygromycin resistant. To maintain the insert containing tetracycline repressor protein we grow cells under blasticidin and hygromycin selection to confirm the presence of the HSP60 cDNA insert.
3. Frozen cells contain DMSO which is highly toxic for the cells. Consequently, when cells start to thaw, **steps** between **2** and **4** should be performed immediately.
4. Cells are fragile at thawing step and they do not survive well with additional antibiotics. For this reason, when we thaw cells, we use DMEM without hygromycin and blasticidin, and culture the cells in T25 flasks. When the cells are attached and growing, we add the antibiotics and keep the cells under selection all times. This is typically after the first week when the cells are passaged for the first time.
5. Image cytometry measurements should be done at 80% confluence to ensure the flask is not too confluent for the staining to be well performed. Since cells will grow more in longer periods, it is important to adjust the amount of the cells to seed for different time points.
6. MitoTracker® Deep Red is susceptible to oxidases in serum. Therefore, working solution is prepared in HBSS. Stock solution of MitoTracker® Deep Red in DMSO is stable at −20 °C for a month.

7. Main stock of antimycin is prepared in 70% ethanol because pure ethanol is a damaging factor for mitochondria. Working solution is prepared in serum-free medium because FBS in the DMEM contains catalase enzyme which converts hydrogen peroxide into water. To inhibit the interference of the enzyme in ROS production, we use serum-free DMEM.
8. HEK cells easily detaching from the plate therefore it is important to wash it very gently and only once. If the washing is causing to lose too many cells, the washing step can be skipped. However, if washing is not performed, it is important to aspirate the remaining medium from the well.
9. MitoSOX Red mitochondrial superoxide indicator is used to determine levels of mitochondrial ROS production. Elevated levels of superoxide anion causes oxidation of MitoSOX reagent which produces red fluorescence and this can be detected with image cytometry. Always use freshly dissolved MitoSOX. MitoSOX is light sensitive. Protect it from light all times. MitoSOX is dissolved in DMSO (13 μL/50 μg). Once it has been dissolved in DMSO, MitoSOX is not reused or stored to obtain the optimal signal with each measurement.
10. Less amounts of cells can be used -down to 500,000 cells. In this case all the reagents should be scaled down accordingly.
11. Mitochondrial membrane potential (MMP) assay was performed with two dyes. JC-1 is a dye which is green as a monomer and red when it aggregated. When JC-1 is in mitochondria, it aggregates at high concentration and becomes red. In cytosol it is at low concentration and green. Red to green fluorescence ratio is used to determine the JC-1 amount in mitochondria compared to cytosol. In healthy cells mitochondria are more negatively charged than the cytosol due to the mitochondrial membrane potential. JC-1 accumulates in the mitochondria because of the charge difference, forms aggregates and becomes red fluorescent. In this way the MMP can be determined by using JC-1. DAPI is used to stain the nuclei of necrotic and late-apoptotic cells. JC-1 and DAPI together gives the information about numbers of healthy, apoptotic, necrotic, and late-apoptotic cells.
12. Selective reaction monitoring (SRM) is a mass spectrometry method that measures specifically targeted peptides in a sample. By using SRM we can detect small amounts of different proteins from one mixture by single analysis and have a relative quantification to the heavy label standard included.
13. For a T25 flask, use 3 mL cold PBS.
14. Use an ice–water bath to keep the sample at 4 °C and prevent overheating from sonication.

15. Coomassie Brilliant Blue R250 is a dye that binds to proteins through ionic interactions. The gel was stained with Coomassie to be able to visualize the region with proteins from the gel.
16. Reduction and alkylation are performed to break disulfide bonds in the protein and prevent them to re-form again. Thus, the proteins are denatured and all the cleavage sites for trypsin are accessible to the enzyme for cut.
17. When the gels pieces are shrunk (they become white), the trypsin in solution is absorbed by the gel pieces faster and all proteins will be in contact with the trypsin solution. This allows all proteins to be cleaved by trypsin. After cleavage of the proteins, the peptides are small enough to go through the pores of the gel and be collected from the liquid.
18. If the gel pieces are not completely covered with the activated trypsin solution, proteins will not be digested completely which will prevent proteins to be released from the gel. For this reason, if some pieces are still partially white, activated trypsin and digestion solution amounts should be adjusted proportionately.
19. At this point peptides are collected. However, eventually remaining impurities are further removed with C-18 spin columns.
20. Maximum loading capacity of columns is 150 μL and 30 μg protein.

Acknowledgments

We acknowledge financial support from the Ludvig and Sara Elsass Foundation and the Eva og Henry Frænkels Mindefond. C.C. was supported by a scholarship from SCITECH, Aarhus University. We thank Anne Bie for introduction to the Flp-in cell model.

References

1. Nunnari J, Suomalainen A (2012) Mitochondria: in sickness and in health. Cell 148(6):1145–1159. https://doi.org/10.1016/j.cell.2012.02.035
2. Picard M, Wallace DC, Burelle Y (2016) The rise of mitochondria in medicine. Mitochondrion 30:105–116. https://doi.org/10.1016/j.mito.2016.07.003
3. Gorman GS, Chinnery PF, DiMauro S, Hirano M, Koga Y, McFarland R, Suomalainen A, Thorburn DR, Zeviani M, Turnbull DM (2016) Mitochondrial diseases. Nat Rev Dis Primers 2:16080. https://doi.org/10.1038/nrdp.2016.80
4. Raimundo N (2014) Mitochondrial pathology: stress signals from the energy factory. Trends Mol Med 20(5):282–292. https://doi.org/10.1016/j.molmed.2014.01.005
5. Magnoni R, Palmfeldt J, Hansen J, Christensen JH, Corydon TJ, Bross P (2014) The Hsp60 folding machinery is crucial for manganese superoxide dismutase folding and function. Free Radic Res 48(2):168–179. https://doi.org/10.3109/10715762.2013.858147
6. Bie AS, Fernandez-Guerra P, Birkler RI, Nisemblat S, Pelnena D, Lu X, Deignan JL, Lee H, Dorrani N, Corydon TJ, Palmfeldt J,

Bivina L, Azem A, Herman K, Bross P (2016) Effects of a mutation in the HSPE1 gene encoding the mitochondrial co-chaperonin HSP10 and its potential association with a neurological and developmental disorder. Front Mol Biosci 3:65. https://doi.org/10.3389/fmolb.2016.00065

7. Nisemblat S, Yaniv O, Parnas A, Frolow F, Azem A (2015) Crystal structure of the human mitochondrial chaperonin symmetrical football complex. Proc Natl Acad Sci U S A 112(19): 6044–6049. https://doi.org/10.1073/pnas.1411718112
8. Bross P, Fernandez-Guerra P (2016) Disease-associated mutations in the HSPD1 gene encoding the large subunit of the mitochondrial HSP60/HSP10 chaperonin complex. Front Mol Biosci 3:49. https://doi.org/10.3389/fmolb.2016.00049
9. Wakabayashi-Nakao K, Tamura A, Koshiba S, Toyoda Y, Nakagawa H, Ishikawa T (2010) Production of cells with targeted integration of gene variants of human ABC transporter for stable and regulated expression using the Flp recombinase system. Methods Mol Biol 648: 139–159.https://doi.org/10.1007/978-1-60761-756-3_9
10. Bie AS, Palmfeldt J, Hansen J, Christensen R, Gregersen N, Corydon TJ, Bross P (2011) A cell model to study different degrees of Hsp60 deficiency in HEK293 cells. Cell Stress Chaperones 16(6):633–640. https://doi.org/10.1007/s12192-011-0275-5

Chapter 15

Superresolution Fluorescence Imaging of Mutant Huntingtin Aggregation in Cells

Steffen J. Sahl and Willianne I. M. Vonk

Abstract

Fluorescence-based nanoscopy methods (also known as "superresolution" microscopy) have substantially expanded our options to examine the distributions of molecules inside cells with nanometer-scale resolution and molecular specificity. In the biophysical analysis of aggregation-prone misfolded proteins and peptides, this has enabled the visualization of distinct populations of aggregated species such as fibrillar assemblies within intact neuronal cells, well below previous limits of sensitivity and resolution. With the Huntington's disease protein, polyglutamine-expanded mutant huntingtin, as an example, we provide sample preparation and imaging protocols for superresolution microscopy down to the ~30 nm-level.

Key words Fluorescence nanoscopy, Cellular superresolution imaging, Huntington's disease, Fibrils, Fibrillar aggregates, Oligomers, Amyloid, Polyglutamine expansion, Trinucleotide repeat disorders, Inclusion bodies

1 Introduction

The seminal insight that the longstanding diffraction resolution limit of optical microscopy can be fundamentally broken [1–5] has changed our outlook on the future of molecular analysis in biological systems. Fluorescence nanoscopy [1, 6] comprises a suite of methods which combine the traditional advantages of fluorescence-based microscopy assays—notably, highest molecular specificity of visualization and sensitivity in principle down to the single-molecule level [2]—with a spatial resolution down to a small fraction of the optical wavelength. At present, resolutions of 20–50 nanometers (nm) are typically achieved.

Among the nanoscopy methods, which are also collectively referred to as "superresolution microscopies", stimulated emission depletion (STED) imaging [3, 5], as well as variants of stochastic optical reconstruction microscopy (STORM) [7] or ground state depletion with individual molecule return (GSDIM) [8–10] nanoscopy in particular are finding increasing application to biological

Cláudio M. Gomes (ed.), *Protein Misfolding Diseases: Methods and Protocols*, Methods in Molecular Biology, vol. 1873, https://doi.org/10.1007/978-1-4939-8820-4_15, © Springer Science+Business Media, LLC, part of Springer Nature 2019

problems [6]. The most commonly encountered implementation of STED nanoscopy retains the main features of laser-scanning confocal microscopy, and can conveniently be used much like a confocal microscope with nanoscale resolution. The STORM/GSDIM methods operate on widefield (epifluorescence or TIRF) microscopy systems and rely on single-molecule imaging with sensitive camera detectors and associated post-processing analysis. In routine practice, both approaches achieve up to a tenfold resolution improvement over conventional diffraction-limited fluorescence microscopy, typically down to ~30 nm. A number of very recent conceptual advances fuel additional excitement about the prospects for routine nanoscopy analyses at yet higher resolutions in the coming years. These include, but are not limited to, the recent development of MINFIELD [11] and DyMIN [12] STED variants with enhanced signal and resolution, the initial demonstrations of molecular (1 nm) resolution with the MINFLUX concept [13], as well as advances in quantitative nanoscopy, notably molecular counting approaches [14, 15].

One important application of these new imaging capabilities is the direct analysis of aggregating mutant huntingtin species within cellular models of Huntington's disease (HD) pathology. In HD, accumulation of polyglutamine (polyQ)-expanded mutant huntingtin (Htt) in inclusion bodies is widely thought to be a neuronal coping response by confining potentially harmful conformations of the mutant protein to protective inclusions. However, knowledge is lacking on the extent to which distinct small oligomeric and other mutant Htt aggregation intermediates exist "en route" or "in parallel" to inclusion body formation. Working toward increased insights into the toxic nature and cellular fate of the distinct pathogenic Htt conformations, and how these can be modulated by quality control activities of chaperones and degradative machineries, is a currently perhaps underappreciated yet promising avenue to advance targeted therapeutic design to halt or delay HD pathology.

At present, no effective treatments to halt, delay, or even cure HD are available. Nanoscopy studies allow for the detailed identification of the abundance and pathogenicity of distinct Htt species at different stages of aggregation. Nanoscopy directly visualizes their presence and, coupled with other readouts, allows to examine the toxic nature of small fibrillar Htt conformers in HD pathogenesis at high spatial resolution. Indeed, in conjunction with dedicated immunodetection of multiple targets, it becomes possible to put cellular quality control in the picture as well. The (potentially selective) association and modes of regulation of various protein homeostasis determinants with the characterized on- and/or off-pathway Htt intermediates can be identified.

The capability to visualize the smallest fibrillar forms of pathogenic Htt in neurons is a recent development [16, 17]. Detailed characterizations of the various Htt species at nanoscale resolution are expected to provide unique insights for therapeutic design targeted at the desired elimination or, alternatively, stabilization of certain defined ((non-)toxic) Htt species in HD pathology [6].

Past HD research has characterized and identified differential parameters as well as effects of pharmacological agents, including purified chaperonin fragments, on large pathogenic Htt aggregates by quantification of "visible aggregates" at low magnification and sensitivity using conventional microscopy techniques [18–20]. Recently however, the perception of the identity of toxic Htt species in HD pathology has shifted and reached the consensus that smaller oligomeric or (pre)fibrillar Htt intermediates induce cellular toxicity, whereas inclusion body formation acts as a neuroprotective mechanism to confine potentially harmful entities [21–23]. Research toward the identity and (toxic) nature of intermediate Htt species has been limited, but very recently we established the visualization of small fibrillar forms of pathogenic Htt in neurons using high-resolution/high-sensitivity nanoscopy [16, 17]. As of now, knowledge is lacking on the pathogenicity and cellular fate of these small Htt conformers, their relevance to pathology, and how their assembly can be modulated by the activities of the various quality control branches. Assessments of the relative abundance of these different species over time require their unambiguous detection, as now provided by the emergent nanoscopy approaches.

Furthermore, a long-term goal in which fluorescence nanoscopy may play a crucial role is the further integration of in cellulo (i.e., mammalian (neuronal) cell model systems) and in vivo (e.g., primary neuronal cells isolated from mouse brain) approaches to produce mechanistic insights into protein quality control in HD. Such increased understanding will also set the stage for reengineering chaperone networks for therapeutic purposes [24, 25].

2 Materials

2.1 Nanoscopy Instrumentation

Instrumentation for optical nanoscopy, such as a stimulated emission depletion (STED) nanoscopy, stochastic optical reconstruction microscopy (STORM), or ground state depletion with individual molecule return (GSDIM) microscopy platform. Leading suppliers of STED nanoscopy platforms include Abberior Instruments, Germany, and Leica Microsystems, Germany. Microscopy systems compatible with PALM/STORM/GSDIM nanoscopy are available from Zeiss, Leica, Nikon, and several other suppliers.

For the huntingtin imaging described in this chapter, with enhanced yellow fluorescent protein (EYFP) as the reporter fluorophore, no special imaging buffers are required. The samples are mounted by standard procedures (*see* Subheading 3.2, **step 6**).

2.2 Cell Line (Example: PC12)

The PC12 cell line, differentiated into sympathetic-neuron-like cells with nerve growth factor (NGF), has served as model for early pathogenic events in Huntington's disease with fluorescent-protein fusions to huntingtin (e.g., [26, 27]). PC12 cells are well suited for imaging experiments in culture due to their moderate thickness of up to ~10 μm when adhered to the microscope slide.

A widely used subclone of the original PC12 neuronal model cell line developed by L. Greene (Columbia University), PC-12 Adh (ATCC CRL-1721.1), had been originally selected for its enhanced attachment to plastics, facilitating culture work, in M. Montminy's laboratory (Salk Institute, La Jolla). These cells are used in the described protocol, but other neuronal cell models may be substituted.

2.3 Expression Constructs of Fluorescently Labeled Huntingtin

The first 94 amino acids of huntingtin, encoding the full exon 1 (Htt-ex1: MATLEKLMKAFESLKSF[Q_n]PPPPPPPPPPPQLPQ PPPQAQPLLPQPQPPPPPPPPPPGPAVAEEPLHRPGS) with a 25Q polyglutamine tract and mutant forms (46Q and 97Q respectively) cloned into a mammalian expression vector (pcDNA3.1(−)) using XhoI and BamHI restriction enzymes, result in CMV promoter-controlled C-terminal fusion constructs to enhanced yellow fluorescent protein (EYFP).

2.4 Culture Media

1. Standard, complete cell culture medium: 10% fetal bovine serum (FBS), 90% phenol-red-free Dulbecco Modified Eagle Medium (DMEM). Store at 4 °C.
2. NGF-supplemented medium: standard complete cell culture medium (Subheading 2.4, **item 1**) with supplement of neural growth factor (NGF 2.5S, ~100 ng/mL of culture medium, Invitrogen) added before cell passaging.

3 Methods

3.1 Cell Culture

1. Grow PC12 cells to confluency at 37 °C in a 5% CO_2, 96% relative humidity incubator in tissue-culture treated flasks (75 cm^2) in standard, complete cell culture medium.
2. Passage cells (with 1:5–1:10 dilution, dependent on the cell density and growth behavior) at least 3 times prior to experiments. A passage consists of a brief yet careful wash with 1× PBS (phosphate buffered saline, pH 7.4) prior to a 3–5 min incubation with 2 mL of trypsin replacement (TrypLE Express), addition of 8 mL cell media to deactivate the enzyme, a brief centrifugation (<2 min, <400 rcf), followed by media replacement to remove any residual trypsin replacement.

3. Add cell suspensions to 6-well tissue culture plates.
4. Grow cells to a monolayer (>90% confluency; typically this is reached within 1–2 days postplating; *see* **Note 1**).
5. Differentiate cells into sympathetic-neuron-like cells by treatment with nerve growth factor (NGF 2.5S, ~100 ng/mL of culture medium) for 48 h.
6. Transfect with the plasmid DNA for fluorescently labeled Htt-ex1 (*see* Subheading 2.3) utilizing the standard protocol for Lipofectamine 2000 or other transfection kit (*see* Subheading 3.2).

3.2 Transfection and Sample Preparation

1. Per sample (i.e., well of 6-well tissue culture plate), prepare complexes using 2 μg of DNA and 6 μL of Lipofectamine 2000 in a total volume of 100 μL of Opti-MEM I media.
2. Add DNA-transfection reagent complexes to cells and incubate for 4–6 h.
3. Aspirate the transfection complex-containing medium and treat each well with 0.5 mL TrypLE Express for 5 min.
4. Suspend cells in 4–8 mL of NGF-supplemented (~100 ng/mL) medium, and distribute cells among 2–4 wells (to achieve appropriate cell densities for imaging) in 6-well plates containing microscope slides. These slides (standard microscopy coverslips) should be precleaned (and optionally ozone-treated) prior to coating with 500 μL each of either ~8 μg/mL fibronectin in 1× PBS (pH 7.4) or ~0.5 mg/mL collagen type-I in 0.02 N acetic acid for 1 h and rinsed with 1× PBS before use.
5. At defined time points posttransfection (e.g., 12 and 48 h after addition of transfection complexes and plating on microscope slides), fix cells in 4% paraformaldehyde in 1× PBS (15 min, RT) and wash carefully with 1× PBS prior to the cell-by-cell examination and superresolution imaging (*see* **Note 2**).
6. For STORM/GSDIM-type nanoscopy, mount (and before store) the microscope slide in 1× PBS in a mechanically stable custom microscopy mount. For STED imaging, mount slide on object slide with a drop of Mowiol.

3.3 Selective Imaging of Small, Dim Aggregates in Close Proximity to Strongly Fluorescent Inclusions (in STORM/GSDIM Nanoscopy)

Highly fluorescent huntingtin inclusion bodies (IBs) frequently hinder visualization of the diverse collection of smaller, dimmer aggregated structures which coexist in many cells in close proximity to the inclusions. Such sizeable brightness disparities of neighboring objects across the sample represent a general problem in wide-field microscopy with an area detector operating at a common gain/pixel. The single, immensely bright IB fully exploits the dynamic range of the detector, while the adjacent dimmer features are not detected, either because the excitation light has to be reduced accordingly or due to bright signal spillover onto the adjacent camera pixels. Consequently, while the inclusions are clearly visible (e.g., [26, 28]), they usually remain the predominant fluorescent feature detected and reported in the cells, especially in work at low magnification and with limited fluorescence dynamic range.

An iris placed centrally in the excitation path before the focusing Köhler ("widefield") lens, can be used to limit excitation to a much smaller region of the sample (Fig. 1b). In this way, the IBs can be selectively photobleached, by illumination over several minutes at ~2 kW/cm^2 intensity (or higher), and their signal magnitude substantially reduced. Many dimmer fluorescent species can be subsequently visualized (even in close proximity to the inclusion) by translating the sample, moving the inclusion outside the iris-patterned excitation beam profile and imaging aggregates with minimal signal interference from IBs (Compare also **Note 3**).

3.4 STORM/GSDIM Nanoscopy: Data Acquisition

Besides commercial platforms for STORM/GSDIM-type nanoscopy, which must be equipped with a laser source to excite fluorescence and induce the dark-state blinking of EYFP (514 nm), a custom optical setup can be used.

In published work [16], 514 nm light (Coherent Sapphire, 50 mW) was focused into the back aperture of a 1.4 NA oil-immersion objective (Olympus UPLSAPO, 100×), achieving standard Köhler illumination for wide-field laser excitation (~30 μm spot size, full width at half maximum) in epifluorescence mode. Fluorescence was selected by a dichroic mirror (Semrock, Di01-R532), 525 nm long pass filter (Semrock, BLP01-514R-25) and a band pass filter for EYFP emission (Semrock, FF01-578/105-25). Neutral density filters were employed in the excitation beam path to switch between a low-intensity (~20 W/cm^2) cell-scanning mode for assessing cellular fluorescence and characterizing phenotypes, and the high-intensity mode (~2 kW/cm^2) required for efficient EYFP blinking [9, 10, 29] and fast image acquisition. The image was formed on a highly sensitive electron-multiplying charge coupled device (EMCCD) camera (Andor iXon DU897) that allowed high frame rates. 35-ms-duration frames are a suitable choice to capture the blinking, with many single-molecule events lasting for about one frame. In [16], appropriate averaging of 100 frames was applied to epifluorescence images (low excitation intensity), to provide a fair comparison of the diffraction-limited epifluorescence mode and the single-molecule-localization-based superresolution reconstructions, of which examples are shown in Fig. 1a.

For the recordings published in [16], no substantial lateral (*xy*) drift was present during the ~4–10 min total acquisition (*see* **Note 4**). Continuous sample drift during the short acquisition window of several minutes would manifest itself as a tendency of points to be systematically offset in an ordered way at later times in the

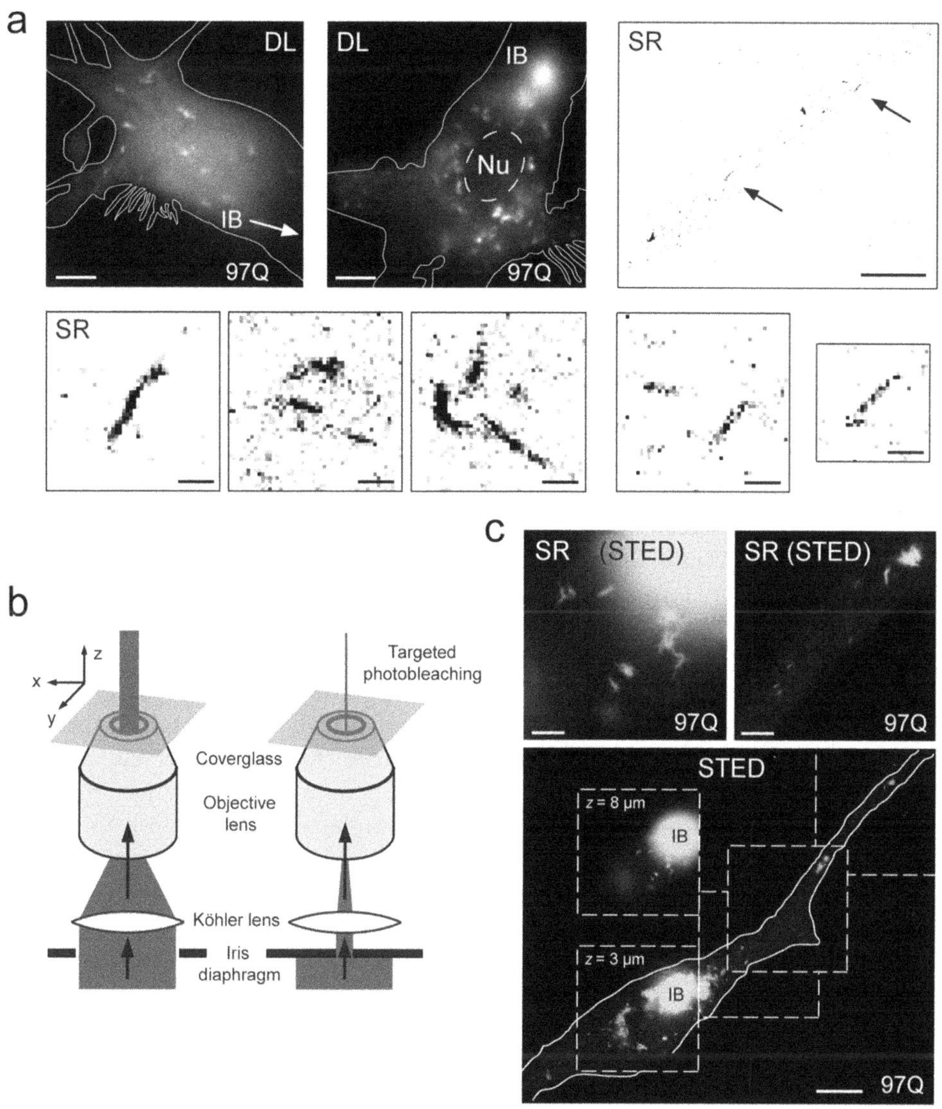

Fig. 1 Fluorescence nanoscopy of intracellular aggregation of mutant huntingtin exon 1. (**a**, **b**) Single-molecule-based STORM/GSDIM and (**c**) STED imaging of huntingtin exon 1 with an expanded polyglutamine tract of 97Q, expressed by transient transfection inside nerve growth factor-differentiated PC12Adh neuronal model cells. The C-terminally fused enhanced yellow fluorescent protein (EYFP) serves as the fluorescent reporter. Distinct from the bright micrometer-sized inclusion bodies (IB), nanoscopy reveals cytosolic linear fibrillar aggregates of sizes well below the diffraction resolution limit. These fibrils stick to each other, forming "coalesced" higher-order assemblies. (**b**) For STORM/GSDIM-type superresolution imaging, a targeted photobleaching protocol (*see* ref. 16) allows to reduce the brightness of IBs and subsequently to visualize the much dimmer fibrils (*see* Subheading 3.3). Data and schematic illustration in (**a**, **b**) adapted from ref. 16, data in (**c**) adapted from ref. 17. *Nu* nucleus, *DL* diffraction-limited fluorescence microscopy (epifluorescence), *SR* superresolved imaging. Scale bars: 5 μm (**a**, top row), 500 nm (**a**, bottom row), 2 μm (**c**, top images), 10 μm (**c**, bottom image)

acquisition compared to earlier localizations. This effect, which would lead to additional broadening of the observed structures (i.e., fiber widths), was not observed. Consistency in the feature sizes of thin linear segments, ~80–120 nm thickness at several orientations within the same fields of view, further supported this conclusion. Localizations contributing to the reconstructed image were obtained after a short (5–15 s) initial bleachdown (30–300 s for denser aggregates), but before all EYFP molecules had been bleached. The useful frame range of single-molecule images typically comprised the first 5 min of a recording.

3.5 STORM/GSDIM Nanoscopy: Data Analysis

The extraction of Htt-ex1 single-molecule coordinates from recordings is performed based on widely established procedures, which are also implemented in commercially available imaging platforms. In brief, single molecules are typically identified by selecting bright pixels which exceed the average background by a multiple of its standard deviation. In [16], positions were then determined by fitting pixelated single-molecule images to symmetric two-dimensional Gaussian functions, which describe the microscope point spread function (PSF), and extracting the center coordinates, (x,y). Besides requiring fit convergence, filters were applied to exclude spurious localizations arising from background noise or overlapping molecules. Reconstructed SR images were computed as histograms, i.e., counts of localization events falling within xy domains (50 × 50 nm bins, displayed with a lookup table from white (no localizations) to black (maximum number of localizations) in the examples of Fig. 1a). The statistical localization precision (imaging resolution) of single molecules was estimated from repetitive localizations of a subset of molecules appearing as multiple isolated clusters [30]. For the data in [16], resolution was ~30 nm. A recommended and easy-to-use solution to extract localizations is the freely available program RapidSTORM [31].

3.6 STED Nanoscopy: Data Acquisition

Examples (Fig. 1c) [17] of STED imaging with Htt Q97-ex1-EYFP at ~80 nm (FWHM, corresponding to a spatial standard deviation σ of ~35 nm) focal-plane resolution were obtained on a Leica platform (TCS STED CW, Leica Microsystems). In this instrument, fluorescence excitation was provided by the 514 nm line of an Argon laser, fluorescence silencing by stimulated emission depletion effected by a fiber laser at 592 nm, both in continuous-wave operation. Excitation and STED laser output powers were set to a maximum of 20% (or 5 mW) and 100% (or 1.5 W), respectively. Fluorescence was detected from 520–580 nm. STED images were scanned at 25.8 nm scanning steps with 8000 Hz linescans over 1024 × 1024 pixel fields. Acquisition with line averaging of 8 or 16 (for confocal) or 32 or 64 (for STED) lines produced the highest image contrast, which is beneficial especially for dim fluorescent features. No image deconvolution needs to be applied to STED raw data (Fig. 2 and **Note 5**).

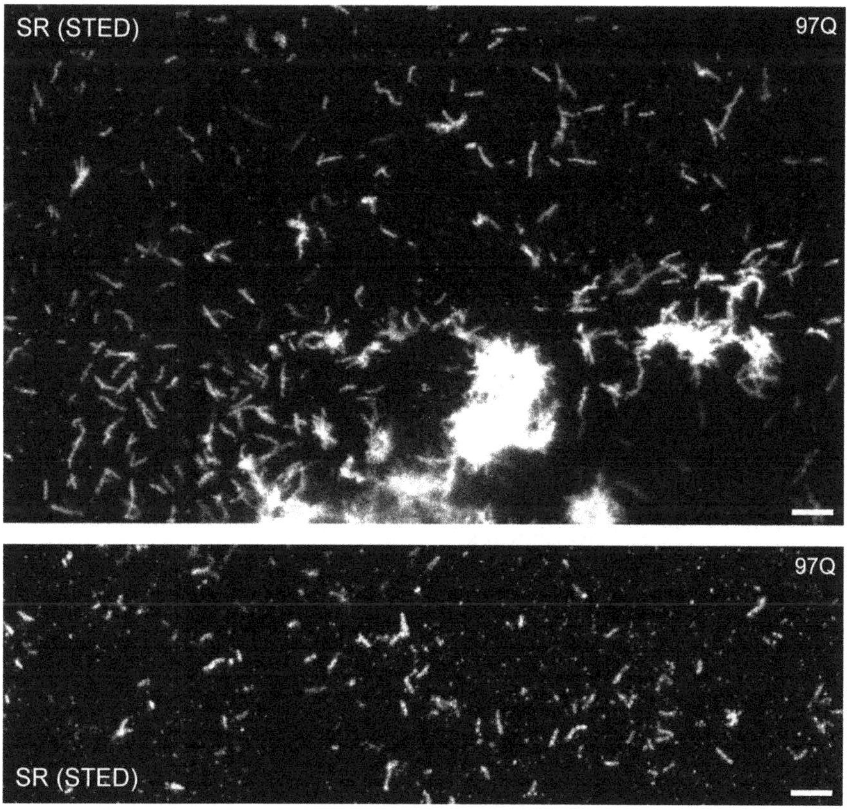

Fig. 2 STED imaging experiments with optimized indirect immunofluorescence labeling (antibody–dye conjugates with the photostable dye Abberior STAR635P) achieve high contrast and reveal isolated and coalesced mutant huntingtin fibrils (97Q), as well as nonelongated aggregate nucleation seeds. These examples were recorded within striatal-derived ST14a cells. The image resolution is approximately 30 nm. Both scale bars: 1 μm

4 Notes

1. Typically, 150,000–200,000 cells are plated into a well of a 6-well plate to reach the desired confluency within 1–2 days. However, it is recommended to optimize the number of cells to be plated as this is highly dependent on the laboratory environment, and on cell growth behavior and passage number.
2. For time points beyond 1 week (e.g., 144 h and 192 h), cells should be transferred to the microscope slides at least 48 h prior to experiments but kept on 6-well tissue culture wells in the interim. This adjustment ensures that the surface treatment of the glass slides is relatively fresh and thus prevents the cell layer from peeling off during fixation and the subsequent washing steps.
3. Above and below the bright inclusion bodies, the strong out-of-plane background fluorescence signals are efficiently blocked by point-scanning nanoscopy modalities with confo-

cal detection, such as most STED nanoscopes. The point-scanning approach, inherently featuring sequential fluorescence detection of adjacent image regions, deals well with brightness disparities of neighboring objects (aggregate seeds and fibrils vs. immensely bright IBs).

4. For best results, the microscope sample slide and the micrometer translation stage should be let to thermally equilibrate before a superresolution recording begins. Avoiding unnecessary temperature fluctuations in the laboratory and allowing the coarse focus knob to rest before recording helps to keep sample drift to a minimum.
5. Excellent results can also be obtained by antibody labeling of epitopes within the Htt protein or its fluorescent-protein fusion tag by standard methods and utilizing secondary antibodies conjugated to STORM- or STED-compatible fluorophores.

Acknowledgments

The research on which the protocols described in this chapter are based was supported by the US National Institutes of Health (NIH) through the National Institute of General Medical Sciences Grant No. R01GM086196 for purchase of equipment in the laboratory of W. E. Moerner, Stanford University, CA, USA, where the first superresolution studies of cellular huntingtin aggregation were carried out jointly by S.J.S. and Lucien E. Weiss [16]. Support for postdoctoral fellowships at Stanford came from the National Eye Institute Grant PN2EY016525 (S.J.S.; to W. E. Moerner) and the Marie Curie International Outgoing Fellowship Program (W.I.M.V.; FP7-PEOPLE-2011-IOF, contract no. 303096 FOLDEG). The European Huntington's Disease Network (EHDN) has generously supported ongoing investigations of S.J.S. and W.I.M.V. with a seed grant (grant no. 797).

References

1. Hell SW (2007) Far-field optical nanoscopy. Science 316:1153–1158
2. Moerner WE (2007) New directions in single-molecule imaging and analysis. Proc Natl Acad Sci U S A 104:12596–12602
3. Hell SW, Wichmann J (1994) Breaking the diffraction resolution limit by stimulated emission: stimulated-emission-depletion fluorescence microscopy. Opt Lett 19:780–782
4. Hell SW, Kroug M (1995) Ground-state-depletion fluorscence microscopy: a concept for breaking the diffraction resolution limit. Appl Phys B Lasers Opt 60:495–497
5. Klar TA, Jakobs S, Dyba M, Egner A, Hell SW (2000) Fluorescence microscopy with diffraction resolution barrier broken by stimulated emission. Proc Natl Acad Sci U S A 97:8206–8210
6. Sahl SJ, Hell SW, Jakobs S (2017) Fluorescence nanoscopy in cell biology. Nat Rev Mol Cell Biol 18:685–701
7. Rust MJ, Bates M, Zhuang X (2006) Sub-diffraction-limit imaging by stochastic optical

reconstruction microscopy (STORM). Nat Methods 3:793–796
8. Bock H et al (2007) Two-color far-field fluorescence nanoscopy based on photoswitchable emitters. Appl Phys B Lasers Opt 88:161–165
9. Fölling J et al (2008) Fluorescence nanoscopy by ground-state depletion and single-molecule return. Nat Methods 5:943–945
10. Biteen JS, Thompson MA, Tselentis NK, Bowman GR, Shapiro L, Moerner WE (2008) Super-resolution imaging in live *Caulobacter crescentus* cells using photoswitchable EYFP. Nat Methods 5:947–949
11. Göttfert F et al (2017) Strong signal increase in STED fluorescence microscopy by imaging regions of subdiffraction extent. Proc Natl Acad Sci U S A 114:2125–2130
12. Heine J, Reuss M, Harke B, D'Este E, Sahl SJ, Hell SW (2017) Adaptive-illumination STED nanoscopy. Proc Natl Acad Sci U S A 114: 9797–9802
13. Balzarotti F et al (2017) Nanometer resolution imaging and tracking of fluorescent molecules with minimal photon fluxes. Science 355: 606–612
14. Ta H et al (2015) Mapping molecules in scanning far-field fluorescence nanoscopy. Nat Commun 6:7977
15. Jungmann R et al (2016) Quantitative super-resolution imaging with qPAINT. Nat Methods 13:439–442
16. Sahl SJ, Weiss LE, Duim WC, Frydman J, Moerner WE (2012) Cellular inclusion bodies of mutant huntingtin exon 1 obscure small fibrillar aggregate species. Sci Rep 2:895
17. Sahl SJ, Lau L, Vonk WIM, Weiss LE, Frydman J, Moerner WE (2016) Delayed emergence of subdiffraction-sized mutant huntingtin fibrils following inclusion body formation. Q Rev Biophys 49:e2
18. Sontag EM, Joachimiak LA, Tan Z, Tomlinson A, Housman DE, Glabe CG et al (2013) Exogenous delivery of chaperonin subunit fragment ApiCCT1 modulates mutant huntingtin cellular phenotypes. Proc Natl Acad Sci U S A 110:3077–3082
19. Tam S, Spiess C, Auyeung W, Joachimiak L, Chen B, Poirier MA et al (2009) The chaperonin TRiC blocks a huntingtin sequence element that promotes the conformational switch to aggregation. Nat Struct Mol Biol 16:1279–1285
20. Nucifora LG, Burke KA, Feng X, Arbez N, Zhu S, Miller J et al (2012) Identification of novel potentially toxic oligomers formed in vitro from mammalian-derived expanded huntingtin exon-1 protein. J Biol Chem 287:16017–16028
21. Kuemmerle S, Gutekunst CA, Klein AM, Li XJ, Li SH, Beal MF et al (1999) Huntington aggregates may not predict neuronal death in huntington's disease. Ann Neurol 46:842–849
22. Slow EJ, Graham RK, Osmand AP, Devon RS, Lu G, Deng Y et al (2005) Absence of behavioral abnormalities and neurodegeneration in vivo despite widespread neuronal huntingtin inclusions. Proc Natl Acad Sci U S A 102: 11402–11407
23. Arrasate M, Mitra S, Schweitzer ES, Segal MR, Finkbeiner S (2004) Inclusion body formation reduces levels of mutant huntingtin and the risk of neuronal death. Nature 431:805–810
24. Escusa-Toret S, Vonk WIM, Frydman J (2013) Spatial sequestration of misfolded proteins by a dynamic chaperone pathway enhances cellular fitness during stress. Nat Cell Biol 15:1231–1243
25. Sontag EM, Vonk W, Frydman J (2014) Sorting out the trash: the spatial nature of eukaryotic protein quality control. Curr Opin Cell Biol 26:139–146
26. van Roon-Mom WMC et al (2008) Mutant huntingtin activates Nrf2-responsive genes and impairs dopamine synthesis in a PC12 model of Huntington's disease. BMC Mol Biol 8:84–96
27. Apostol BL et al (2006) Mutant huntingtin alters MAPK signaling pathways in PC12 and striatal cells: ERK1/2 protects against mutant huntingtin-associated toxicity. Hum Mol Gen 15:273–285
28. Legleiter J et al (2009) Monoclonal antibodies recognize distinct conformational epitopes formed by polyglutamine in a mutant huntingtin fragment. J Biol Chem 284:21647–21658
29. Dickson RM, Cubitt AB, Tsien RY, Moerner WE (1997) On/off blinking and switching behavior of single green fluorescent protein molecules. Nature 388:355–358
30. Lee HD, Sahl SJ, Lew MD, Moerner WE (2012) The double-helix microscope super-resolves extended biological structures by localizing single blinking molecules in three dimensions with nanoscale precision. Appl Phys Lett 100:153701
31. Wolter S, Schüttpelz M, Tscherepanow M, van de Linde S, Heilemann M, Sauer M (2010) Real-time computation of subdiffraction-resolution fluorescence images. J Microsc 237:12–22

Part III

Protein Folding Recovery and Correction Assays

Chapter 16

Thermal Shift and Stability Assays of Disease-Related Misfolded Proteins Using Differential Scanning Fluorimetry

Tânia G. Lucas, Cláudio M. Gomes, and Bárbara J. Henriques

Abstract

Systematic identification of buffer formulations and small molecule chaperones that improve the expression, stability, and storage of proteins with therapeutic interest has gained enormous importance in biochemical research as well as in biotechnology and biomedical applications. In particular, the biochemical characterization of disease-related proteins and their genetic variants that result in misfolding requires systematic determination of protein stability, screening of optimal buffer conditions for biophysical and structural studies, and in some cases, the identification of small molecule chaperones with the potential to ameliorate folding defects. Among the several techniques available, differential scanning fluorimetry (DSF) is currently an extensively employed screening and analysis method for thermal shift and protein stability assays. Here we describe a step-by-step generic protocol for fast characterization of protein thermal stability and analysis of stabilization in thermal-shift assays by additives, ligands and chemical chaperones using β-oxidation mitochondrial dehydrogenases as model. These enzymes are associated to inborn errors of metabolism caused by mutant variants with folding and stability defects for which we previously established folding correction afforded by their cognate cofactors and substrates. With this example we thus illustrate the potential applications of the method in screening small molecule folding correctors among metabolites, ligands, cofactors or candidate drugs with therapeutic potential in protein folding diseases.

Key words Differential scanning fluorimetry, Thermal stability, Melting temperature, Protein stability, Pharmacological chaperones, Chemical chaperones, Enzyme, Cofactors

1 Introduction

Protein stabilization by small molecules may significantly improve protein expression and purification [1], the quality of purified proteins [2], crystallization process [3], or have a potential therapeutic application to recover and maintain protein structure and function [4–7]. Small molecules with the potential to rescue misfolded proteins by stimulating their correct folding, stabilizing native-like conformations, and/or correct trafficking and function in vivo, are known as pharmacological or chemical chaperones, respectively if the ligand binds directly to the protein or not [8, 9].

Cláudio M. Gomes (ed.), *Protein Misfolding Diseases: Methods and Protocols*, Methods in Molecular Biology, vol. 1873, https://doi.org/10.1007/978-1-4939-8820-4_16, © Springer Science+Business Media, LLC, part of Springer Nature 2019

Protein stability can be experimentally determined in vitro by manipulating environmental conditions (e.g., temperature) to promote protein denaturation. The unfolding reaction can be followed by different techniques and the midpoint of the unfolding transition will be characteristic for a specific system. For thermal unfolding, a melting temperature (T_m), that corresponds to the midpoint of the unfolding curve, can be determined and the effect of a given condition over protein stabilization evaluated by an effect over the T_m value. A stabilizing compound or buffer formulation will thus result in an increase in the denaturation temperature and the resulting thermal shift (ΔT_m) is a relative measure of protein stabilization under the tested conditions.

Differential scanning fluorimetry (DSF) is among the most attractive screening methods to determine changes in the T_m value of a protein in medium high-throughput conditions. This method is based in the use of a fluorophore that will emit only upon interacting with an unfolding protein during thermal denaturation, and from which the fraction of folded versus unfolded protein can be determined. Its advantages include the fact that DSF is typically performed in a conventional real-time PCR (RT-PCR) instrument allowing testing 96 different conditions simultaneously, without the need for a specialized instrument. Typical assays are fast, inexpensive and relatively precise, providing transition temperatures with reported precision of ±0.05 °C [10–12]. Another enormous advantage is that it monitors thermal protein unfolding in the presence of a fluorescent dye, being therefore independent of protein intrinsic fluorescence. Also it can be applied to a wide range of proteins since small amounts and low concentration are required [10–13]. As potential caveats, one should point out that care must be taken in reporting melting temperatures from a given protein exclusively from DSF. Frequent technical issues in published DSF profiles relate to misinterpretation of events involving unfolding of oligomeric proteins (for which the dependence of the T_m with protein concentration is frequently neglected) or running thermal melts at inadequate melting rates (frequently higher than the ideal rate of 1 °C.min^{-1}, thus resulting in nonequilibrium measurements). One other frequently observed flaw results from the neglecting the fact that when protein precipitation occurs with thermal denaturation the system is no longer in thermodynamic equilibrium and what is actually being measured is an *apparent* T_m which should be explicitly referred to as such. The formalism to correct execution and interpretation of protein thermal melts is well established for the protein folding expert but this is not always the case of many DSF users which should refer to specialized literature to avoid the specific problems referred above (e.g., [14, 15]). Ideally, the thermal stability profile of a protein determined by DSF needs always to be compared to that obtained by an independent method such as differential scanning calorimetry (DSC) or circular dichroism (CD).

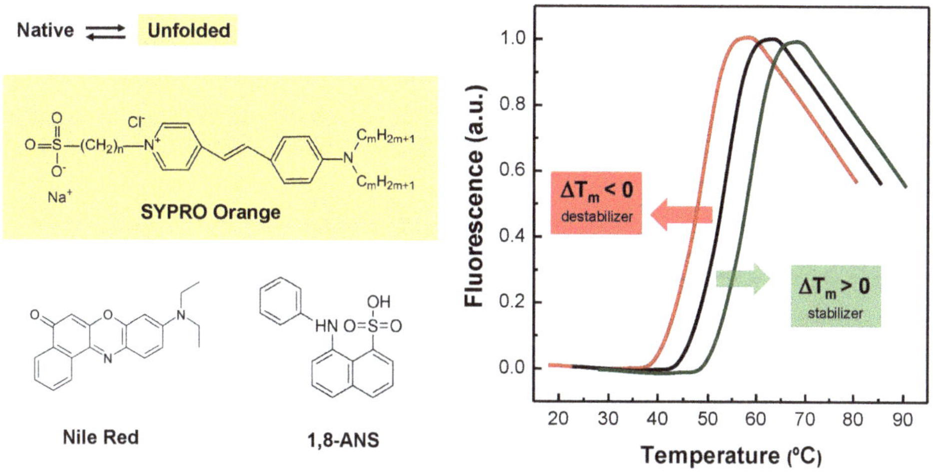

Fig. 1 Differential scanning fluorimetry dyes and thermal shift assay. SYPRO Orange, Nile Red and 1,8-ANS (left) are commonly used dyes in DSF assays whose fluorescence increases significantly upon binding to hydrophobic protein moieties exposed in a protein during thermal denaturation (right). In DSF assays a post-transition decrease in fluorescence is frequently observed due to quenching effects and changes in quantum yield with increasing temperature along with protein aggregation/precipitation and probe dissociation. The represented curves show stabilizing (green) and destabilizing (red) conditions in respect to the control (black), which result respectively in positive and negative thermal shifts (ΔT_m)

Several different fluorescent dyes have been used to monitor the unfolding process such as Nile Red, SYPRO Orange, and 8-anilino-1-naphthalenesulfonic acid (1,8-ANS) [10, 11]. The fluorescence of these dyes is quenched in aqueous solutions, but they become highly fluorescent in nonpolar environments, such as the hydrophobic regions of unfolded proteins. The difference in fluorescence emission permits to follow protein unfolding, as an increase in fluorescence emission of the dye will occur upon interaction with the exposed hydrophobic protein moieties (Fig. 1). SYPRO Orange has proven to have the most favorable properties for DSF experiments due to its high signal-to-noise ratio and the lower probability of interference of small molecule with its optical properties, which would cause quenching of the fluorescence intensity since its excitation maximum is near 500 nm [10, 11].

DSF is best used to compare the stability of a particular protein in the presence of different compounds (Fig. 1). These include generic additives such as salts, amino acids, sugars and buffers, or compounds such as inhibitors, substrates, or cofactors [16, 17]. These strategies are useful in protein stabilization strategies allowing the identification of additives that could improve protein expression and purification, thereby reducing protein propensity to unfold, aggregate, or be susceptible to proteolysis [2, 3, 10]. DSF has also been very useful to evaluate and optimize stabilizers for protein crystallization, contributing significantly to the ability to generate crystal structures of proteins [3].

Here we present a step-by-step protocol using DSF for screening small molecules that can be used as therapeutic drugs in misfolding enzyme-related disorders, using β-oxidation flavoenzyme dehydrogenases as models. Mitochondrial β-oxidation disorders are associated with misfolding of different enzymes of the metabolic pathway. In particular, short-chain acyl-CoA deficiency (SCADD) and medium-chain acyl-CoA deficiency (MCADD) are associated with impaired fatty acid metabolism, and are caused by mutations on the *SCAD* and *MCAD* genes, respectively [18]. Glutaric aciduria type I (GA-I), caused by mutation on *GCDH* gene, is a disease of the lysine, hydroxylysine and tryptophan metabolism [18]. Mild cases of these disorders are usually associated to the presence of misfolded proteins that have an impaired stability and thus become more prone to degradation, resulting in loss of activity [17, 19]. In many instances, the stabilization of even a small fraction of the misfolded variant will afford catalytic recovery that may be sufficient to overcome disease threshold. In recent years a substantial effort has been placed in the identification of the potential stabilization of these affected enzymes by their flavin cofactors and by substrates. While flavins can be directly enriched in mitochondria through oral intake of riboflavin [20], substrates and their derivatives tend to accumulate as unprocessed metabolites and are therefore locally available at increased levels [20]. Therefore, these approaches constitute promising starting points toward the development of tailored pharmacological chaperones to ameliorate these mitochondrial pathologies.

2 Materials

2.1 DSF for Screening of Small Molecules as Protein Stabilizers

1. DSF can be typically performed in a conventional RT-PCR instrument, e.g., an iCycler iQ RT-PCR Detection System (Bio-Rad).
2. White 96-well thin-wall PCR plates are used for the iCycler iQ RT-PCR Detection System (*see* **Note 1**).
3. Optical quality sealing tape is used to seal the plates to prevent evaporation and cross-contamination. A smooth rubber roller is used for sealing.
4. Protein buffer: 10 mM hepes, pH 7.8.
5. All compounds, such as sugars, amino acids, ligands, and cofactors, were of the highest purity grade commercially available. Prepare the different compound solutions to be tested. Prepare 2× concentrated compound stock solutions for compounds dissolved in protein buffer (*see* **Note 2**), or stock solutions at higher concentrations for compounds dissolved in other solutions, such as DMSO. In this case it is recommended to apply lower volumes to avoid possible interferences, (*see* **Note 3**).

6. Prepare SYPRO Orange fluorescence dye (Invitrogen) solution (*see* **Note 4**). SYPRO Orange is delivered as 5000× solution in 100% (vol/vol) DMSO, diluted to 20× solution in protein buffer (*see* **Note 5**).
7. Protein samples to be analyzed, e.g., recombinant human short-chain acyl-CoA dehydrogenase wild type and clinical variant (SCAD-WT and SCAD-p.Gly185Ser), medium-chain acyl-CoA dehydrogenase (MCAD) and glutaryl-CoA dehydrogenase (GCDH), expressed in *Escherichia coli* and purified as described [17, 21].
8. Prepare protein–dye solution, 2.4 μM protein solution (SCAD, MCAD, or GCDH), in the presence of 10× the dye, in protein buffer (*see* **Note 6**).
9. Worksheet "DSF Analysis v3.0.2.xlsx", available from ftp://ftp.sgc.ox.ac.uk/pub/biophysics

3 Methods

3.1 DSF for Screening of Small Molecules as Protein Stabilizers

1. High-throughput screening of protein stabilizers is performed in an iCycler iQ RT-PCR instrument by monitoring protein thermal unfolding in the presence of an extrinsic fluorescent probe, SYPRO Orange.
2. Distribute 10 μL per well of the different compound solutions, and different compound concentrations, into the 96-well microplate with multichannel pipettes or using other robotic instrumentation. For control samples add 10 μL of the solution in which the compounds have been dissolved.
3. Add 10 μL of the protein–dye solution to each well plate with multichannel pipette. Final plate concentrations are 1.2 μM protein and 5× dye.
4. Seal the microplates with optical quality sealing tape using the rubber roller. This is a very important step to prevent evaporation and cross-contamination, ensure that the microplate is well sealed. The microplate should be protected from light to reduce photobleaching and immediately loaded into the iCycler iQ RT-PCR instrument.
5. Set instrument excitation filter from 530 to 560 nm and the emission filter from 575 to 595 nm. (SYPRO Orange excitation and emission wavelengths are, respectively, 490 and 575 nm.) Monitor thermal denaturation following the increase of SYPRO Orange fluorescence intensity associated with protein unfolding process.
6. Run the assay over a temperature range from 20 to 90 °C, at a linear rate of 1 °C.min^{-1}.

7. Fluorescence intensity is plotted as a function of temperature, resulting in a sigmoidal curve that can be described by a two-state transition. Thus, the inflection point of the transition curve, which is assumed to be equal to the melting temperature (T_m) under equilibrium conditions, could be calculated for each unfolding curve obtained for the different condition tested. The simplest way to calculate T_m values is to determine the maximum of the first derivative, which is a feature offered by most PCR software packages.
8. Data visualization, and comparison of the effects of the different compounds under various conditions, could be easily performed through custom calculation software based on Microsoft Excel. Open the "DSF Analysis v3.0.2.xlsx" file and import the first derivative results from the RT-PCR software into the "paste in transformed data" sheet (*see* **Note 7**). Further fitting analysis is also possible (*see* **Note 8**).
9. Compound concentration dependence is determined for the different compounds tested. Figure 2 depicts a representative assay of the thermal shift variations obtained, for the three proteins, in the presence of different FAD: protein ratios, showing

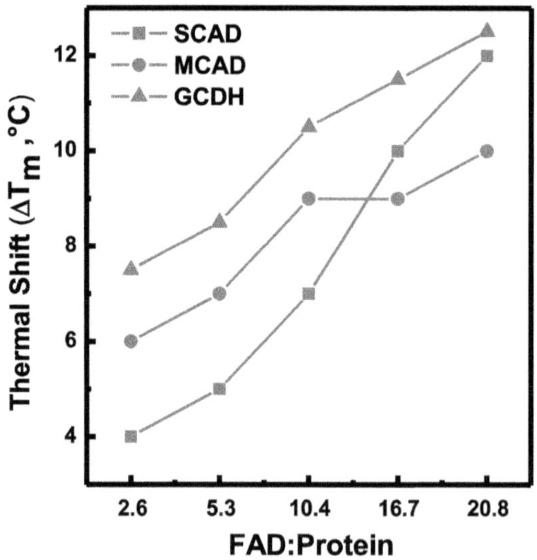

Fig. 2 Protein thermal shift at increasing cofactor concentrations determined by DSF. The midpoints of thermal unfolding curves were determined for SCAD (T_m = 50 °C), MCAD (T_m = 54 °C) and GCDH (T_m = 54 °C) at 1.2 μM in 10 mM Hepes, pH 7.8 and at increasing molar ratios of FAD: protein. The melting temperatures determined in the presence of the FAD cofactor (T_m^{FAD}) resulted in positive thermal shifts ($\Delta T_m = T_m^{FAD} - T_m$) which are represented in the graph. The order of stabilizing effect of FAD at the highest ratio is GCDH > SCAD > MCAD. Adapted from [17] with modifications

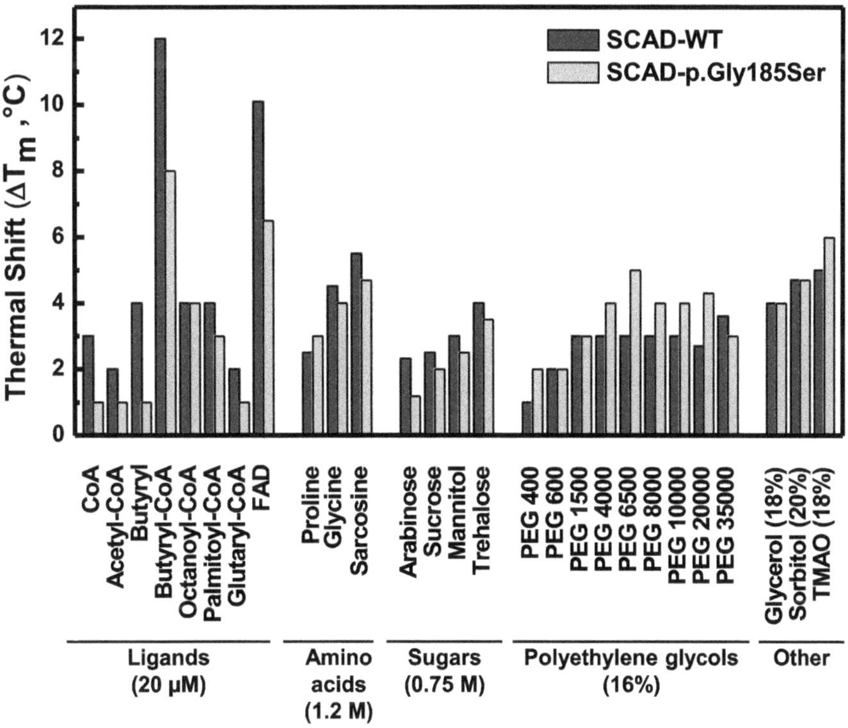

Fig. 3 DSF screening of stabilizers for disease-related misfolded proteins. The effect on protein stability of different classes of compounds known to have stabilizing effects (ligands, amino acids, sugars, and polyethylene glycols) was evaluated for SCAD-WT (T_m = 50 °C) and for the SCAD-p.Gly185Ser clinical variant (T_m = 49 °C). The midpoints of thermal unfolding were determined from at least three independent DSF curves, in absence and presence of compounds, and thermal shifts determined as described. All experiments were performed in 10 mM Hepes, pH 7.8 with 1.2 μM protein and 5× dye. Adapted from [17] with modifications

that FAD stabilizing effect is concentration dependent. A good practice is to always perform a compound concentration dependence test.

10. After determining the melting temperature of the protein in the absence (control, T_0) and in the presence of a specific compound concentration (T_m), the effects on protein stability are evaluated calculating ΔT_m ($T_m - T_0$) for each well. Figure 3 shows that SCAD-WT protein is stabilized by different compounds, such as ligands analogous to the enzyme substrate, glycine, sarcosine, glycerol, sorbitol and TMAO. A higher stabilization was achieved in the presence of butyryl-CoA and FAD, the two natural ligands of the enzyme. These two compounds also have a high effect in the clinical mutant. Polyethylene glycols, known to be general protein stabilizers, have a higher stabilizing effect on the clinical variant probably because the enzyme is conformationally destabilized in relation to wild type.

4 Notes

1. It is important to use white PCR plates due to its consistent optical properties and decreased refraction, thus enhancing the signal.
2. Maintenance of the pH can be achieved using a buffer at 10–100 mM concentration. For compound screening, lower ionic strengths are preferred to prevent any interference in the binding of the ligand to the protein.
3. Compounds prepared in 100% (vol/vol) DMSO must be diluted in final screening conditions to avoid a DMSO final concentration higher than 2% (vol/vol).
4. SYPRO Orange is one of the most common dyes used in DSF high-throughput screenings for proteins stabilizers [10, 22]. Although it is one of the best choices, not all proteins can be analyzed using this dye, and other dyes should be tested for cases where no unfolding transition is observed. One example is 1-anilino-8-naphthalene sulfate (1,8-ANS) dye, whose excitation maximum occurs at ~350 nm. Use this probe with caution once many compounds interfere with dye spectral properties and cause, for example, quenching of the fluorescence intensity.
5. SYPRO Orange solution should be diluted, typically 10× or 20×, in protein buffer before adding the protein (to prevent protein damage due to the high concentrations of DMSO). Final plate concentration should be determined for each protein, although a 5× dilution is usually appropriated.
6. Protein concentration on the assay is protein dependent, but 0.05–0.1 mg.mL^{-1} are normally sufficient to accurately measure T_m values.
7. Raw data could be used directly on the "DSF Analysis v3.0.2.xlsx" file since it was determined that fluorescent background correction has little effect on the fitted values of T_m [23]. To help filling the "DSF Analysis v3.0.2.xlsx" see example file provided at ftp://ftp.sgc.ox.ac.uk/pub/biophysics.
8. Further analysis of the most promising compounds could be performed individually using a two-state thermal unfolding fitting equation, by using a simple equation such as the Boltzmann equation

$$y = \min + \frac{(\max - \min)}{1 + \exp\left(\frac{T_m - x}{m}\right)}$$

where min and max are, respectively, the values of minimum and maximum intensities, *m* denotes the slope of the curve which reflects the transition cooperativity, and T_m is the transition midpoint. These fittings can be generated for example using Origin software (http://www.originlab.com).

Acknowledgments

This work was supported by the Fundação para a Ciência e Tecnologia (FCT/MCTES, Portugal) through fellowships to T.G.L. (SFRH/BD/89690/2012) and B.J.H. (SFRH/BPD/74475/2010) and a grant PTDC/BBB-BQB/5366/2014 (to B.J.H.) and PTDC/NEU-NMC/2138/2014 (to C.M.G.). The Gomes laboratory is partly supported by grant UID/MULTI/04046/2013 from FCT/MCTES/PIDDAC (to BioISI).

References

1. Boivin S, Kozak S, Meijers R (2013) Optimization of protein purification and characterization using thermofluor screens. Protein Expr Purif 91(2):192–206. https://doi.org/10.1016/j.pep.2013.08.002
2. Rodrigues JV, Prosinecki V, Marrucho I, Rebelo LP, Gomes CM (2011) Protein stability in an ionic liquid milieu: on the use of differential scanning fluorimetry. Phys Chem Chem Phys 13(30):13614–13616. https://doi.org/10.1039/c1cp21187k
3. Vedadi M, Niesen FH, Allali-Hassani A, Fedorov OY, Finerty PJ Jr, Wasney GA, Yeung R, Arrowsmith C, Ball LJ, Berglund H, Hui R, Marsden BD, Nordlund P, Sundstrom M, Weigelt J, Edwards AM (2006) Chemical screening methods to identify ligands that promote protein stability, protein crystallization, and structure determination. Proc Natl Acad Sci U S A 103(43):15835–15840. https://doi.org/10.1073/pnas.0605224103
4. Ulloa-Aguirre A, Janovick JA, Brothers SP, Conn PM (2004) Pharmacologic rescue of conformationally-defective proteins: implications for the treatment of human disease. Traffic 5(11):821–837.https://doi.org/10.1111/j.1600-0854.2004.00232.x
5. Loo TW, Clarke DM (2007) Chemical and pharmacological chaperones as new therapeutic agents. Expert Rev Mol Med 9(16):1–18. https://doi.org/10.1017/S1462399407000361
6. Tropak MB, Blanchard JE, Withers SG, Brown ED, Mahuran D (2007) High-throughput screening for human lysosomal beta-N-Acetyl hexosaminidase inhibitors acting as pharmacological chaperones. Chem Biol 14(2):153–164. https://doi.org/10.1016/j.chembiol.2006.12.006
7. Farinha CM, King-Underwood J, Sousa M, Correia AR, Henriques BJ, Roxo-Rosa M, Da Paula AC, Williams J, Hirst S, Gomes CM, Amaral MD (2013) Revertants, low temperature, and correctors reveal the mechanism of F508del-CFTR rescue by VX-809 and suggest multiple agents for full correction. Chem Biol 20(7):943–955. https://doi.org/10.1016/j.chembiol.2013.06.004
8. Leandro P, Gomes CM (2008) Protein misfolding in conformational disorders: rescue of folding defects and chemical chaperoning. Mini Rev Med Chem 8(9):901–911
9. Gomes CM (2012) Protein misfolding in disease and small molecule therapies. Curr Top Med Chem 12(22):2460–2469. https://doi.org/10.2174/1568026611212220002
10. Niesen FH, Berglund H, Vedadi M (2007) The use of differential scanning fluorimetry to detect ligand interactions that promote protein stability. Nat Protoc 2(9):2212–2221. https://doi.org/10.1038/nprot.2007.321
11. Senisterra GA, Finerty PJ Jr (2009) High throughput methods of assessing protein stability and aggregation. Mol BioSyst 5(3):217–223. https://doi.org/10.1039/b814377c
12. Johnson RJ, Savas CJ, Kartje Z, Hoops GC (2014) Rapid and adaptable measurement of pro-

tein thermal stability by differential scanning fluorimetry: updating a common biochemical laboratory experiment. J Chem Educ 91(7):1077–1080. https://doi.org/10.1021/ed400783e

13. Pantoliano MW, Petrella EC, Kwasnoski JD, Lobanov VS, Myslik J, Graf E, Carver T, Asel E, Springer BA, Lane P, Salemme FR (2001) High-density miniaturized thermal shift assays as a general strategy for drug discovery. J Biomol Screen 6(6):429–440. https://doi.org/10.1177/108705710100600609
14. Ibarra-Molero B, Naganathan AN, Sanchez-Ruiz JM, Muñoz V (2016) Chapter twelve—modern analysis of protein folding by differential scanning calorimetry. In: Feig AL (ed) Methods in enzymology, vol 567. Academic Press, pp 281–318. https://doi.org/10.1016/bs.mie.2015.08.027
15. Jaenicke R, Lilie H (2000) Folding and association of oligomeric and multimeric proteins. In: Advances in protein chemistry, vol 53. Academic Press, pp 329–401. https://doi.org/10.1016/S0065-3233(00)53007-1
16. Ericsson UB, Hallberg BM, Detitta GT, Dekker N, Nordlund P (2006) Thermofluor-based high-throughput stability optimization of proteins for structural studies. Anal Biochem 357(2):289–298. https://doi.org/10.1016/j.ab.2006.07.027
17. Lucas TG, Henriques BJ, Rodrigues JV, Bross P, Gregersen N, Gomes CM (2011) Cofactors and metabolites as potential stabilizers of mitochondrial acyl-CoA dehydrogenases. Biochim Biophys Acta 1812(12):1658–1663. https://doi.org/10.1016/j.bbadis.2011.09.009
18. Henriques BJ, Lucas TG, Gomes CM (2016) Therapeutic approaches using riboflavin in mitochondrial energy metabolism disorders. Curr Drug Targets 17(13):1527–1534
19. Henriques BJ, Olsen RK, Bross P, Gomes CM (2010) Emerging roles for riboflavin in functional rescue of mitochondrial beta-oxidation flavoenzymes. Curr Med Chem 17(32):3842–3854
20. Rodrigues JV, Henriques BJ, Lucas TG, Gomes CM (2012) Cofactors and metabolites as protein folding helpers in metabolic diseases. Curr Top Med Chem 12(22):2546–2559
21. Nguyen TV, Riggs C, Babovic-Vuksanovic D, Kim YS, Carpenter JF, Burghardt TP, Gregersen N, Vockley J (2002) Purification and characterization of two polymorphic variants of short chain acyl-CoA dehydrogenase reveal reduction of catalytic activity and stability of the Gly185Ser enzyme. Biochemistry 41(37):11126–11133
22. Simeonov A (2013) Recent developments in the use of differential scanning fluorometry in protein and small molecule discovery and characterization. Expert Opin Drug Discovery 8(9):1071–1082. https://doi.org/10.1517/17460441.2013.806479
23. Stevens RC, Sancho J, Martinez A (2010) Rescue of misfolded proteins and stabilization by small molecules. Methods Mol Biol 648:313–324. https://doi.org/10.1007/978-1-60761-756-3_22

Chapter 17

Methods to Screen Compounds Against Mutant p53 Misfolding and Aggregation for Cancer Therapeutics

Giulia Diniz da Silva Ferretti, Danielly C. Ferraz da Costa, Jerson L. Silva, and Luciana Pereira Rangel

Abstract

p53 is a critical tumor suppressor that functions as a transcription factor. Mutations in the *TP53* gene are observed in more than 50% of cancer cases worldwide. Several of these mutations lead to a less stable, aggregation-prone protein that accumulates in cancer cells. These mutations are associated with a gain of oncogenic function, which leads to cancer progression. p53 amyloid aggregation is a common feature in most of these mutants; thus, it can be used as a druggable target to reactivate or induce the degradation of p53 and promote a retraction in the aggressive pattern of mutant p53-containing cells. We show here a series of experiments for the screening and validation of new p53 antiamyloid compounds.

Key words p53, Aggregation, Amyloid, Anticancer therapy

1 Introduction

The *TP53* gene is widely considered a "guardian of the genome" due to its protective properties against different types of cell damage [1]. Under normal conditions, it has the ability to suppress tumor progression by different mechanisms, particularly by inducing apoptosis [2]. This gene encodes the protein p53, which is composed of four chains of 393 amino acids that form a stable and active tetramer, which develops its functions by inducing the transcription of target genes and interacting with other proteins or transcription factors [3]. Mutations in *TP53* have been detected in more than 50% of cancer cases. They induce gain-of-function phenotypes that are associated with cancer progression, including metastasis, drug resistance, and others [4]. Additionally, when altered by missense mutations in its gene, p53 acquires a less stable conformation with a greater tendency to form intracellular amyloid aggregates with prion-like properties [5–10].

Cláudio M. Gomes (ed.), *Protein Misfolding Diseases: Methods and Protocols*, Methods in Molecular Biology, vol. 1873,
https://doi.org/10.1007/978-1-4939-8820-4_17,

It has also been shown that p53 may accumulate and form amyloid aggregates when incubated with cholesterol secosterol aldehydes, which are lipids related to inflammatory processes [11]. In addition, studies have demonstrated the coaggregation of p53 and its paralogues p63 and p73 [12, 13] and that p53 can interact with the HSP70 chaperone and its regulatory protein MDM2 to form structures with amyloid *characteristics* described as a stabilized form of the p53 mutant [14].

Given the recent description of this new pharmacological target, new efforts are necessary to develop drugs that can clear mutant aggregates or stabilize the structure of p53 to regain its native p53 functions [9]. The clearing of aggregates could help disable gain-of-function effects, leading to a more sensitive status of the cell to chemotherapy; the restoration of a wild-type conformation, however, could lead to programmed cell death [15]. Thus, although p53 may be an interesting target for cancer due to its high mutation rate, the behavioral complexity of p53 mutants makes understanding their aggregation, which is a common feature among many of these mutants, necessary for the development of new drugs.

Here, we detail methodological approaches to evaluate the presence of aggregates in tumor cells and tissues and explore the effects of new molecules targeting mutant p53 aggregation.

1.1 Experimental Approach

In this chapter, we present a few techniques that are applicable for the discovery of p53 antiamyloid compounds. These experiments have been established with a series of compounds tested in our lab and include both in vitro protein aggregation assays and intracellular p53 aggregation techniques. We show a key technique for the screening of compounds (purified *E. coli* p53 DNA-binding domain temperature aggregation), two approaches for confirming the antiamyloid effect in cells and two simple assays for validating the positive effect presented by these compounds on tumor cells. In Fig. 1, we show an example of what we expect to observe in each experiment described here, using a fictitious compound named "Candidate", which is evaluated as a good example of a p53 aggregation inhibitor when subjected to all tests.

2 Materials

2.1 p53 DNA-Binding Domain (p53C) Purification

Prepare all solutions using ultrapure water and analytical grade reagents. All solutions must be freshly prepared and filtered through a 0.22-μm membrane. All p53 purification steps must be performed at 4 °C.

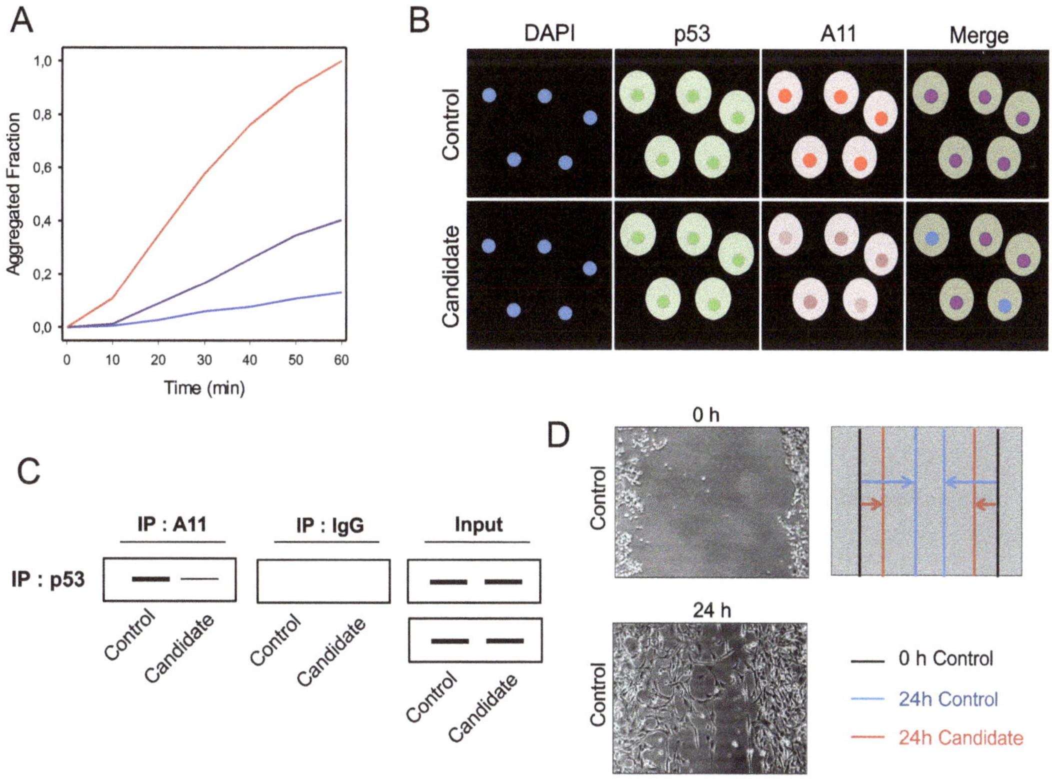

Fig. 1 Representative schemes of expected results obtained in the experiments described in this chapter. "Candidate" was used as an example of one compound to be evaluated. (**A**) p53C aggregation kinetics. Red line: p53C control; purple line: p53C in the presence of candidate less concentrated; Blue line: p53C in the presence of candidate more concentrated showing less p53C aggregation compared with the other conditions. (**B**) Immunofluorescence assay. Cell line expressing mutant p53 treated with the candidate demonstrating lower fluorescence intensity for anti-amyloid oligomer staining compared with control. First column: nuclear marker DAPI; second column: anti-p53 staining; third column: anti-amyloid oligomers antibody and the last one the merged fields. (**C**) Immunoprecipitation assay. Cell line expressing mutant p53 was treated with candidate overnight, immunoprecipitated with A11 antibody and then immunoblotted for p53 showing lower p53 levels after treatment in the amyloid fraction of the cell extract. Inputs are the total extracts used to prepare immunoprecipitation samples. (**D**) Cell migration assay, using candidate. Cells expressing mutant p53 had their migration partially inhibited by the candidate compound, compared to control. Taken together, the results shown indicate that this candidate could be a good compound for p53 aggregation inhibition

2.1.1 BL21DE3 Escherichia coli Cell Transformation and Protein Expression

1. PET11a plasmid containing the R248Q mutant p53 DNA-binding domain sequence (p53CR248Q), obtained as described by Ishimaru et al. [5, 16].
2. Liquid Luria–Bertani (LB) medium.
3. 100 mg/mL ampicillin solution in sterile water.
4. 1 M IPTG (isopropyl β-D-1-thiogalactopyranoside) solution in sterile water.

2.1.2 Cell Lysis and Protein Purification

1. Protease inhibitor cocktail.
2. SP-Sepharose Fast Flow chromatography resin, packed in a 100-mL column (GE Healthcare Life Sciences, Buckinghamshire, UK).
3. Superdex 75 10/300 GL gel filtration column (GE Healthcare Life Sciences, Buckinghamshire, UK).
4. Buffer A: 50 mM Tris, 50 mM NaCl, and 2.5 mM DTT, pH 7.2.
5. Buffer B: 50 mM Tris, 400 mM NaCl, and 2.5 mM DTT, pH 7.2.
6. Buffer C: 50 mM Tris, 150 mM NaCl, 5 mM DTT, and 5% (v/v) glycerol, pH 7.2.

2.2 p53 Aggregation Kinetics

1. Buffer C.
2. 10 mM thioflavin T solution in water
3. Purified p53C solution.
4. Compound stock solutions at 100 mM in DMSO or water.

2.3 Coimmunoprecipitation Followed by Western Blotting

1. Phosphate-buffered saline (PBS): 137 mM NaCl, 2.7 mM KCl, 10 mM Na_2HPO_4, 1.8 mM KH_2PO_4, pH 7.4.
2. Lysis buffer: 10 mM Tris–HCl, 150 mM NaCl, 1 mM EDTA, 1% Triton, and 1× protease Cocktail Inhibitor, pH 7.4.
3. Antibodies mouse IgG, anti-amyloid oligomers (anti-rabbit, A11), and anti-p53 (anti-mouse, DO-1).
4. Protein A/G-Agarose beads.
5. Electrophoresis sample buffer (1×): Tris–HCl 50 mM, SDS 2%, bromophenol blue 0.1%, and 10% glycerol.
6. Low fluorescence PVDF membrane.
7. Filter paper.
8. Transfer buffer: 25 mM Tris–HCl, 190 mM glycine, and 20% methanol.
9. Transfer system.
10. Blocking buffer (Odyssey® Blocking Buffer–TBS, LI-COR Biotechnology, Lincoln, USA).
11. Tris buffered saline (TBS): Tris–HCl 20 mM, NaCl 385 mM.
12. TBS containing 0.1% Tween 20 (TBS-T).
13. Primary antibody anti-p53 (anti-mouse, DO-1).
14. Secondary antibody (anti-mouse) conjugated with IRDyes® (IRDye® 800CW) (LI-COR Biotechnology, Lincoln, USA).

2.4 Immunofluorescence Assay for the Detection of Amyloid Oligomers Colocalized with p53

1. 24-well plates for cell culture.
2. Microscope glass coverslips (13 mm diameter).
3. PBS.
4. Acetone–methanol (1:1) fixation solution at −20 °C.
5. 2% Bovine serum albumin (BSA), 0.2% Tween 20, 10% (v/v) glycerol, solubilized in PBS.
6. Filter paper.
7. Anti-p53 primary antibody (anti-mouse, DO-1).
8. Anti-oligomers primary antibody (anti-rabbit, A11).
9. Anti-mouse Alexa Fluor 488 and anti-rabbit Alexa Fluor 546 secondary antibodies (Molecular Probes, Life Technologies, Oregon, USA).
10. Mounting medium with DAPI.

2.5 Cell Proliferation Assay

1. Microtubes.
2. PBS.
3. 0.4% trypan blue dye solution in PBS.
4. Neubauer chamber and coverslip.
5. Cell counter.
6. Bright-field microscope.

2.6 Wound Healing Assay

1. 12- or 24-well plates for cell culture.
2. Sterile plastic p10 pipette tip.
3. PBS.
4. Fresh cell culture media.
5. Mitomycin C.
6. Bright-field inverted microscope.

3 Methods

3.1 p53 Purification

The protocol for the purification of the p53 DNA-binding domain (p53C) was first described by Bullock et al. [17]. Subsequently, we described this protocol with a few modifications in Kovachev et al. [18]. The protocol detailed here is suitable for the purification of both WT and mutant p53 DNA binding domains and full-length p53.

3.1.1 Competent BL21-DE3 E. coli Cells Transformation

1. Place cell samples on ice for 15 min (~50 μL of cells suspension in conic tubes).
2. Add 1 μg of plasmid solution to the bacterial samples (control samples receive the same treatment, without plasmid addition).

3. Heat cells to 42 °C for 2 min.
4. Place cells on ice for 5 min.
5. Plate cells in LB-agar medium containing 100 μg/mL ampicillin. One of the control samples should be plated on LB-agar medium for cell viability control.
6. Incubate plates at 37 °C overnight.
7. Assess transformation efficiency on the next day by comparing the controls with the transformed sample (*see* **Note 1**).

3.1.2 Transformed Bacterial Cell Growth for Protein Purification

1. Prepare a sterile preinoculum vial with ~25 mL of LB medium containing 100 μg/mL ampicillin. Add an isolated colony from the transformed plate and culture overnight at 37 °C with shaking.
2. The next morning, add ~8 mL of the bacterial solution to a 2 L flask containing 800 mL of fresh LB with 100 μg/mL ampicillin. Incubate at 37°C with shaking until an OD_{600} of 1.2 is reached.
3. Lower the incubator temperature to 15 °C and add IPTG to a 1 mM final concentration. Incubate overnight (*see* **Note 2**).

3.1.3 Cell Harvesting and Sonication

1. The next morning, the cells are harvested through centrifugation at 10,000 × *g* for 10 min.
2. Add 30 mL of lysis buffer with protease inhibitor cocktail to the pellet. Homogenize thoroughly.
3. Sonicate for 15 min (10 s ON/10 s OFF) with 70% amplitude (total of 30 min) on ice (*see* **Note 3**).
4. Centrifuge at 10,000 × *g* for 10 min.
5. Collect the supernatant for protein purification.

3.1.4 Protein Purification

1. Add the lysed bacterial supernatant to the resin using a peristaltic pump at low speed at 4 °C.
2. Wash the column with three volumes of buffer 1.
3. Apply a gradient containing 40 mL of buffer 1 vs. buffer 2 using a peristaltic pump at medium speed.
4. Collect samples of ~3 mL. Pool the fractions with an $OD_{280\text{ nm}}$ of >0.6.
5. Perform size-exclusion chromatography by using Superdex 75 on HPLC and buffer 3 with a 1 mL/min flow. Follow peaks at 280 nm absorbance (*see* **Note 4**).
6. Dose samples at 280 nm (p53C molar extinction coefficient = 17,130 M^{-1} cm^{-1}). Store the 200 μL aliquots at −80 °C.

3.2 p53 Aggregation Kinetics

In this experiment, the main objective is to follow p53C aggregation and compare it to samples containing the experimental compounds to identify p53 aggregation inhibitors. After screening, the best aggregation inhibitors should be selected for subsequent experiments (*see* **Note 5**).

1. Mix the appropriate amounts of p53C to a final concentration of 5 μM, ThT to a final concentration of 25 μM, and the experimental compounds at different concentrations for each sample. Add the solution to a fluorescence-appropriate 96-well microplate (*see* **Notes 6, 7** and **8**).
2. Incubate the reactions in a microplate reader with temperature control at 37 °C for 2 h. Follow the ThT fluorescence increase due to p53 amyloid aggregation at 440 nm (excitation) and 482 nm (emission).
3. Compare the resulting curve profiles after 2 h to evaluate the inhibition of p53C aggregation by the experimental compounds (*see* **Note 9**).

3.3 Coimmunoprecipitation Followed by Western Blotting

1. Incubate the cells from the mutant p53 cell line (e.g., MDA-MB231) in a 75-cm^2 flask until 80–90% confluence.
2. Wash the cells three times with 2 mL of PBS.
3. For adherent cells, add 200 μL of liquid nitrogen. When the cells begin to thaw, add cold lysis buffer (*see* **Note 10**).
4. Scrape and disrupt the cells by repeated aspiration. Subsequently, transfer the cells to a 1.5 mL microcentrifuge tube.
5. Pellet the cellular debris by centrifugation at 3000 × *g* for 5 min at 4 °C and transfer the supernatant to a new 1.5 mL microcentrifuge tube and discard the pellet.
6. Quantify the amount of total protein. We typically use the Lowry et al. [19] method. Transfer 500 μg of protein to a 1.5 mL microcentrifuge tube. Add 10 μL of protease cocktail inhibitor and 2 μg of anti-amyloid oligomers antibody (A11), and bring the volume to 1 mL with PBS. Next, incubate the mixture for 1 h at 4 °C under stirring. In this step, incubate the control with mouse IgG.
7. Add 20 μL of Protein A/G agarose beads to each tube and incubate overnight at 4 °C under stirring.
8. Centrifuge at 1000 × *g* for 5 min at 4 °C and discard the supernatant. Then, wash the beads four times with cold PBS. Repeat centrifugation.
9. After washing, add 40 μL of 1× electrophoresis sample buffer and boil the samples for 3 min.

10. Centrifuge at 1000 × *g* for 5 min at room temperature and collect the supernatant.
11. Run a 12.5% SDS-PAGE gel with 20 μL of each A11 coimmunoprecipitation sample.
12. For western blotting, immerse the filter paper in transfer buffer and activate the low-florescence PVDF membrane for 1 min in 100% methanol.
13. Mount a sandwich: place sandwich components in a transfer system in the following sequence: filter paper, the activated membrane, the gel, and another filter paper.
14. Perform protein transfer.
15. Remove the membrane and incubate in blocking buffer. Block the membrane for 1 h in a 4 °C chamber under mild agitation.
16. Dilute the anti-p53 primary antibody (1:1000) in blocking buffer.
17. Remove the blocking buffer from the case and add the diluted anti-p53 primary antibody. Incubate overnight at 4 °C under mild agitation.
18. Wash the membrane five times with TBS-T.
19. Dilute the secondary antibody (1:10,000) in blocking buffer.
20. Incubate the membrane for 2 h in room temperature under mild agitation and protected from light.
21. Wash the membrane three times with TBS-T and 2 times with TBS.
22. Perform immunodetection by using appropriate equipment with fluorescence detection.

3.4 Immunofluorescence Assay for the Detection of Amyloid Oligomers Colocalized with p53

1. Grow cells on glass coverslips until 60–70% confluence is reached.
2. Remove the cell culture media and immediately wash the cells twice with PBS (200 μL/well).
3. Incubate the cells with acetone–methanol (1:1) fixation solution (200 μL/well) for 20 min at −20 °C (*see* **Note 11**).
4. Incubate the cells with PBS (200 μL/well) for 30 min at room temperature.
5. Dilute the anti-p53 (1:200) and anti-amyloid oligomers (1:1000) primary antibodies in a solution containing 2% bovine serum albumin (BSA), 0.2% Tween 20, and 10% (v/v) glycerol in PBS.

6. Incubate the cells with primary antibodies (150 μL/well) for 2 h at room temperature.
7. Remove the antibodies, and gently wash the cells three times with PBS.
8. Dilute the anti-mouse Alexa Fluor 488 and anti-rabbit Alexa Fluor 546 (1:2000) secondary antibodies in the same solution used for diluting the primary antibodies.
9. Incubate the cells with secondary antibodies (150 μL/well) for 1 h at room temperature in the dark.
10. Wash the cells three times with PBS and remove the coverslips from the wells.
11. Plate the coverslips on a smooth surface and gently dry the edges using slices of filter paper.
12. Mount each coverslip with a drop of mounting medium with DAPI (*see* **Note 12**).
13. Store the slides in the dark at −20 °C.

3.5 Cell Proliferation Assay

1. Place 50–100 μL of the cell suspension in a microtube.
2. Add an equal volume of 0.4% trypan blue dye solution to the cell suspension (*see* **Note 13**).
3. Incubate the mixture for less than 3 min at room temperature (*see* **Note 14**).
4. With the coverslip in place, pipette 10–20 μL of this mix onto the Neubauer chamber.
5. Place the Neubauer chamber on the stage of a light microscope and focus on the cells.
6. Count all cells (clear and blue) in each large square of the Neubauer chamber using a cell counter (blue cells are the nonviable cells).
7. Calculate the percentage of viable cells, by dividing the number of viable cells by the number of total cells and multiplying by 100.

3.6 Wound Healing Assay

1. Grow cells on 12- or 24-well plates for cell culture until 60–70% confluence is reached.
2. Wash the cells twice with PBS (200–500 μL/well).
3. Scratch the cell monolayer with a sterile plastic p10 pipette tip.
4. Gently remove the peeled cells with two PBS washes, and add fresh cell culture media containing mitomycin C (0.5 μg/mL) and the compound of interest to the wells (*see* **Note 15**).

5. Incubate the cells for 24–48 h to allow wound coverage.
6. Using a bright field microscope, acquire images at zero and final time points.
7. Use ImageJ software to measure wound width in each condition.

4 Notes

1. If transformation is not successful, then add 200 μL of LB medium after the second ice step and incubate the cells at 37 °C for 2 h, with shaking, followed by plating.
2. You can add a 25 °C step between the 37 °C and 15 °C steps. Maintain this temperature for 2 h and then lower the temperature to 15 °C.
3. During cell sonication, resuspend the cells every 5 min.
4. During HPLC size-exclusion chromatography, p53C should be eluted approximately in the middle of the run using a Superdex 75 column.
5. Before this assay, analyze the absorption and emission profiles of the experimental compounds at the wavelengths used in the experiment. Quenching might be present and incur false-positive results.
6. Always assess the p53 structure and undesired aggregation using the tryptophan and tyrosine intrinsic fluorescence of this protein at 278 nm (excitation) and 300–415 nm (emission). Additionally, assess p53 previous aggregation using light scattering.
7. To prevent the seeding of the aggregation process from previous aggregates present in the p53C solution, centrifuge samples at 15,000 × *g* for 15 min before preparing the reaction mixture. Dose the protein content again.
8. The p53C solution must be the last thing added to the mixture, just before the initiation of the reaction.
9. If the experimental compounds fluoresce, absorb or quench fluorescence in the range used for ThT, then perform the aggregation kinetics assay using light scattering at 420 nm (excitation and emission).
10. Avoid using a lysis buffer containing sodium dodecyl sulphate (SDS). The p53 oligomers are susceptible to this detergent and may not bind the anti-amyloid oligomers antibody.

11. Incubate the 24-well plate inside the freezer.
12. Seal the coverslip with nail polish to prevent movement under the microscope.
13. Mix by pipetting up and down to avoid cell clumping.
14. If the cells are counted after more than 5 min, then the viability might be inaccurate due to an increase in cell death.
15. Mitomycin is used as an inhibitor of cell proliferation.

5 Concluding Remarks

Conformational changes in p53 caused by mutations can lead to an accumulation of p53 in the cell. During the past 15 years, studies have shown that p53 aggregation is likely related to both dominant-negative and gain-of-function effects in cancer [6, 7, 9, 20, 21]. The formation of p53 aggregates was observed in breast tumor biopsies [22] and was reported in the transactivation, central, and tetramerization domains [5, 12]. These aggregates can lead to an inactive conformation of the wild-type protein, making the cell susceptible to malignant transformation [7, 23]. Evaluation of the aggregation of mutant p53 proteins might have applications in the diagnosis and prognosis of different types of cancers as well as in the search for new therapeutic compounds. p53 aggregation has been detected in malignant tumors or cells lines of the breast [6, 22, 23], ovaries [24, 25], skin [26], prostate [27], and lungs [12, 23], among others. Thus, the aggregation of mutant p53 may affect the development of cancer through loss-of-function, dominant-negative, and gain-of-function effects. Our research group has also demonstrated the prion-like amyloid nature of p53 through several spectroscopic techniques using both recombinant proteins and cells expressing the R280K mutant [6]. This observation has been confirmed by different groups [23, 26, 28].

The prion-like behavior of mutant p53 could explain both the dominant-negative and gain-of-function effects [6, 7]. However, it is still necessary to confirm that this intracellular prionic conversion can be transmitted between cells (acting as a "real prion"). In any case, we aimed to identify compounds that inhibit mutant protein aggregation and prionic conversion as potential antitumor drugs. The main goal of this chapter is to show a roadmap for screening compounds that target p53 aggregation in different tissues (Fig. 2).

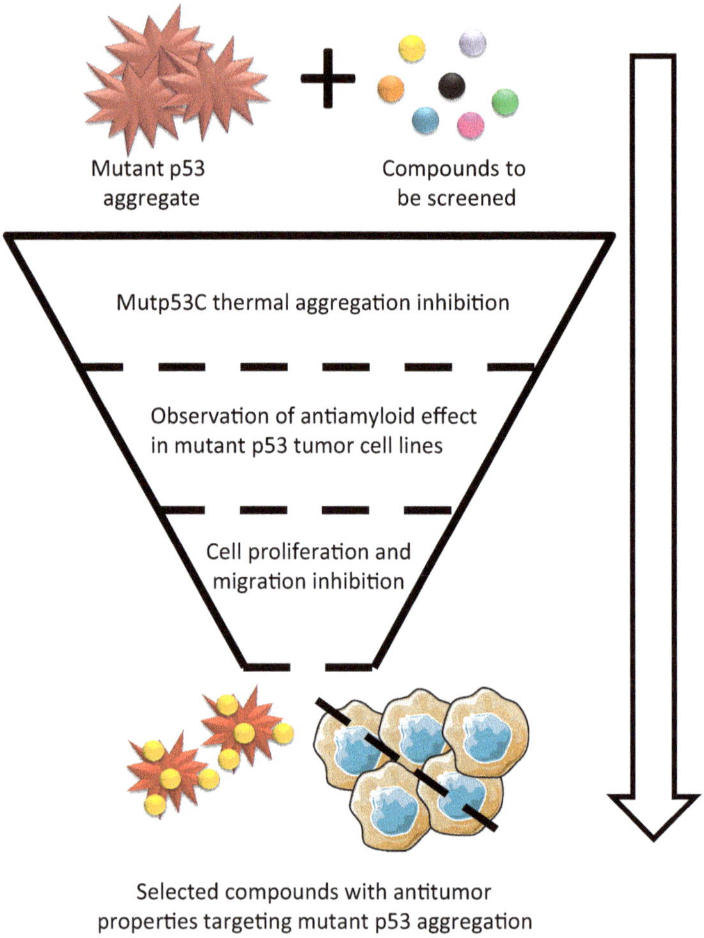

Fig. 2 Schematic route for the screening of p53 antiamyloid inhibitors

Acknowledgments

We thank the Brazilian funding agencies CAPES, CNPq, FAPERJ, Fundação do Câncer and Serrapilheira Institute for the support.

References

1. Lane DP (1992) Cancer. p53, guardian of the genome. Nature 358:15–16
2. Kastan MB, Onyekwere O, Sidransky D, Volgestain B, Craig RW (1991) Participation of p53 protein in the cellular response to DNA damage. Cancer Res 51:6304–6311
3. Vousden KH, Lane DP (2007) p53 in health and disease. Nat Rev Mol Cell Biol 8:275–283
4. Freed-Pastor WA, Prives C (2012) Mutant p53: one name, many proteins. Genes Dev 26: 1268–1286
5. Ishimaru D, Andrade LR, Teixeira LS, Quesado PA, Maiolino LM, Lopez PM et al (2003) Fibrillar aggregates of the tumor suppressor p53 core domain. Biochemistry 42:9022–9027
6. AnoBom AP, Rangel LP, Costa DC, De Oliveira GA, Sanches D, Braga C et al (2012)

Mutant p53 aggregates into prion-like amyloid oligomers and fibrils: implications for cancer. J Biol Chem 287:28152–28162
7. Silva JL, De Moura Gallo CV, Costa DC, Rangel LP (2014) Prion-like aggregation of mutant p53 in cancer. Trends Biochem Sci 39:260–267
8. Wang G, Fersht AR (2015) Mechanism of initiation of aggregation of p53 revealed by Φ-value analysis. Proc Natl Acad Sci U S A 112:2437–2442
9. Silva JL, Cino EA, Soares IN, Ferreira VF, De Oliveira GAP (2018) Targeting the prion-like aggregation of mutant p53 to combat cancer. Acc Chem Res 51:181–190
10. De Smet F, Saiz Rubio M, Hompes D, De Baets G, Langenberg T et al (2017) Nuclear inclusion bodies of mutant and wild-type p53 in cancer: a hallmark of p53 inactivation and proteostasis remodelling by p53 aggregation. J Pathol 242:24–38
11. Nieva J, Song BD, Rogel JK, Kujawara D, Altobel L 3rd, Izharrudin A et al (2011) Cholesterol secosterol aldehydes induce amyloidogenesis and dysfunction of wild-type tumor protein p53. Chem Biol 29:920–927
12. Xu J, Reumers J, Couceiro JR, De Smet F, Gallardo R, Rudyak S et al (2011) Gain of function of mutant p53 by coaggregation with multiple tumor suppressors. Nat Chem Biol 7:285–295
13. Cino EA, Soares IN, Pedrote MM, De Oliveira GA, Silva JL (2016) Aggregation tendencies in the p53 family are modulated by backbone hydrogen bonds. Sci Rep 6:32535
14. Milena W, Maciej BO, Zuzanna TG, Bartosz W, Maciej Z, Alicja Z (2012) Molecular mechanism of mutant p53 stabilization: the role of HSP70 and MDM2. PLoS One 7:e51426
15. Bykov VJ, Wiman KG (2014) Mutant p53 reactivation by small molecules makes its way to the clinic. FEBS Lett 588:2622–2627
16. Ishimaru D, AnoBom AP, Lima LM, Quesado PA, Oyama MF, De Moura Gallo CV et al (2009) Cognate DNA stabilizes the tumor suppressor p53 and prevents misfolding and aggregation. Biochemistry 48:6126–6135
17. Bullock AN, Henckel J, Dedecker BS, Johnson CM, Nikolova PV, Proctor MR et al (1997) Thermodynamic stability of wild-type and mutant p53 core domain. Proc Natl Acad Sci U S A 94:14338–14242
18. Kovachev PS, Banerjee D, Rangel LP, Eriksson J, Pedrote MM, Martins-Dinis MMDC et al (2017) Distinct modulatory role of RNA in the aggregation of the tumor suppressor protein p53 core domain. J Biol Chem 292:9345–9357
19. Lowry OH, Rosebrough NJ, Farr AL, Randall RJ (1951) Protein measurement with the Folin phenol reagent. J Biol Chem 193:265–275
20. Silva JL, Vieira TCRG, Gomes MPB, AnoBom AP, Lima LMTR, Freitas MS et al (2010) Ligand binding and hydration in protein misfolding: insights from studies of prion and p53 tumor suppressor proteins. Acc Chem Res 43:271–279
21. Silva JL, Rangel LP, Costa DC, Cordeiro Y, De Moura Gallo CV (2013) Expanding the prion concept to cancer biology: dominant-negative effect of aggregates of mutant p53 tumour suppressor. Biosci Rep 33(4):e00054
22. Levy CB, Stumbo AC, AnoBom AP, Portari EA, Carneiro Y, Silva JL et al (2011) Co-localization of mutant p53 and amyloid-like protein aggregates in breast tumors. Int J Biochem Cell Biol 43:60–64
23. Ghosh S, Salot S, Sengupta S, Navalkar A, Ghosh D, Jacob R et al (2017) p53 amyloid formation leading to its loss of function: implications in cancer pathogenesis. Cell Death Differ 24:1784–1798
24. Soragni A, Janzen DM, Johnson LM, Lindgren AG, Thai-Quynh Nguyen A, Tiourin E et al (2016) Designed inhibitor of p53 aggregation rescues p53 tumor suppression in ovarian carcinomas. Cancer Cell 29:90–103
25. Yang-Hartwich Y, Soteras MG, Lin ZP, Holmberg J, Sumi N, Craveiro V et al (2015) p53 protein aggregation promotes platinum resistance in ovarian cancer. Oncogene 34:3605–3616
26. Lasagna-Reeves CA, Clos AL, Castillo-Carranza D, Sengupta U, Guerrero-Muñoz M, Kelly B et al (2013) Dual role of p53 amyloid formation in cancer; loss of function and gain of toxicity. Biochem Biophys Res Commun 430:963–968
27. Kluth M, Harasimowicz S, Burkhardt L, Grupp K, Krohn A, Prien K et al (2014) Clinical significance of different types of p53 gene alteration in surgically treated prostate cancer. Int J Cancer 135:1369–1380
28. Forget KJ, Tremblay G, Roucou X (2013) p53 aggregates penetrate cells and induce the co-aggregation of intracellular p53. PLoS One 8:e69242
29. Costa DC, De Oliveira GA, Cino EA, Soares IN, Rangel LP, Silva JL (2016) Aggregation and prion-like properties of misfolded tumor suppressors: is cancer a prion disease? Cold Spring Harb Perspect Biol 8. https://doi.org/10.1101/cshperspect.a023614

Chapter 18

Early Stage Discovery and Validation of Pharmacological Chaperones for the Correction of Protein Misfolding Diseases

Oscar Aubi, Per M. Knappskog, and Aurora Martinez

Abstract

Pharmacological chaperones are small molecular weight molecules that bind specifically to protein targets and stabilize unstable and misfolded conformations. In particular, there is an increasing interest in the application of this type of compounds for the correction of genetic conformational disorders, which are caused by mutations leading to protein instability. The discovery of compounds with pharmacological chaperone ability is customarily initiated by a high-throughput screening of chemical libraries searching for stabilizing binders. However, there is no established consensus for the subsequent steps. Therefore, here, we introduce an example of a successful protocol directed to the discovery of pharmacological chaperones with potential for the therapeutic correction of phenylketonuria, a defect caused by mutations in the enzyme phenylalanine hydroxylase.

Key words Drug discovery and development, Orthogonal binding assays, Functional assays

1 Introduction

The folding of a protein in the interior of a cell is a complex process that requires the action of quality control systems in order to preserve cellular proteostasis—i.e., balance between correct protein folding, protein degradation, and, in some cases, aggregation [1, 2]. The key components of this cellular protein quality control comprise molecular chaperones, co-chaperones and associated regulatory proteins, as well as components of the ubiquitin-dependent proteasome and selective autophagy systems [1, 3]. Despite the efficacy of the proteostasis machinery, diverse factors—notably genetic mutations—can lead to instability and misfolding of a target protein, resulting in conformational diseases caused by increased degradation (loss-of-function) and/or aggregation (gain-of-function) of aberrant proteins [4–6].

Phenylketonuria (PKU) is an inborn error of metabolism originated from mutations in phenylalanine hydroxylase (PAH)

Cláudio M. Gomes (ed.), *Protein Misfolding Diseases: Methods and Protocols*, Methods in Molecular Biology, vol. 1873, https://doi.org/10.1007/978-1-4939-8820-4_18, © Springer Science+Business Media, LLC, part of Springer Nature 2019

that prompt the accumulation of the aromatic amino acid L-phenylalanine (L-Phe) in blood and, ultimately, in brain [7]. PAH is a tetrameric, highly regulated enzyme that catalyzes the catabolic degradation of L-Phe using (6*R*)-5,6,7,8-tetrahydrobiopterin (BH_4) as cofactor and O_2 as additional substrate [8]. Remarkably, PKU is a prototypic genetic conformational disease, where the main pathogenic mechanism and determinant of the phenotypic severity is the degree of instability caused by the specific mutations [9].

Novel therapies for this type of disorders include proteostasis regulators and pharmacological chaperones (PCs) [4]. PCs are small molecular weight compounds that bind specifically to unstable proteins, recovering partially or totally the structure, intracellular half-life, and function of the normal wild-type protein [10, 11]. PCs are customarily discovered through derivatization of natural ligands of a target protein or by high-throughput screening (HTS) campaigns [11–13].

The classical modality of HTS within drug discovery has been the application of empirical phenotypic screenings where, using cells and whole organisms, the apparent desired biological responses are monitored. Interestingly, in the recent years, there has been a revival of renewed phenotypic assays by the use of new technological advances, such as quantitative proteomics and metabolomic profiling [14], as they seem advantageous for the discovery of novel targets as well as multitarget drug candidates potentially active in complex diseases [15]. On the other hand, target-based mechanistic screenings investigate the molecular interactions between a selected target and ligands. This rational approach has been greatly developed methodologically during the last two decades [16] and has resulted very successful when applied to genetic disorders for which the target is well established, as is the case with PKU. In particular, differential scanning fluorimetry (DSF) has proven to be a suitable method for mechanistic screenings of compound libraries searching for compounds with PC potential through stabilization of a target protein, such as PAH variants associated with PKU [10, 13].

Following the primary HTS by DSF, a series of analyses is required in order to validate and select hit compounds with capacity to stabilize a target protein. In our case, these steps include the confirmation of binding by an orthogonal assay such as surface plasmon resonance (SPR) [17], enzymatic activity measurements to eliminate inhibitory hit compounds and efficacy tests in eukaryotic cells to evaluate the increased activity and steady state levels of the expressed protein.

We thus present here the protocols of the battery of assays conducted on the enzyme PAH (excluding the DSF-based HTS, which is covered elsewhere in this volume) with the aim of obtaining hit compounds with PC potential as corrective therapeutics for PKU.

2 Materials

2.1 Surface Plasmon Resonance

1. A Biacore 3000 (GE Healthcare) instrument (*see* **Note 1**).
2. Plastic vials of 7 mm diameter and ventilated rubber caps.
3. Target protein sample, e.g., recombinant wild-type human PAH [18], in a concentration of 0.025 mg/mL in 10 mM sodium acetate, pH 5.0 (approx. 150 μL).
4. Sensor chip CM5 (GE Healthcare).
5. 1-ethyl-3-(3-dimethylaminopropyl) carbodiimide hydrochloride.
6. N-hydroxysuccinimide.
7. 1 M ethanolamine–HCl, pH 8.5.
8. HBS buffer: 0.01 M HEPES, pH 7.4, 0.15 M NaCl.
9. Hit compound(s) solution: 50 μM in 1% dimethyl sulfoxide (DMSO)—final assay concentration in HBS buffer, from the MyriaScreen Diversity Collection (Sigma-Aldrich/TimTec) (*see* **Note 2**).
10. Negative control: 1% DMSO—final assay concentration in HBS buffer.
11. Positive control, e.g., BH_4 (Schircks Laboratories): 50 μM, 1.25 mM dithiothreitol (DTT), 31.25 μM HCl—final assay concentration in HBS buffer.

2.2 Enzymatic Activity Measurements with Purified PAH

1. An automated dispenser system, e.g., BRAVO liquid handling platform (Agilent Technologies).
2. Standard mixture: 0.005 mg/mL protein (recombinant wild-type human PAH), 100 mM HEPES, pH 7.0, 20 mM NaCl, 0.04 mg/mL catalase, 1 mM L-Phe, 0.05% bovine albumin serum, 2% DMSO, 0.04 mg/mL (115 μM average concentration) of each hit compound (*see* **Note 3**).
3. Blank solutions with analogous mixture composition but in the absence of compound.
4. 96-well PCR plates or other type of plate configuration, a benchtop centrifuge, and a thermoblock.
5. BH_4 solution: 75 μM, 1.875 mM DTT, 46.87 μM HCl, and a 1% (v/v) acetic acid–ethanol solution (*see* **Note 3**).
6. High pressure liquid chromatography (HPLC) instrument equipped with a Zorbax 300 strong cation exchange column (4.6 × 250 mm and 5 μm particle size) as stationary phase (Agilent Technologies) and a 1 mM acetic acid solution (2% propanol) as mobile phase.

2.3 Measurement of Enzymatic Activity and Protein Levels

2.3.1 Generation of the Cell Line PAH-Flp-In™-293

1. Wild-type human PAH cDNA (NM_000277).
2. AmpliTaq Gold® DNA polymerase (1.25 U) in 1× PCR Gold® buffer—final reaction concentrations provided.
3. 0.2 μM forward primer (5′-CTAGCTAGCATGTCCAC TGCGGTCCTGGAA-3′) and 0.2 μM reverse primer (5′-CGGGGTACCTTACTTTATTTTCTGGAGG GCACT-3′)—final reaction concentrations provided.
4. 1.5 mM $MgCl_2$, 0.2 mM dNTP mix, and a thermal cycler—final reaction concentrations provided.
5. Expression vector pcDNA5/FRT (Invitrogen).
6. Restriction endonucleases KpnI (10 U) and NheI (10 U) in 1× NEBuffer-1 (New England BioLabs)—final concentrations provided.
7. T4 DNA ligase (5 U) in 1× ligase reaction buffer (New England BioLabs)—final concentration provided.
8. QIAquick PCR purification kit and QIAEX II gel extraction kit.
9. Agarose gels in 1× Tris–acetate–EDTA buffer.
10. A temperature-controlled and water-jacketed CO_2 incubator.
11. Cell line Flp-In™-293 (ThermoFisher Scientific).
12. Dulbecco's Modified Eagle's Medium (DMEM) supplemented with 10% fetal bovine serum and 2 mM L-glutamine.
13. Selection antibiotics gentamycin (final concentration 100 μg/mL) and hygromycin B (final concentration 25–100 μg/mL).
14. Expression vector pOG44 (ThermoFisher Scientific).
15. Lipofectamine™ 2000 reagent and Opti-MEM™ reduced serum medium.

2.3.2 Cell Culture and Preparation of Lysates

1. Cell line Flp-In™-293 constitutively expressing wild-type human PAH.
2. DMEM supplemented with 10% fetal bovine serum, 2 mM L-glutamine, 20 μg/mL gentamycin, and 50 μg/mL hygromycin B.
3. Water-jacketed CO_2 incubator.
4. Hit compound(s) solution: 0.02 mg/mL final concentration (average of 55.6 μM), 1% DMSO, or the corresponding negative control: no compound, 1% DMSO.
5. Resuspension buffer I: 1× phosphate-buffered saline containing EDTA-free protease inhibitor cocktail (Roche Applied Science).
6. Benchtop centrifuge.

7. Detergent Triton X-100 solution: 0.1% final assay concentration.
8. Resuspension buffer II: 1× Tris-buffered saline (TBS) containing EDTA-free protease inhibitor cocktail.
9. Zeba-Spin desalting columns, 0.5 mL (ThermoFisher Scientific).
10. Direct Detect spectrometer (Millipore) or an equivalent method (e.g., Bradford) for total protein quantitation.

2.3.3 Enzymatic Activity Measurements with PAH-Flp-In™-293 Cell Extracts

1. Assay solution: 100 mM HEPES, pH 7.0, 0.04 mg/mL catalase, 1 mM L-Phe, 50 μM ammonium iron (II) sulfate (*see* **Note 3**).
2. PAH-Flp-In™-293 lysates generated in the previous section.
3. BH_4 solution: 200 μM, 5 mM DTT, 125 μM HCl, and a 1% (v/v) acetic acid–ethanol solution (*see* **Note 3**).
4. An identical HPLC instrument and configuration as outlined in Subheading 2.2, a benchtop centrifuge, and a thermoblock.

2.3.4 Immuno-quantification of PAH-Flp-In™-293 Cell Extracts

1. Electrophoresis running buffer (1×): 24.7 mM Tris base, 191 mM glycine, 0.1% sodium dodecyl sulfate (SDS).
2. Loading sample buffer (2×): 167 mM Tris base, pH 6.8, 0.83 M sucrose, 0.008% bromophenol blue, 4.17 mM EDTA, 16.7 mM DTT, 3% SDS (*see* **Note 4**).
3. PAH-Flp-In™-293 lysates from Subheading 2.3.2.
4. Western protein standard (e.g., MagicMark™ XP from ThermoFisher Scientific) and prestained protein standard (e.g., Precision Plus™ Dual Color from Bio-Rad Laboratories).
5. A benchtop centrifuge and a thermoblock.
6. 10% SDS-PAGE precast electrophoresis gels, 0.2 μm PVDF membranes and a Trans-Blot Turbo instrument (Bio-Rad Laboratories) (*see* **Note 5**).
7. 1× TBS buffer.
8. Blocking solution: 5% nonfat dried milk, 1% Tween 20 in TBS.
9. Washing solution: 0.1% Tween 20 in TBS.
10. Primary antibodies: α-PAH (Millipore-MAB5278) and α-glyceraldehyde 3-phosphate dehydrogenase (GAPDH) (Abcam-ab9485).
11. Secondary antibodies: goat anti-mouse (Bio-Rad Laboratories) and goat anti-rabbit (Bio-Rad Laboratories), conjugated to horseradish peroxidase.

12. Enhanced chemiluminescence Western Blotting buffer (ECL): 0.1 M Tris base, pH 8.5, 0.23 mg/mL luminol, 0.03 mg/mL *p*-coumaric acid, 0.7% DMSO, 0.03% hydrogen peroxide.
13. ChemiDoc XRS imaging system (Bio-Rad Laboratories) or an equivalent platform.

3 Methods

3.1 Surface Plasmon Resonance

1. Blank immobilization of flow cell 1 and covalent protein immobilization with a flow rate of 5 μL/min and 5 min injection time of flow cell 2, using the surface preparation method template included in the Biacore 3000 control software (*see* **Note 6**).
2. Washing of the sensor surface with HBS buffer (20 μL/min) for approximately 2 h to eliminate unbound molecules and obtain a stable baseline (*see* **Note 7**).
3. Conduct the experiments on the hit compound(s), as well as the positive and negative controls, at 25 °C and flow rate of 20 μL/min HBS, with each individual run consisting of three phases: 2 min injection, 2.5 min equilibration and 2 min regeneration by dissociation in HBS buffer (30 μL/min) (*see* **Note 8**).
4. Analysis of the data within the BIAevaluation software. The sensorgrams are normalized respect to the negative 1% DMSO control and the reference surface to account for possible fluctuations in the bulk refractive index and nonspecific binding to the biosensor, correspondingly. Meanwhile, the response of the positive control—BH_4 in this case—and the calculated theoretical maximum response (R_{max} in Eq. (1) define the threshold for hit compound selection in this binding validation assay by SPR (Fig. 1).

$$R_{max} = \frac{MW_{ref.comp.}}{MW_{prot.}} \times R_{prot.} \tag{1}$$

$MW_{ref.comp.}$ and $MW_{prot.}$ are the molecular weight of a reference compound and protein, respectively, while $R_{prot.}$ is the experimental immobilized amount of protein.

3.2 Enzymatic Activity Measurements with Purified PAH

1. Dispense, robotically, 45 μL of the standard mixture or blank solution into 96-well plates and incubate them at 42 °C for 40 min (*see* **Note 9**).
2. After a 3 min equilibration period of the plates at 37 °C, initiate the enzymatic reaction by adding 5 μL of BH_4 solution

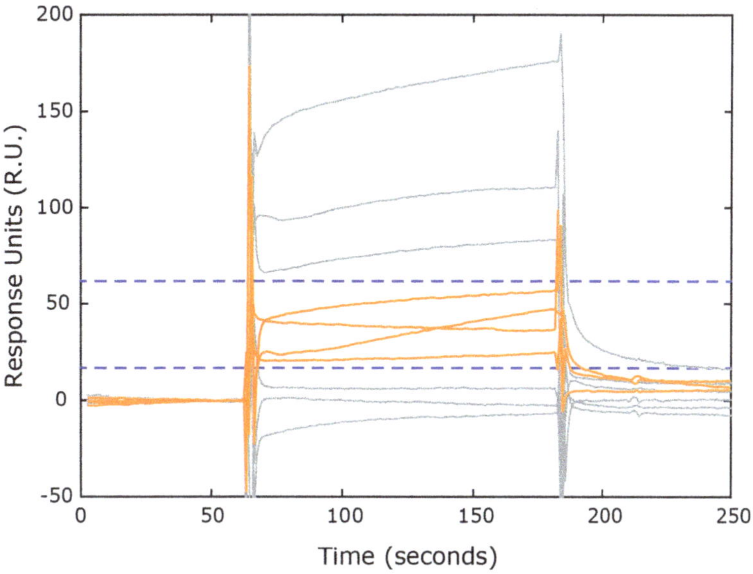

Fig. 1 Validation of hit compound binding by surface plasmon resonance. Representative sensorgrams depicting nonspecific binding compounds (grey trace) above the saturation response (upper dashed line), low affinity binders (grey trace) below the positive control BH_4 reference (lower dashed line), and selected hit compound binders (highlighted in orange). All hit compounds and BH_4 are assayed at 50 μM and 1% DMSO

and stop it after 1 min with 50 μL of the acetic acid–ethanol mix.

3. Place the sample plates at −20 °C for 90 min and, thereafter, centrifuge (3200 × *g*, 10 min) to precipitate and eliminate the protein, in this order.
4. Load and inject (10 μL) the supernatants into the HPLC system (under a controlled temperature of 4 °C) using a mobile phase flow rate of 2.5 mL/min. The amount of L-tyrosine (L-Tyr)—the product of the reaction—is measured by fluorimetric detection (λ_{ex} = 274 nm; λ_{em} = 304 nm) after chromatographic separation from the substrate L-Phe, whereas the enzymatic activity is quantified by calibration with an external L-Tyr standard.
5. Analysis of the data, which includes the evaluation of the different compound samples applying the values from the blank solutions as reference. Accordingly, the compounds can be classified in strong inhibitors (residual activity ≤70%), weak inhibitors (residual activity between 70 and 99%) and potential pharmacological chaperones (residual activity ≥100%) (Fig. 2).

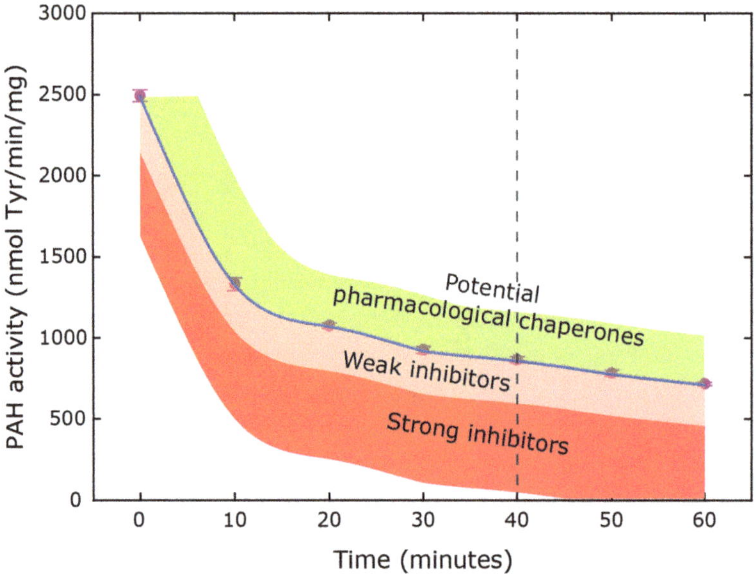

Fig. 2 Effect of hit compounds on the thermal deactivation of PAH. Decay in PAH activity over time upon thermal incubation at 42 °C is represented by the experimental scatter plot (●). This time-course assay is fundamental for the selection of the ideal single-point conditions necessary to handle the amount of hit compounds coming out from the screening stages. After 40 min, a residual activity of approx. 40% is measured, which is an adequate value for discriminating inhibitors from potential pharmacological chaperones (illustrated by the different colored theoretically defined regions in the graph). The enzymatic activity measurements are thus conducted at this particular individual temperature and time (42 °C and 40 min; dashed line)

3.3 Measurement of Enzymatic Activity and Protein Levels in Eukaryotic Cells with Stable Expression of Human PAH

3.3.1 Generation of the Cell Line PAH-Flp-In™-293

1. Amplification of the PAH cDNA construct (0.5 μg in a total reaction volume of 50 μL) with the primers containing tail recognition sequences for the restriction enzymes KpnI and NheI—initial denaturation at 95 °C/10 min followed by 30 cycles of 95 °C/50 s for denaturation, 58 °C/40 s for annealing and 72 °C/2 min for extension, with a final extension at 37 °C for 10 min.
2. Purification of the PCR products according to the instructions from the QIAquick PCR purification kit.
3. Digestion of pcDNA5/FRT plasmid (1 μg) and PCR product (40 μL) by restriction endonucleases KpnI and NheI (37 °C/2 h).
4. Loading and resolution of the two digested products in 1% agarose gel (70 V for approx. 2 h) followed by a purification step as described in the QIAEX II gel extraction kit and a posterior ligation of the two fragments with T4 DNA ligase (16 °C/overnight) to render the vector pcDNA5/FRT-PAH (*see* **Note 10**).

5. Culture the Flp-In™-293 cell line in supplemented DMEM medium containing the first selection antibiotic gentamycin.
6. Transfect the cells with 4 μg of the expression vector pcDNA5/FRT-PAH (inclosing the inserted PAH construct) and 4 μg of the plasmid pOG44 (harboring the Flp recombinase) using Opti-MEM™ reduced serum medium and Lipofectamine™ 2000 transfection reagent, following manufacturer's recommendations.
7. Substitute the transfection elements by the aforementioned supplemented DMEM growth medium containing gentamycin and incubate the cells 4 days at 37 °C in 5% CO_2.
8. Split the cells at day 5 and add the second selection antibiotic hygromycin B at 25 μg/mL.
9. Increase this concentration to 100 μg/mL on day 7 and maintain it the rest of the culturing time (*see* **Note 11**).
10. Select the antibiotic resistant clones—i.e., positive integration of the plasmid into the genome, split them and allow for growth for an additional period of approx. 20 days replacing the supplemented growth medium periodically.
11. Confirm the PAH expression of the newly generated PAH-Flp-In™-293 cell line (e.g., by qPCR).

3.3.2 Cell Culture and Preparation of Lysates

1. Seed 800,000 PAH-Flp-In™-293 cells onto 35 × 10 mm petri dishes in 2 mL supplemented DMEM growth medium and incubate at 37 °C in 5% CO_2 (*see* **Note 12**).
2. Replace the old medium 24 h later with 2 mL of fresh one containing either the hit compound(s) or the negative control solutions.
3. After a new incubation period of 24 h (i.e., 48 h from the start of the experiment), harvest and resuspend the cells in 500 μL resuspension buffer I (*see* **Note 13**).
4. Centrifuge the harvested cells (2700 × *g*, 5 min).
5. Resuspend the pellets in 200 μL of resuspension buffer II.
6. Incubate for 15 min with Triton X-100, 0.1% final concentration.
7. Additional centrifugation stage (20,000 × *g*, 15 min).
8. Collect and filter the supernatants (approx. 150 μL) through a Zeba-Spin desalting column (*see* **Note 14**).
9. Measure the protein concentration of the lysates.

3.3.3 Enzymatic Activity Measurements with PAH-Flp-In™-293 Cell Extracts

1. Add, manually, 20 μL of the PAH-Flp-In™-293 lysate into a microcentrifuge tube containing 25 μL of the assay solution equilibrated at 37 °C.
2. Incubate the mix for 5 min, start the enzymatic reaction by the addition of 5 μL of BH_4 solution and stop it after 10 min with 50 μL of the acetic acid–ethanol mix.

3. The next steps leading to the quantification of L-Tyr, namely precipitation, centrifugation (minor adjustment: 10,000 × *g*, 10 min), chromatographic separation and fluorimetric detection are equivalent to Subheading 3.2. In this case, the results of the different compound treatments are normalized respect to the negative control (1% DMSO) and expressed as the total enzymatic activity of the filtered cellular lysates (Fig. 3a).

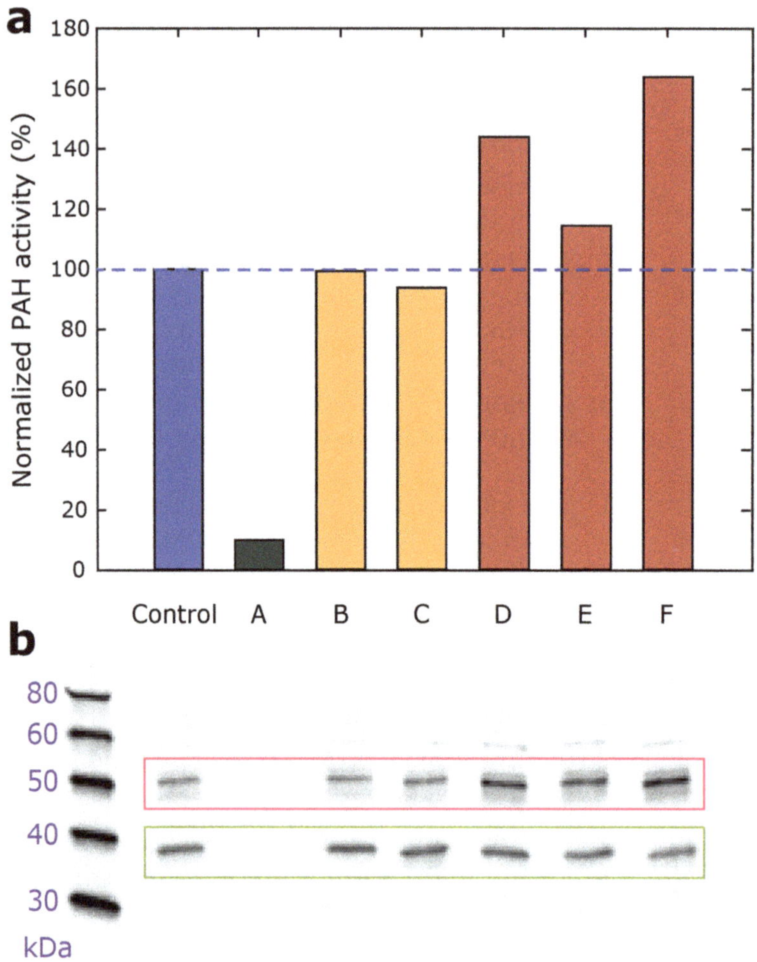

Fig. 3 Pharmacological chaperone effect of hit compounds in the cell line PAH-Flp-In™-293. Representative examples of the analyzed stabilizing effect of six hit compounds (average concentration of 55.6 μM and 1% DMSO) in terms of intracellular (**a**) PAH activity and (**b**) PAH protein levels. (**a**) Results are normalized respect to the 1% DMSO control treatment (blue bar/dashed line) and illustrate the different outcomes observed: A (green bar) is a toxic hit compound; B and C (yellow bars) are hit compounds with no significant impact in PAH activity, whereas D, E, and F (red bars) are hit compounds that increased PAH activity (to different extents) and are, thus, potential PCs. (**b**) Immunoblot image analysis of the same hit compounds. The intensity of the PAH band (red rectangles), at approx. 52 kDa, is normalized respect to the GAPDH band (green rectangles), at approx. 37 kDa, used as loading control. A good correlation between enzymatic activity and protein levels for each compound is observed

3.3.4 Immunoquantification of PAH-Flp-In™-293 Cell Extracts

1. Mix at a 1:1 (v/v) ratio 10 μg (total protein) of lysate and 2× loading sample buffer—the exact relative volumes will depend on the previous measurements of protein concentration (*see* **Note 15**).
2. Heat the lysate samples and protein standards at 95 °C for 5 min and centrifuge at room temperature (10,000 × *g*, 5 min).
3. Fill the electrophoretic chamber with 1× running buffer, load the prepared samples/standards into the precast SDS-PAGE gel and apply a constant current of 20 mA for approx. 1 h (*see* **Note 16**).
4. Transference of the proteins from the gel matrix to the PVDF membrane is carried out in a high-throughput blotting system according to the manufacturer instructions (constant current 2.5 A for 5 min, with a voltage limit of 25 V).
5. Block the membrane overnight at 4 °C with blocking solution.
6. Incubate 1 h with primary antibodies: α-PH8 (1:1000) and α-GAPDH (1:1000), both in blocking solution, followed by a washout for 30 min using washing solution.
7. Incubate 1 h with secondary antibodies: goat anti-mouse (1:2000) and goat anti-rabbit (1:2000), both in blocking solution, preceding a new washout step of 30 min with washing solution (*see* **Note 17**).
8. Detection and quantification of the bands intensity (Fig. 3b) using the ImageLab software after development of the membrane by a 1 min treatment with ECL solution (*see* **Note 18**).

4 Notes

1. The protocol presented here can be adapted to other instruments/applications following some general guidelines: (https://www.biacore.com/lifesciences/Application_Support/laboratory-guidelines/index.html).
2. Other compound libraries can be used subjected to the purpose of the screening campaign.
3. All concentrations refer to the final reaction volume of 50 μL.
4. It is advised that the components DTT and SDS are only added to the 2× loading sample buffer prior to each individual experiment.
5. Another valid option would be to cast the SDS-PAGE gels manually and employ a different system of sample transference to the membrane.

6. Alternatively, it is also possible to set the desired surface immobilization levels although the amount of protein required would be higher. In the same manner, different flow cell combinations can be selected.
7. The injection of a 10 mM DTT solution (40 μL and 120 s contact time) at the beginning of the washing step allows for a reduction in the total time of this stage. This procedure is limited to proteins whose integrity is not affected by the presence of this reducing agent, as is the case of PAH.
8. The total number of compounds assayed per experiment will depend on the stability of the protein once immobilized.
9. The incubation temperature and time will depend on the protein of study. Therefore, it is recommended to carry out preliminary time course experiments in the absence of compound at different temperatures in order to select the right conditions.
10. It is important, at this point, to sequence the produced plasmid to confirm the incorporation of the cDNA construct and exclude possible mutations introduced during the PCR stage.
11. In order to determine the ideal concentration of hygromycin B for selection, it is necessary to perform beforehand a sensitivity experiment with varying antibiotic concentrations.
12. The final volume and amount of cells chosen is the result of a previous optimization phase. In the case that these parameters shall be modified, we encourage to include such a stage prior to the main experiments.
13. The harvesting of the cells and the subsequent steps of this protocol must be conducted at 4 °C.
14. The Zeba-Spin desalting columns serve the purpose of removing small molecular weight molecules that could interfere in succeeding workflow assays and they are customarily preequilibrated (3 × 300 μL) in the experimental buffer of choice prior to the lysate loading and filtration. This practice could be dispensable if there is sufficient knowledge of no matrix effects in the selected method for the analysis of the lysate sample.
15. In order to achieve uniform lane migrations in the electrophoresis gel, it is optimal to correct the discrepancies in final volume with 1× loading sample buffer (1:1 dilution from 2× stock solution).
16. Apart from the prestained and Western protein ladders, it can be useful to add a control lane with the purified protein of interest (i.e., PAH) for band analysis and quantification.
17. It is important to avoid dryness in the membrane. Thus, the TBS buffer can be utilized as solvent for that purpose in intermediate steps, if necessary.

18. To ensure a high quality in the development of the bands, the elements of the ECL solution, luminol/*p*-coumaric acid and hydrogen peroxide, respectively, are prepared in separate individual tubes of equal volume and mixed (1:1) right before starting the 1 min incubation treatment of the membrane.

Acknowledgments

This work was supported by The Research Council of Norway (RCN; Program Forny2020 248889/O30 and FRIMEDBIO 214012), the infrastructure project Nor-Openscreen (RCN), and The K.G. Jebsen Foundation. We thank Guri E. Matre and Ming Ying for expert technical assistance.

References

1. Chen B, Retzlaff M, Roos T, Frydman J (2011) Cellular strategies of protein quality control. Cold Spring Harb Perspect Biol 3(8):a004374
2. Sontag EM, Samant RS, Frydman J (2017) Mechanisms and functions of spatial protein quality control. Annu Rev Biochem 86:97–122
3. Kevei E, Pokrzywa W, Hoppe T (2017) Repair or destruction-an intimate liaison between ubiquitin ligases and molecular chaperones in proteostasis. FEBS Lett 591(17):2616–2635
4. Muntau AC, Leandro J, Staudigl M, Mayer F, Gersting SW (2014) Innovative strategies to treat protein misfolding in inborn errors of metabolism: pharmacological chaperones and proteostasis regulators. J Inherit Metab Dis 37(4):505–523
5. Labbadia J, Morimoto RI (2015) The biology of proteostasis in aging and disease. Annu Rev Biochem 84:435–464
6. Hartl FU (2017) Protein misfolding diseases. Annu Rev Biochem 86:21–26
7. Blau N, van Spronsen FJ, Levy HL (2010) Phenylketonuria. Lancet 376(9750):1417–1427
8. Flydal MI, Martinez A (2013) Phenylalanine hydroxylase: function, structure, and regulation. IUBMB Life 65(4):341–349
9. Wettstein S, Underhaug J, Perez B, Marsden BD, Yue WW, Martinez A, Blau N (2015) Linking genotypes database with locus-specific database and genotype-phenotype correlation in phenylketonuria. Eur J Hum Genet 23(3):302–309
10. Underhaug J, Aubi O, Martinez A (2012) Phenylalanine hydroxylase misfolding and pharmacological chaperones. Curr Top Med Chem 12(22):2534–2545
11. Convertino M, Das J, Dokholyan NV (2016) Pharmacological chaperones: design and development of new therapeutic strategies for the treatment of conformational diseases. ACS Chem Biol 11(6):1471–1489
12. Pey AL, Ying M, Cremades N, Velazquez-Campoy A, Scherer T, Thony B, Sancho J, Martinez A (2008) Identification of pharmacological chaperones as potential therapeutic agents to treat phenylketonuria. J Clin Invest 118(8):2858–2867
13. Hole M, Jorge-Finnigan A, Underhaug J, Teigen K, Martinez A (2016) Pharmacological chaperones that protect Tetrahydrobiopterin dependent aromatic amino acid hydroxylases through different mechanisms. Curr Drug Targets 17(13):1515–1526
14. Lee JA, Carragher NO, Berg EL (2017) Empirical drug discovery: a view from the proteome. Drug Discov Today Technol 23:1–5
15. Wang J, Zhang X, Fang Y, Liang X (2016) Label-free cell phenotypic assays for assessing drug polypharmacology. Curr Pharm Des 22(21):3190–3200
16. Schenone M, Dancik V, Wagner BK, Clemons PA (2013) Target identification and mechanism of action in chemical biology and drug discovery. Nat Chem Biol 9(4):232–240
17. Flatmark T, Stokka AJ, Berge SV (2001) Use of surface plasmon resonance for real-time

measurements of the global conformational transition in human phenylalanine hydroxylase in response to substrate binding and catalytic activation. Anal Biochem 294(2):95–101

18. Martínez A, Knappskog PM, Olafsdottir S, Døskeland AP, Eiken HG, Svebak RM, Bozzini M, Apold J, Flatmark T (1995) Expression of recombinant human phenylalanine hydroxylase as fusion protein in Escherichia coli circumvents proteolytic degradation by host cell proteases. Isolation and characterization of the wild-type enzyme. Biochem J 306:589–597

Chapter 19

Constructing Kinetically Controlled Denaturation Isotherms of Folded Proteins Using Denaturant-Pulse Chaperonin Binding

Pierce T. O'Neil, Alexandra J. Machen, Jackie A. Thompson, Wei Wang, Quyen Q. Hoang, Michael R. Baldwin, Karen R. Khar, John Karanicolas, and Mark T. Fisher

Abstract

Methods to assess the kinetic stability of proteins, particularly those that are aggregation prone, are very useful in establishing ligand induced stabilizing effects. Because aggregation prone proteins are by nature difficult to work with, most solution based methods are compromised by this inherent instability. Here, we describe a label-free method that examines the denaturation of immobilized proteins where the dynamic unfolded protein populations are captured and detected by chaperonin binding.

Key words Denaturant pulse, Biolayer interferometry, Protein folding, GroEL, Denaturation, Aggregation, Chaperonin

1 Introduction

The use of protein denaturation procedures coupled with an appropriate detection output (e.g., spectroscopic; calorimetric thermophoresis) allows researchers to assess the impact of ligand binding to proteins both stabilizing and destabilizing and to compare the stability of mutant proteins with their wild-type counterparts. These methods using temperature or chaotrope perturbation work well for well-behaved proteins where their denaturation profiles are not complicated by extensive competing aggregation side reactions. Unfortunately, a large number of proteins for which protein denaturation profiles are used to assess protein stability undergo extensive aggregation under physiologically relevant conditions. This is particularly problematic for assessing the validity of potentially therapeutic stabilizing ligands of less stable disease

In memoriam: Mark T. Fisher (1954-2018)

Cláudio M. Gomes (ed.), *Protein Misfolding Diseases: Methods and Protocols*, Methods in Molecular Biology, vol. 1873,
https://doi.org/10.1007/978-1-4939-8820-4_19,

proteins that undergo pathological unfolding or aggregation. These aggregation side reactions interfere with the reproducibility of isotherms. If one simply relies on scoring those candidate ligands that prevent the off-pathway aggregation, there is the potential for identifying false positive candidate ligands when the small molecule simply inhibits the aggregation reaction rather than truly stabilizing the "native"-like fold.

In this methods monograph, expanded and detailed protocols are provided to easily construct denaturant isotherms using a label-free platform. The methods involve immobilizing target proteins on a biosensor surface which has a twofold purpose. First, immobilization of the less stable proteins prevents aggregation side reactions which are a confounding factor for solution-based assays. Second, the protein is oriented in a manner to allow for the adequate readout detection of exposed hydrophobic surfaces formed during a timed denaturant pulse using the large nonspecific chaperonin GroEL to amplify the signal. In this instance, the readout is the detection of chaperonin binding, which serves as a "promiscuous antibody" to partially folded proteins. Due its large size (802 kDa), GroEL binding is easily discerned using label-free approaches. Reproducibility in the binding readout is obtained using automated, precisely timed GroEL interaction and wash phases. In this work, new examples of utility and process development are presented, including preruns of an alpha-synuclein construct, demonstrating cooperative unfolding, supporting the notion that alpha-synuclein contains a folded region (alpha helical interacting regions). In addition, vastly improved reproducible isotherms are generated with increasing the purity of the protein of interest, strengthening the assessment that these methods can distinguish between destabilized and ligand-stabilized forms during small molecule development for protein stabilization.

The label-free technique utilized in this monograph is Biolayer Interferometry (BLI), which is similar to Surface Plasmon Resonance (SPR). The biosensor contains both a reference layer and the biocompatible surface layer. White light is directed down the biosensor and is reflected back up by both layers. The difference in the reflection from both layers creates an interference pattern. As the mass on the biocompatible surface layer changes, the interference pattern maxima change. Monitoring the nanometer shifts in maxima is recorded as the response versus time by the BLI instrument.

2 Materials

1. Octet RED96 (ForteBio).
2. Software: Data Acquisition 8.2 and Data Analysis 8.2 (ForteBio). The exact programming may change with later

versions of the programs. The screenshots and descriptions given in this monograph are specific for version 8.2.

3. Ni-NTA Biosensors (FortéBio).
4. 96-well nonbinding microplates, nonsterile (Greiner Bio-One #655900). Note that the plates MUST be black and opaque so as to not reflect projected light or transmit outside light. Additionally, this nonbinding property is required so that the *protein of interest (POI)* does not adhere to the well itself, effectively changing the functional concentration.
5. p53 Buffer: 50 mM TRIS pH 7.5, 300 mM NaCl, 1 mM DTT. p53 Buffer is used as an example of a POI buffer. It is generally prepared as a 10× stock solution without DTT and upon dilution with ultrapure (>18.2 MΩ) water to working concentration DTT is added. This stock solution is especially useful when preparing solutions containing denaturant.
6. GroEL Buffer: 50 mM Tris pH 7.5, 50 mM KCl, 10 mM $MgCl_2$, 0.5 mM EDTA. GroEL Buffer is used during the denaturant pulse assay as a readout of a protein's hydrophobic exposed nature. This buffer is generally prepared as a 10× stock solution and then diluted with ultrapure water.
7. 8 M Urea Stock: 8 M urea, 1× POI buffer. The initial denaturant is urea as it is both chaotropic and nonionic. The urea stock solution can be made no longer than 12 h before its use as it degrades in solution into ammonia and isocyanic acid. Storage on ice is recommended to limit degradation, but stock solutions will reversibly precipitate under 20 °C.
8. Regeneration Solution: 10 mM glycine, pH 1.7. Biosensor regeneration solution is generally prepared as a dilution of a 10× stock solution. The stock solution is prepared as 100 mM glycine in ultrapure water with its pH adjusted to 1.7. Dilution into ultrapure water does not significantly impact pH.
9. Recharge Solution: 10 mM $NiCl_2$.Biosensor recharge solution is generally prepared as a dilution of a 10× stock solution in ultrapure water.

3 Methods

3.1 Programming the Octet

1. Check the power state of the Octet RED96 and the light source. If off, turn both on now as the light source requires at least 1 h to warm up, preferably 4 h. By the time the system has been programmed, the biosensors readied, and the assay plate pipetted, at least 1 h should have passed.
2. Open the Data Acquisition software provided with the Octet. In the Experiment Wizard window, select New Kinetics

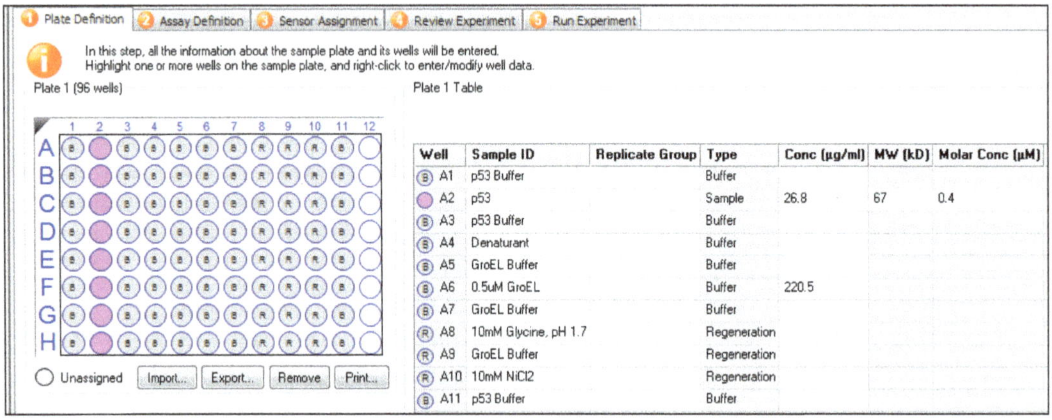

Fig. 1 Tab 1—Plate definition. Screenshot displaying the setup for the standard denaturant pulse assay plate definition with plate on the left and descriptors on the right

Experiment and click the arrow. In the first tab of the new window, program the "Plate Definition" for the denaturant pulse automated assay into the computer according to Fig. 1. This is accomplished by right clicking on the column number and selecting the proper column type then typing the in the sample ID in the right pane. This example protocol is for His-tagged p53 and biosensor regeneration. Adjust the buffer composition in Columns 1–4 for your POI. However, the composition of Columns 6–8 should not be adjusted as this is the best buffer for GroEL binding.

3. In the next tab, program the "Assay Definition" using the Plate Definition created in Subheading 3.1, **step 2**. Add the steps shown in the top right pane using the proper time in seconds and step type as shown in Fig. 2. To program the assay steps, click on the step and then double click on the appropriate column on the plate. Once all 15 steps shown have been programed, select all of them and click replicate button. In the new window selected "Append to current assay" radio option and deselect, if necessary, the checkbox for "Offset steps". Repeat these steps so that the assay is repeated three times in a row which will give an $n = 3$ for statistics and error bars. The example is for p53 denaturant pulse but is appropriate for most proteins. The *Denaturant Pulse* step length can be adjusted (*see* **Note 1**). The short baseline step (step 5 in Fig. 1) is a brief wash time to remove the denaturant but may permit partial refolding and can be adjusted (*see* **Note 2**).

4. In the third tab, ensure the correct *Sensor Assignment* settings are selected (Fig. 3). In the top left pane, only the first row of squares (representing biosensors) is selected as tips will be regenerated between assays. In the bottom left pane, ensure all wells programmed in the first tab are highlighted blue indicating

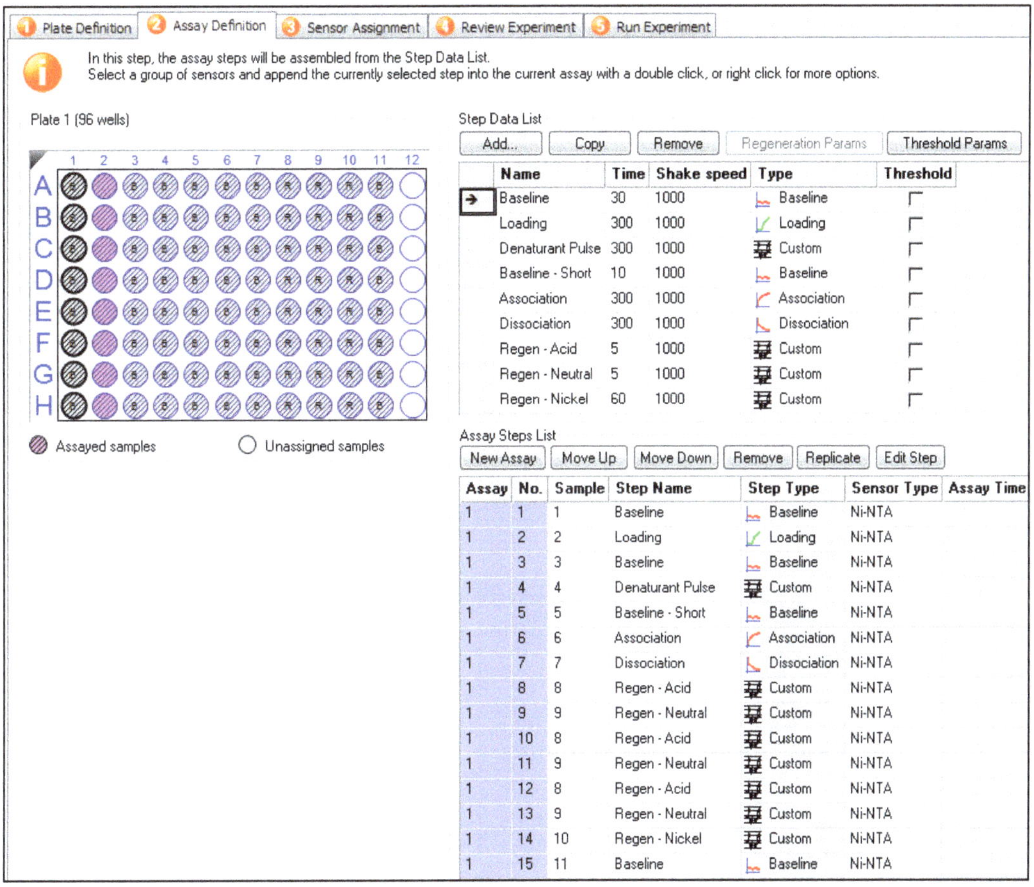

Fig. 2 Tab 2—Assay definition. Screenshot displaying the setup for programming the denaturant pulse protocol with needed steps at the top right and the assigned steps at the bottom right. To program a step click on the step at the top right and double-click on the well column to assign that step to the well. Only the first 15 steps are shown, which constitutes one full assay

they will be used in the assay and that the assay was properly programmed in the second tab. In the right pane, ensure the proper tip type is selected. In the example assay, the protein is His-tagged so Ni-NTA tips are selected. Tip regeneration may not be possible with all proteins. If tip regeneration appears to be not working properly, alterations to the method need to be carried out (*see* **Notes 3–6**).

5. In the fourth tab, use the arrows at the top right to step through and review the assay programming. If anything doesn't look correct, return to the appropriate tab and make corrections. Items to pay close attention to are the following: step time, step position, and step type.
6. In the fifth tab, use the top left pane to select a save destination for your data. In the middle pane, enter 25 °C in the temperature box and ensure the plate delay is set to 600 s to

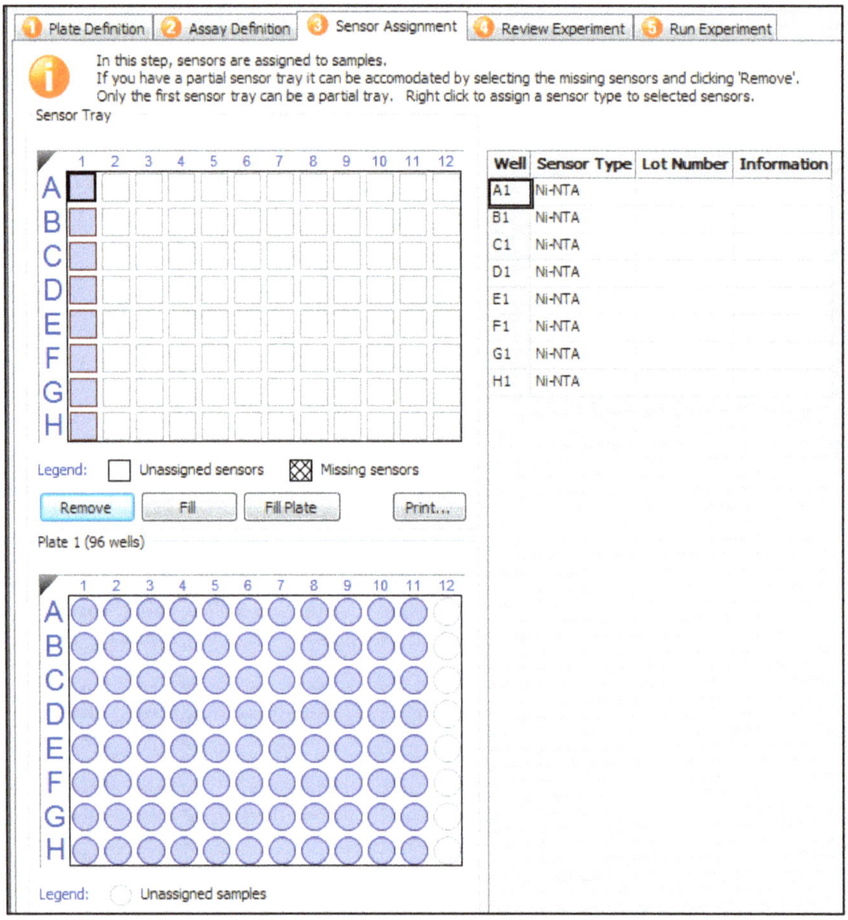

Fig. 3 Tab 3—Sensor assignment. Screenshot displaying the sensor assignment tab. Biosensors are represented as squares in the top right pane. The plate assigned in tab 1 is in the bottom left pane. Sensor type is selected at the top right pane

permit the solutions plate to have sufficient time to equilibrate to the specified temperature. The temperature can be adjusted from 25 °C (*see* **Note 7**).

7. Do not press the GO button at this time. The plates and biosensors are programmed but still need to be physically prepared.

3.2 Hydrate the Biosensors

1. Using a spare biosensor box, remove the green tip rack and place a black 96 well plate into the bottom of the box. Fill Column 1 using 250 μL of your POI buffer. In this example, it would be p53 buffer. In all remaining wells add 250 μL ultrapure water to greatly reduce evaporation losses from your assay plate during the experiment. If you do not have a spare biosensor box, omit the addition of water from the extra wells.

2. Replace the green biosensor tray. Transfer a set of tips into the first column. Ensure the end of biosensor tip does not touch anything including the green tip tray. Any contact directly with the end of the tip will irreparably damage the biosensor. Discard any compromised tips in glass discard. After all tips have been transferred, either note the time or start a timer counting up. Additionally, return the unused biosensor back into their foil pouch with desiccant as soon as possible to protect them from moisture and light. The biosensors *MUST* hydrate in buffer for at least 10 min and be used within 24 h of hydration. Immediately place the hydrating biosensor box into the left side of the Octet and close the door to protect the biosensors from light. Do NOT leave the lid on the hydrating tips as this can break the optics box of the Octet.

3.3 Filling the 96 Well Plate

1. Dilute the protein stocks. Dilute GroEL to 0.5 μM in GroEL buffer for a total of 2 mL and place tube on ice. Dilute your POI to an appropriate dilution and place on ice. In our example assay, p53 is diluted to 0.4 μM which was selected as it produces a protein loading signal of 1 nm as well as being within the protein's linear response range, which is a critical parameter for normalization across the small variability during loading of tips (*see* **Note 8**).
2. Prepare dilutions of denaturant. Dilute the stock denaturant in protein buffer. In this example, the denaturant is urea. For urea, the dilutions range 0 M to 7 M by 1 M steps. For discussion of stronger denaturants (*see* **Note 9**).
3. Fill the columns of the black 96 well assay plate with 250 μL of the appropriate solution as planned in Subheading 3.1, **step 2**. The order of addition should be solutions, denaturant, and finally proteins. Specifically, add POI buffer (this example p53 Buffer) to well Columns 1, 3, & 11. Add GroEL Buffer to well Columns 5, 7, and 9. Add 10 mM glycine, pH 1.7 to well Column 8. Add 10 mM $NiCl_2$ to column 10. Add ultrapure water to unused column 12. Next, add the denaturant range with 0 M urea in Row A through 7 M urea in Row H. Finally, add diluted POI in well Column 2 and diluted GroEL to well Column 6.
4. Transfer the filled assay plate to the right side of the Octet, orienting the A1 position at the back right. Immediately close the door after.

3.4 Running the Experiment

1. Once the machine has been programmed, the biosensors hydrated, and the assay plate filled, click the GO button in the fifth tab to initiate the experiment following the programmed delay. Example data output is shown in Fig. 4.

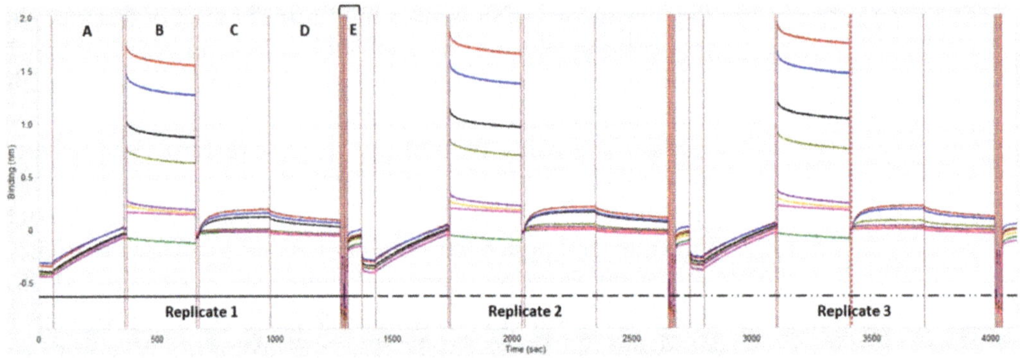

Fig. 4 Sample data. Sensorgram trace of three assay replicates using acid/neutral regeneration. Key steps are labeled: A—POI loading, B—Denaturant pulse, C—GroEL Binding, D—Dissociation, E—Regeneration steps. Each replicate assay is identified by bars at the bottom of the sensorgram

3.5 Data Analysis

1. Open the Data Analysis program provided with the Octet RED96. In tab 1, select your data folder to load you .frd data files. In tab 2, select the radio button option for "All Steps Aligned." Click the "Export Aligned Steps (.csv files) to output the raw data in a file which can be read with Microsoft Excel. Open the exported data with Excel. To normalize the GroEL association against the p53 (or POI) loading, the endpoint of GroEL association is divided by p53 (or POI) binding, as long as it is within the linear range (*see* **Note 8**).
2. Plot the denaturant concentration on the *x*-axis and normalized GroEL binding on the *y*-axis to yield the kinetically controlled denaturant isotherm. Comparison between 50% GroEL binding and denaturant concentration will yield information about a compound's ability to stabilize or destabilize the POI, corresponding to a right-shift or left-shift respectively of the 50% GroEL binding denaturant concentration. The compounds should be added to the baseline before denaturant and during the denaturant solutions. Before testing compounds, it is important to establish if stabilization is possible. To test this, a general osmotic stabilizer (sucrose or glycerol) should be included during the denaturant step. Figure 5 demonstrates stabilization of a protein using either 4 M glycerol or 1 M sucrose.
3. Modeling of the data can permit quantitative comparisons between conditions, as was presented in the previous point. Additionally, if the model is a sigmoidal curve, this indicates cooperative denaturation. Most proteins fold and unfold in a cooperative manner. If the protein lacks native structure, this cooperativity will be not present. For alpha-synuclein, which is generally thought to be unfolded and lacking structure in solution, the sigmoidal response to guanidine denaturation in Fig. 6 indicates cooperative unfolding, suggesting at least partially folded structure in the native state.

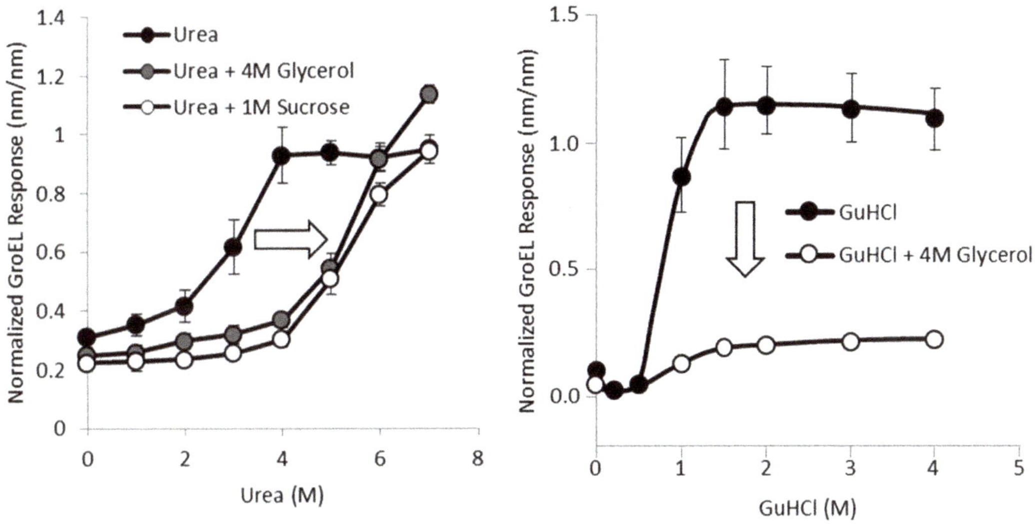

Fig. 5 Denaturant pulse stabilization. Denaturant pulse isotherms demonstrating how inclusion of 1 M sucrose or 4 M glycerol in the urea (denaturant) solution can stabilize the POI. This stabilization is seen as a right-shift in the curve as indicated by the arrow for MalZ (left). Additionally, the GroEL binding may decrease upon stabilization as is seen for p53 using glycerol (right). Error bars are present on all data points of standard deviation for sample $n = 3$. Adapted with permission from [1]. Copyright 2016 American Chemical Society

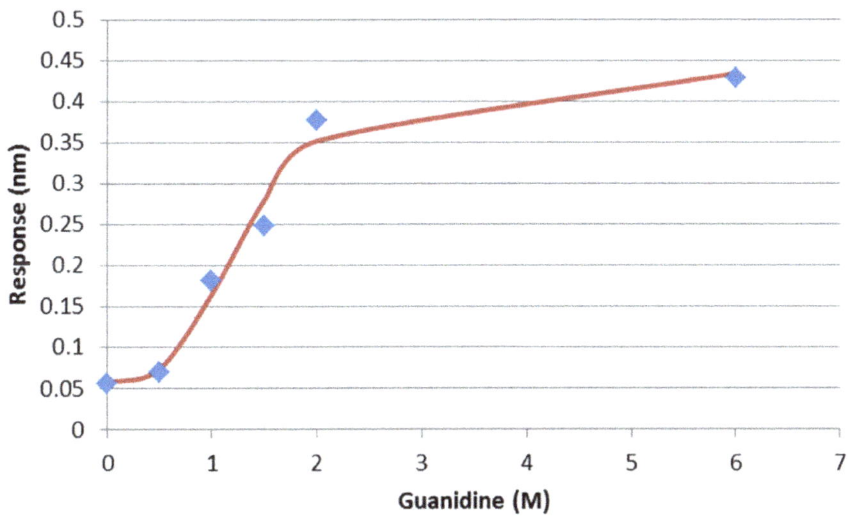

Fig. 6 Sigmoidal model of denaturant pulse. Plot of alpha-synuclein guanidine hydrochloride denaturant pulse with sigmoidal curve fit. Sigmoidal shape indicates a cooperative unfolding

4 Notes

1. Denaturant Pulse Length. A denaturant pulse length of 5 min (300 s) is applicable for most proteins. If the normalized GroEL binding response is at maximum at low denaturant, a shorter denaturant time (as short as 1 min) will right-shift the

isotherm such that the transition is observed at lower denaturant concentrations. Additionally, if the protein does not appear to have any GroEL binding even at maximal denaturant concentration, the denaturant pulse length can be increased to left-shift the isotherm enabling observation of the protein unfolding. We have tested pulses up to 10 min and found these to yield useful responses.

2. Wash Step Length. If refolding of the protein during the short (10s) wash prior to GroEL association, the length can be shortened. Wash steps from 0s to 60s have been examined in our lab. We found very little differences between 10s and 60s for anti-GST antibodies [1]. In addition, the Chaudhuri group has examined refolding on the tip using GroEL as the readout concluding the refolding rate is slow with only 40% refolding after 15 m at 25 °C [2]. Therefore, it is not recommended to adjust this step as it is very unlikely your POI is refolding during the short wash step.
3. Protein Purity. The first item to check if inconsistent loading or GroEL binding using regeneration is protein purity. Impure protein samples have been observed by our lab to produce large variation between replicates.
4. Protein Stability. If protein purity has been checked and is not the confounding factor, the second item to check is the aggregation state of the POI. Because the triplicate assay requires over an hour of time at 25 °C, aggregation can occur particularly for proteins with marginal stability at this temperature. In order to prevent aggregation, include either 1 M sucrose or 1 M glycerol into the loading solution. This may prevent/slow down aggregation during the triplicate assay; however, the inclusion of either osmolyte will kinetically slow the loading rate, which requires a higher concentration of protein to achieve the same signal. In some special cases, an endogenous stabilizer can be used to stabilize the protein in solution without affecting the kinetics of loading (e.g., GTP stabilizes the nucleotide binding domain of CFTR protein [1] or p53 stabilization by CDB3, a peptide derived from the endogenous p53 binding protein 2 [3]).
5. Alternative Biosensor Regeneration Methods. If acid induced aggregation is suspected, regeneration using imidazole as a His-tag competitor can be attempted. In place of the acid/neutral and nickel steps, program a single 2-min incubation in 300 mM imidazole. Alternatively, inclusion of a detergent (up to 0.05% Tween 20) in the neutralization buffer during regeneration can assist in regeneration.
6. Using Multiple Tips Instead of Biosensor Regeneration. If the loading response is not consistent after regeneration, one

likely cause is incomplete regeneration or acid induced aggregation, which we have observed in some cases. To avoid regeneration complications, multiple tips can be used for obtaining triplicate runs to quantitatively evaluate the variation between runs. To use multiple tips, three modifications need to be made. First, when programming the assay in Subheading 3.1, **step 3**, select the radio button for "Add as a new assay" when replicating the assay. This forces the machine to use a new set of tips each assay run. Second, when hydrating biosensors, fill the first three columns with protein buffer in the 96 well plate and with biosensors. Third, the regeneration steps can be omitted from the assay steps.

7. Temperature of the Denaturant Pulse Assay. The temperature of the denaturant pulse assay may be raised from 25 °C to a more physiological temperature of 37 °C. However, the protein linear response range, the concentration for 1 nm loading, and the aggregation propensity will need to be reassessed at this higher temperature as any kinetic rate will be increased with the increase in temperature. Additionally, the higher temperature will promote a greater amount of evaporation from the assay plate. For extended experiments this will change the functional concentrations. At 25 °C, only 5 μL evaporation has been observed for the Octet RED96. The newer model Octet RED96e has greatly reduced evaporation at all temperatures and would be more appropriate to use for higher temperature experiments.

8. Linear Response Range. By loading the protein onto the tip within the linear response range, the GroEL binding response can be normalized to the POI loading. This accounts for minor variation in loading. However, outside this range normalization is not valid. For most proteins there are two linear ranges: one at very low concentrations and a second at higher concentrations, *see* Fig. 7. It is preferable it use the lower concentration linear range to limit any crowding or protein-protein interactions on the biosensor.

9. Other Denaturants. For proteins which are particularly stable as indicated by low GroEL binding at high denaturant concentrations, stronger denaturants can be used during the assay. In the same manner, a weaker denaturant can be used if the protein is denatured too easily. The denaturants we have tested are the following: urea < guanidine hydrochloride < guanidine thiocyanate. All the denaturants should be made fresh due to decomposition and/or evaporation, both of which will change the effective concentration.

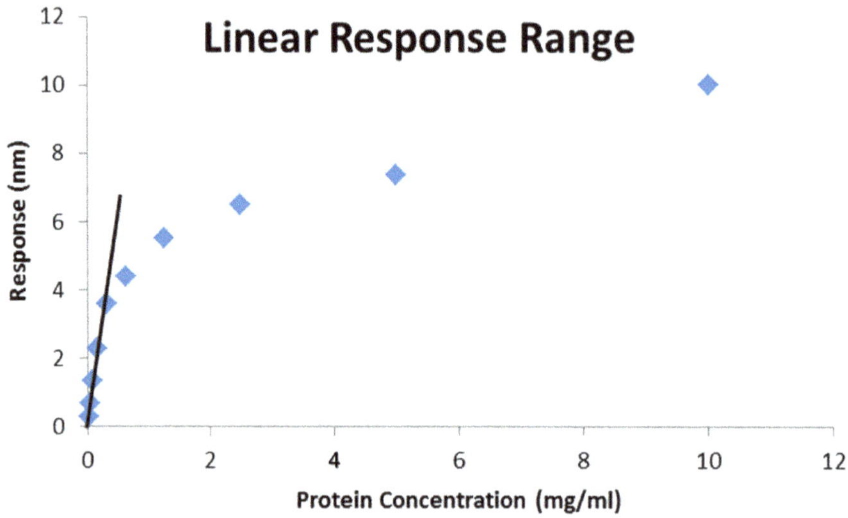

Fig. 7 Linear response range plot. Plot of response versus protein concentration. For tetanus neurotoxin, the linear range can be seen at both very low concentration (solid line; <0.5 mg/mL) and at high concentrations (>2 mg/mL). These values are specific to every protein

Acknowledgments

This work was supported by NIH RO1AI090085 (M.T.F.), University of Kansas Innovation Center Grant (M.T.F.), 5R01GM112736 (J.K.), NSF EPScoR Kansas Board of Regents grant (M.T.F. and J.K.), NIH R21NS079881 (Q.Q.H.), the KBOR grant, T32GM008359 (P.T.O.), and Madison & Lila Self Graduate Fellowship (A.J.M).

References

1. Lea WA, O'Neil PT, Machen AJ, Naik S, Chaudhri T, McGinn-Straub W, Tischer A, Auton MT, Burns JR, Baldwin MR, Khar KR, Karanicolas J, Fisher MT (2016) Chaperonin-based biolayer interferometry to assess the kinetic stability of metastable, aggregation-prone proteins. Biochemistry 55(35):4885–4908. https://doi.org/10.1021/acs.biochem.6b00293
2. Pastor A, Singh AK, Fisher MT, Chaudhuri TK (2016) Protein folding on biosensor tips: folding of maltodextrin glucosidase monitored by its interactions with GroEL. FEBS J 283(16):3103–3114. https://doi.org/10.1111/febs.13796
3. Friedler A, Hansson LO, Veprintsev DB, Freund SM, Rippin TM, Nikolova PV, Proctor MR, Rudiger S, Fersht AR (2002) A peptide that binds and stabilizes p53 core domain: chaperone strategy for rescue of oncogenic mutants. Proc Natl Acad Sci U S A 99(2):937–942. https://doi.org/10.1073/pnas.241629998

Chapter 20

In Vitro Prion Amplification Methodology for Inhibitor Screening

Tuane Cristine R. G. Vieira and Jerson L. Silva

Abstract

Prion (PrP^C) is an endogenous protein found mainly in the nervous system, and its misfolded isoform (PrP^{Sc}) is associated with a group of neurodegenerative disorders known as transmissible spongiform encephalopathies, or simply prion diseases. The PrP^{Sc} isoform shows an intriguing ability to self-perpetuate, acting as template for PrP^C misfolding and consequent aggregation. Aggregation in vitro and in vivo follows a fibrillation processes that is associated with neurodegeneration. Therefore, it is important to investigate and understand the molecular mechanisms involved in this process; such understanding also allows investigation of the action of possible candidate molecules to inhibit this process. Here, we highlight useful in vitro methodologies and analyses that were developed using PrP as a protein model but that, as other amyloid proteins also exhibit the same behavior, may be applied to understand other "prion-like" diseases such as Alzheimer's and Parkinson's disease.

Key words Prion, Aggregation, Amplification, RT-QuIC

1 Introduction

The term prion was coined in the 1980s by Stanley Prusiner to designate a protein responsible for the development of transmissible spongiform encephalopathies (TSEs) [1]. TSEs are rare and fatal diseases that affect humans and other mammals and are caused by the conversion and aggregation of this protein. A short time later, it was discovered that this protein was constitutive and could therefore present itself in two forms, a nonpathogenic form, called cellular PrP (PrP^C), and a pathogenic form, called PrP *scrapie* (PrP^{Sc}) [1, 2]. These two proteins differ structurally, which causes them to have distinct physicochemical properties, including toxicity and self-perpetuating aggregation by PrP^{Sc} [3]. Recently, the term prion has spread to other proteins, and we also use the terms "Prionoid," "Prion-like," and "Prion-related" to group them, since the molecular mechanisms that were originally described for the PrP protein have been

Cláudio M. Gomes (ed.), *Protein Misfolding Diseases: Methods and Protocols*, Methods in Molecular Biology, vol. 1873, https://doi.org/10.1007/978-1-4939-8820-4_20, © Springer Science+Business Media, LLC, part of Springer Nature 2019

described for other proteins associated with different types of amyloidoses, including Parkinson's and Alzheimer's diseases [3–5].

PrP^{Sc} is a beta-sheet rich protein that serves as a template for PrP^{C} conversion [2, 3]. After conversion, PrP polymerizes into amyloid fibrils in vitro or in vivo. Fibrillation follows a general pathway that includes the following: (a) the formation of a misfolded/partially folded monomer with a high propensity to aggregate; (b) oligomer formation following a nucleation phase (polymerization is thermodynamically unfavorable up to the formation of a nucleus); and (c) a growth phase with the formation of protofibrils and fibrils (thermodynamically favored after nucleus formation) [6]. The evaluation of protein fibrillation generates a sigmoidal-shaped curve, explained by the nucleation and growth phases (Fig. 1). This ability to change protein folding through protein–protein interaction is what allows prion propagation.

All amyloidoses remain incurable diseases. Finding treatment that leads to a cure, or at least postponing the disease has been the focus of several investigations [7–10]. Molecules capable of modulating fibrillation thermodynamics and kinetics or those interfering with the nucleation and/or growth steps are interesting candidates

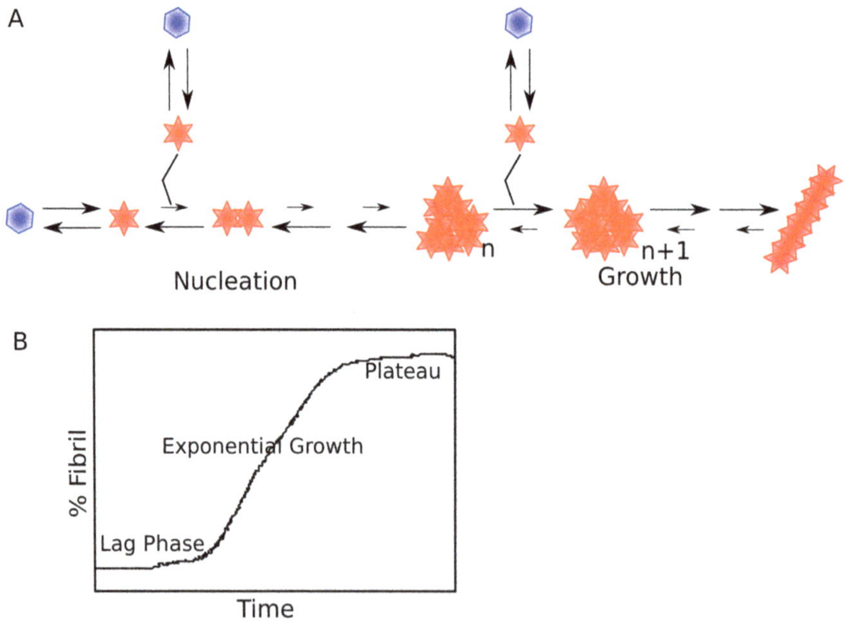

Fig. 1 (**a**) PrP fibrillation follows a nucleated conformational conversion model. PrP^{Sc} acts as a template for PrP^{C} conversion. Converted PrP^{Sc} aggregates form small oligomers up to the formation of a nucleus. This nucleation phase is thermodynamically unfavorable, but once the nucleolus is formed, protofibrils are formed following the growth phase (thermodynamically favored), which will produce mature fibrils. (**b**) Fibril formation can be monitored by ThT fluorescence. ThT fluorescence increases following a sigmoidal curve. Nucleation occurs during the Lag Phase. ThT binds to protofibrils and fibrils during exponential growth until the monomers are exhausted from the solution, reaching a plateau

for therapeutic intervention [11]. Assays involving cells are widely used to investigate the effect of inhibitors; however, they do not allow the molecular mechanism to be elucidated.

1.1 Experimental Approach

In this chapter, we present in vitro protein amplification methodologies to observe the process of aggregation of amyloidogenic proteins, as well as the effect of inhibitors using mathematical models to explain the mechanisms of action of these molecules. The approaches described here were obtained in our studies on the PrP aggregation process and its interaction with low-molecular-weight heparin [7], which proved to be a true inhibitor, increasing the stability of the monomeric protein, leading to an effect on both the thermodynamics and kinetics of the fibrillation process. This same approach was performed to investigate the interaction of the PrP protein with other inhibitors, as well as the fibrillation process of other proteins [8, 9, 12–14].

2 Materials

2.1 PrP Purification

All solutions have to be prepared using ultrapure water and analytical-grade reagents. All solutions must be freshly prepared and filtered through a 0.22 μm membrane. All purification steps are done at room temperature, unless otherwise stated.

2.1.1 BL21DE3 Escherichia coli Cell Transformation and Protein Expression

1. PET24a plasmid (EMD Biosciences) containing full-length PrP (23-231) or truncated PrP (90-231) with a mouse or hamster amino acid sequence (can be applied for other species sequence).
2. Liquid Luria–Bertani (LB) medium.
3. Kanamycin 50 mg/mL in sterile water.
4. 1 M IPTG (isopropyl β-D-1-thiogalactopyranoside) solution in sterile water.

2.1.2 Cell Lysis and Protein Purification

1. Bug Buster Master Mix 1× (Merck-Millipore, Massachusetts, cat. no. 71456).
2. Ni-NTA beads (Qiagen, Hilden, cat. no. 1018401).
3. SnakeSkin Dialysis Tubing 7 K (Thermo Fisher Scientific, Waltham, cat. no. 68700).
4. Filter 0.22 μM.
5. Denaturing Buffer 1: pH 8.0, 0.1 M sodium phosphate, 8 M GdnHCl, 10 mM Tris Base.
6. Denaturing Buffer 2: pH 8.0, 0.1 M sodium phosphate, 6 M GdnHCl, 10 mM Tris Base.
7. Refolding Buffer: pH 8.0, 0.1 M sodium phosphate, 10 mM Tris Base.

8. Elution Buffer: pH 5.8, 0.1 M sodium phosphate, 10 mM Tris base, 0.5 M imidazole.
9. Dialysis buffer: pH 5.8, 10 mM sodium phosphate.

2.2 Brain Homogenate Preparation

1. Brain Homogenization Buffer (BHB): 1 mM EDTA, 0.15 M NaCl, 0.5% Triton X-100, 1 tablet of cOmplete Protease Inhibitor w/o EDTA.

2.3 RT-QuIC Reagents

1. N2 media supplement (Thermo Fisher Scientific, Waltham, cat. no. 17502-048).
2. Sodium dodecyl sulfate.
3. Phosphate buffered saline (PBS).
4. EDTA.
5. Thioflavin T.
6. Sodium chloride.
7. Black, transparent, flat bottom 96-well plates and sealers (Thermo Fisher Scientific, Waltham, cat. no. 232702).
8. 100 KDa filters (Pall Life Sciences, cat. no. OD100C35).
9. SDS–PBS Buffer: 0.1% (wt/vol) SDS prepared in PBS. Keep the buffer stock at 4 °C.
10. 5× PBS, pH 6.9, buffer. Keep the buffer stock at 4 °C.
11. 100 mM EDTA. Keep the buffer stock at 4 °C.
12. 2 M NaCl.
13. 1 mM ThT. Keep the 1 mM stock in the dark at room temperature.

2.4 In Vitro-Produced Fibril Reagents

1. Stock solution of recombinant murine PrP (23–231) in 6 M GdnHCl.
2. Diluting Buffer: pH 6.8, 10 mM phosphate buffer, 1 M GdnHCl, 3 M urea, 150 mM NaCl.
3. Analysis Buffer: pH 5.5, 10 mM sodium acetate buffer, 10 μM ThT, PrP substrate at 0.5 μM.
4. Reaction mixture: pH 7.4, 10 mM phosphate buffer, 300 mM NaCl, 10 μM ThT, PrP substrate at 2 μM, and in vitro-produced fibrils at 1% (mol/mol).

3 Methods

3.1 PrP Purification

The recombinant PrP is prepared according to the previously published method with minor modification as described in [15]. The protocol detailed here is suitable for the purification of both full-length and truncated PrP.

3.1.1 Competent BL21-DE3 E. coli Cell Transformation

1. Place cell samples on ice for 15 min (~50 μL of cells suspension in conic tubes).
2. Add 1 μL of plasmid solution to the bacterial samples (control samples receive the same treatment, without plasmid addition).
3. Heat cells to 42 °C for 2 min.
4. Place cells on ice for 5 min.
5. Plate cells in LB-agar medium containing 50 μg/mL kanamycin. One of the control samples should be plated on LB-agar medium for cell viability control.
6. Incubate plates at 37 °C overnight.
7. Assess transformation efficiency the next day by comparing the controls with the transformed sample.

3.1.2 Transformed Bacterial Cell Growth for Protein Purification

1. Prepare a sterile preinoculum vial with ~25 mL of LB medium containing 50 μg/mL kanamycin. Add an isolated colony from the transformed plate and culture overnight at 37 °C with ~200 rpm shaking (*see* **Note 1**).
2. The next morning, add ~3 mL of the bacterial solution to each of four 2 L flask containing 1 L of fresh LB with 50 μg/mL kanamycin. Incubate at 37 °C with ~200 rpm shaking until an OD 600 nm of 0.8 is reached.
3. Centrifuge the contents of the four flasks at 2500 × *g*, collect the pellets and resuspend in 1 L of LB medium containing 50 μg/mL kanamycin and 1 mM IPTG (final concentration). Incubate overnight at 37 °C with ~200 rpm shaking.
4. Spin at 5000 rpm for 10 min and decant supernatant.
5. Weigh the cell paste and freeze in NL2 (*see* **Note 2**).
6. Transfer the frozen pellet to a 50 mL conical tube and thaw the pellet in a 37 °C bead bath (*see* **Note 3**).

3.1.3 Inclusion Body Preparation

1. Thaw the cell paste in a 37 °C bead bath.
2. Resuspend in a 1× Bug Buster Master Mix (add 5 mL reagent for each 1 g of cell pellet) (*see* **Note 4**).
3. Incubate on a rotating mixer for 20 min at RT.
4. Centrifuge at 13,000 × *g* for 20 min at 4 °C.
5. Discard the supernatant (*see* **Note 5**).
6. Resuspend the pellet in the same volume of 1× Bug Buster reagent used in **step 2** and put on the rotator for 15 min (*see* **Note 6**).
7. Q.S. to 40 mL with 1:10 (0.1×) Bug Buster reagent and mix by inversion.

8. Centrifuge at 7900 × *g* for 15 min at 4 °C to collect the inclusion bodies.
9. Discard the supernatant and resuspend the pellet in 20 mL of 0.1× Bug Buster reagent using a homogenizer.
10. Q.S. to 40 mL with 0.1× Bug Buster reagent and centrifuge at 13,000 × *g* for 15 min at 4 °C.
11. Discard the supernatant and freeze the inclusion body pellet for later use.

3.1.4 Protein Purification

1. Dissolve the inclusion body pellet in 14 mL Denaturing Buffer 1 using the Tissue Homogenizer and then put it on a rotator for ~50 min.
2. Rinse 30 mL of Ni-NTA superflow resin beads with ~100 mL of water using the in-house vacuum and a 0.22-μm bottle top filter (*see* **Note 7**). Transfer the beads into a 50 mL tube. Q.S. the tube to 30 mL with Denaturing Buffer 2 and equilibrate for ~50 min on a rotator at room temperature.
3. Spin the solubilized inclusion body pellet at 13,000 × *g* for 5 min to remove the insoluble debris and add the supernatant to the equilibrated beads.
4. Put on the rotator for 50 min to allow the protein to bind to the beads.
5. Load the resin onto an empty column and connect to an ÄKTA Explorer System (GE Healthcare).
6. Wash the resin with three column volumes of Denaturing Buffer 2 and then run a refolding gradient at 0.8 mL/min for 240 min (from 0% B to 100% B). Pump A1 (Denaturing Buffer) and Pump B (Refolding Buffer).
7. Let the Refolding Buffer flow through column for an additional 30 min at 0.8 mL/min after gradient.
8. To elute protein, run the Elution Buffer (100% Pump A2) at 2 mL/min.
9. Watch the UV 280 signal increase, then collect the center of the large peak. Dilute this peak immediately with approximately one-third of its volume of Dialysis Buffer (*see* **Note 8**).
10. Filter the protein with a 0.2 μM filter.
11. Put into Snakeskin dialysis tubing and dialyze in 4 L of prechilled dialysis buffer for 2 h and then transfer it to a fresh 4 L dialysis buffer for further dialysis overnight (*see* **Note 9**).
12. Filter the protein again following dialysis with a prewashed 0.2 μM filter.
13. Quantify the protein, watching the absorbance at 280 nm (use appropriate extinction coefficient) (*see* **Note 10**).

14. Immediately aliquot the protein and place in a −80 °C freezer.
15. When the aliquots are thawed, filter with a 100 KDa filter to remove the oligomeric PrP.
16. Check the protein on a gel for contaminants or fragmentation, and analyze by circular dichroism to check the secondary structure.

3.2 Brain Homogenate Preparation

1. Use 10% weight per volume of Brain Homogenate Buffer.
2. Dounce ten times with loose Dounce and ten times with tight Dounce.
3. Spin at 2000 × *g* for 2 min to partially clarify and transfer the supernatant to another tube.
4. Aliquot into small aliquots and place at −80 °C for storage.

3.3 RT-QuIC Protocol

1. Dilute N2 media supplement 100× in 0.1% SDS–PBS.
2. Dilute normal and infected 10% Brain Homogenate in 0.1% SDS–PBS–N2 at the desired dilution (*see* **Note 11**).
3. Filter recombinant PrP with a 100 kD microtube filter (PALL): 7000 × *g* for 5 min. Check the PrP concentration and calculate the volume to be pipetted (0.1 mg/mL rPrP in 100 μL of reaction) (*see* **Note 12**).
4. Prepare RT-QuIC cocktails as follows (for 1 well with 100 μL of final volume): 20 μL of 5× PBS Buffer, 1 μL of 100 mM EDTA, 1 μL 1 mM ThT, 8.5 μL 2 M NaCl, calculated rPrP volume. Bring to the final volume with Milli-Q water (*see* **Notes 13** and **14**).
5. Load RT-QuIC into the wells of a black 96-well plate (98 μL).
6. Seed each well with 2 μL of Brain Homogenate (desired dilution).
7. Compounds to be tested may be preincubated with either the recombinant protein or the seed or may be added at different reaction times (for example at the start of the reaction, at the end of the nucleation phase, or during the growth phase).
8. Seal the plate with a plate sealer.
9. Run RT-QuIC in a FluoStar OMEGA microplate reader (BMG LabTech) with double orbital shaking, at 42 °C for ~40 h.
10. Shaking program: 1 min shaking at 700 rpm Double Orbital/1 min resting.
11. Fluorescence measurement settings: excitation at 450 nm; emission at 480 nm; bottom read; 20 flashes; 20 μs integration time. Measure ThT fluorescence every ~15 min with a shaking kinetic cycle (*see* **Note 15**).

3.4 In Vitro-Produced Fibril Preparation and Amplification

1. Dilute the murine rPrP 23–231 in 6 M GdnHCl stock solution to a final concentration of 10 μM with Diluting Buffer and incubate this solution at 37 °C with continuous shaking at 600 rpm in conical plastic tubes (Eppendorf) in a 0.75 mL reaction volume.
2. To monitor PrP fibrillation, collect aliquots from the sample during the time course of the process (~70 h), dilute them in Analysis Buffer, and record the ThT fluorescence emission. Using a spectrofluorometer, excite the samples at 450 nm and record the emission spectra from 460 to 520 nm (maximum approximately 485 nm).
3. The subsequent steps follow **step 4** from RT-QuIC protocol with a different Reaction Mixture (described in Subheading 2).

3.5 Curve Fitting

Measured sigmoidal curves can be fit using scientific data analysis and graphing software. As shown by [12], a sigmoidal curve can be described by the following equation:

$$\Upsilon = y_{\mathrm{i}} + m_{\mathrm{i}}x + \frac{y_{\mathrm{f}} + m_{\mathrm{f}}x}{1 + \mathrm{e}^{-[(x - x_0)/\tau]}}$$

where Υ is the fluorescence intensity, x is time, and x_0 is the time to 50% of the maximal fluorescence. The apparent rate constant k_{app} for the growth of the fibrils is given by $1/\tau$, and the lag time is given by $x_0 - 2\tau$.

4 Notes

1. It is possible to maintain a glycerol culture at −80 °C. Aliquot this preinoculum with 25% glycerol (final concentration) and transfer to a freezer at −80 °C.
2. This growth method generates ~8 g cell paste.
3. You can freeze a second time in NL2 and place at −20 °C until further use.
4. Be sure to fully resuspend the pellet; use a tissue homogenizer.
5. Be sure to fully pellet all the material.
6. Completely dissolving the pellet is critical to obtain a high yield. Use a homogenizer or a sonicator if needed.
7. Make sure not to dry out the beads.
8. Diluting the protein immediately in dialysis buffer helps reduce protein loss.
9. Avoid bubbles.
10. Typical yield is ~20–30 mg.

11. 10^{-7} usually provides a good signal.
12. Keep on ice.
13. Reaction must contain a final concentration of 0.002% SDS.
14. Prepare a single cocktail to sufficiently fill all the wells to be analyzed.
15. Adjust the gain the first time the negative control is used.

5 Concluding Remarks

The prion protein fibrillation process associated with the establishment of TSEs involves the conversion of the protein using a seed that acts as a template. This conversion leads the protein to an aggregation pathway that can form fibrillar aggregates. Visualizing this process in vitro can provide important information about the development of the disease and has been the approach used by several researchers. One approach that has been used extensively to diagnose the disease is the RT-QuIC assay. As a prion amplification method, it can also be used to evaluate the effect of inhibitor candidates on the aggregation process, and extract information about the molecular details of inhibition.

The RT-QuIC assay and the in vitro-produced fibril amplification assay have been shown to be very reproducible and easy to perform [16] (Fig. 2). Because recombinant PrP is used as a substrate, the assays can be repeated many times with different approaches. PrP with

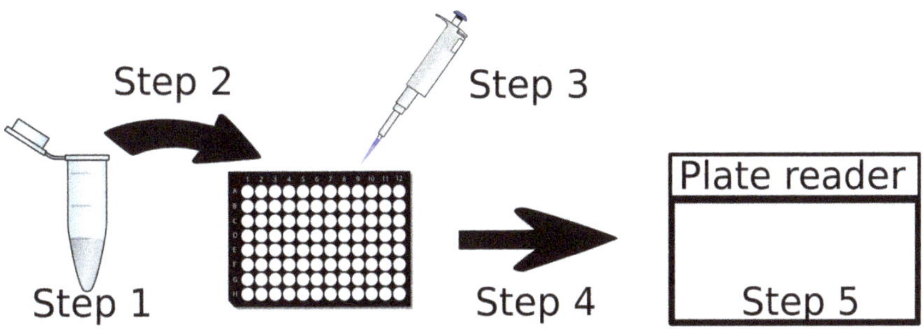

Fig. 2 RT-QuIC and in vitro-produced fibril amplification assays follow a very similar step-by-step procedure. Step 1 is to prepare the RT-QuIC Cocktail or Reaction Mixture. In both methods this step consists in subjecting the recombinant Prion protein that will be converted to the experimental buffer. Unlike RT-QuIC, in the in vitro-produced fibril amplification assay the fibers that will seed the aggregation are added in this step, not Step 3. Compounds can be added in this step in order to investigate their effect. 100 μL of this sample are transferred to each well of the 96-well black plate with transparent bottom (Step 2). In Step 3 2 μL of normal (negative control) or infectious brain homogenate are added during RT-QuIC protocol. Compounds can also be added during Step 3. After the plate is prepared and sealed, it is taken to a plate reader with double orbital shaking, at 42 °C for ~40 h (Step 4). The analysis is done following the increase in ThT fluorescence (described in Fig. 1)

different sequences (hamster, mouse and others) can be used as the full-length or truncated form. This is an advantage because it is possible to investigate if the inhibitory effect of a molecule is broad or restricted to a specific sequence, and if it needs the unstructured N-terminal or globular C-terminal domain.

We used the RT-QuIC assay to investigate the effect of LMWHeparin on PrP conversion and aggregation [7]. The assay enabled us to show that LMWHep inhibits the conversion and aggregation of PrP induced by infected brain homogenate seeds through interaction with the N-terminus of the protein. Furthermore, the interaction with both the N- and C-terminus dramatically reduced the formation of fibers, interfering with the nucleation process. The same approach was used to search a library of aromatic organic compounds [9], and identify trimethoxychalcone as a promising anti-prion compound, reducing fibrillation though interaction with the globular domain [8]. Acridine, dextran sulfate sodium, and tannic acid have been shown to inhibit PrP conversion in the RT-QuIC assay [14].

The RT-QuIC assay can be performed using the brain, cerebrospinal fluid, nasal brushings, and plasma to seed the reaction. In the absence of such samples, fibers produced with the recombinant protein in vitro may be used. In 2005, Baskakov described [17, 18] a method using shaking and chemical denaturants to produce PrP fibrils in vitro. An adaptation of the RT-QuIC method using these fibrils as seeds that were shown to be efficient and corroborated the results found for RT-QuIC seeded with brain homogenates [8].

The assays showed here use the fluorescent amyloid-specific dye thioflavin T (ThT). ThT fluoresces at 490 nm after it binds to amyloid fibrils, and the fluorescence is proportional to the fibril concentration over a wide range. It does not affect the aggregation process, and can be monitored in real-time [19].

ThT fluorescence kinetics show a sigmoidal curve (Fig. 1), and the data can be fit to different equations depending on the aggregation model [6]. PrP fibrillation is believed to follow a nucleated conformational conversion model (Fig. 1) [20], and the data can be fit by a logistic equation for comparison of the kinetics of fibrillation [6, 12]. At least three important parameters can be mathematically measured and compared: (a) the amount of fibril (relative to ThT intensity; (b) the lag time of the nucleation phase; and (c) the rate constant of the exponential growth (k_{app}). Inhibitors that show a thermodynamic effect reduce the fibril mass at the end of the reaction (Fig. 3). Inhibitors that show this effect or not can also affect the kinetics, affecting the nucleation when delaying the lag time, and/or altering the k_{app} when affecting fibril growth (Fig. 3). An inhibitor that affects both the thermodynamics and

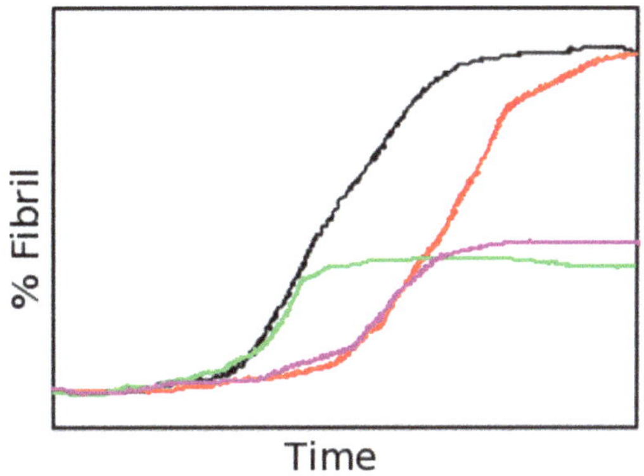

Fig. 3 Sigmoidal fibrillation curve of a protein in the absence (black) and in the presence of inhibitors (colored traces). Some inhibitors can affect fibrillation kinetics, changing the lag time and/or k_{app} without changing the fibril content (red curve). Others can affect the fibril thermodynamics, without altering the kinetics (green curve). A true inhibitor changes both the kinetics and thermodynamics (pink curve)

kinetics of the fibrils would be the best option for a therapeutic candidate (Fig. 3).

Several other proteins follow a nucleated conformational conversion model [21, 22]. These proteins are associated with amyloidoses that have been considered as prion-like diseases [3, 4]. As these proteins also show self-propagation, the assays disclosed herein are promising for adaptation to other amyloid proteins and can be used for the diagnosis of such diseases and the screening of therapeutic molecules.

Acknowledgments

We are thankful to Lucas M. Ascari (Faculty of Pharmacy-UFRJ) for revision of the in vitro-produced fibril assay.

References

1. Prusiner SB (1998) Prions. Proc Natl Acad Sci U S A 95(23):13363–13383. https://doi.org/10.1073/pnas.95.23.13363
2. Silva JL, Cordeiro Y (2016) The "Jekyll and Hyde" actions of nucleic acids on the prion-like aggregation of proteins. J Biol Chem 291(30):15482–15490. https://doi.org/10.1074/jbc.R116.733428
3. Eraña H, Venegas V, Moreno J, Castilla J (2017) Prion-like disorders and transmissible spongiform Encephalopathies: an overview of the mechanistic features that are shared by the various disease-related misfolded proteins. Biochem Biophys Res Commun 483(4):1125–1136. https://doi.org/10.1016/j.bbrc.2016.08.166
4. Soto C (2012) Transmissible proteins: expanding the prion heresy. Cell 149(5):968–977. https://doi.org/10.1016/j.cell.2012.05.007
5. Rangel LP, Costa DCF, Vieira TCRG, Silva JL (2014) The aggregation of mutant p53 produces prion-like properties in cancer. Prion 8(1):75–84. https://doi.org/10.4161/pri.27776
6. Morris AM, Watzky MA, Finke RG (2009) Protein aggregation kinetics, mechanism, and

curve-fitting: a review of the literature. Biochim Biophys Acta 1794(3):375–397. https://doi.org/10.1016/j.bbapap.2008.10.016

7. Vieira TC, Cordeiro Y, Caughey B, Silva JL (2014) Heparin binding confers prion stability and impairs its aggregation. FASEB J 28(6): 2667–2676. https://doi.org/10.1096/fj.13-246777
8. Ferreira NC, Ascari LM, Hughson AG, Cavalheiro GR, Góes CF, Fernandes PN, Hollister JR, da Conceição RA, Silva DS, Souza AMT, Barbosa MLC, Lara FA, Martins RAP, Caughey B, Cordeiro Y (2017) A promising anti-prion trimethoxychalcone binds to the globular domain of PrPC and changes its cellular location. Antimicrob Agents Chemother. https://doi.org/10.1128/AAC.01441-17
9. Ferreira NC, Marques IA, Conceição WA, Macedo B, Machado CS, Mascarello A, Chiaradia-Delatorre LD, Yunes RA, Nunes RJ, Hughson AG, Raymond LD, Pascutti PG, Caughey B, Cordeiro Y (2014) Anti-prion activity of a panel of aromatic chemical compounds: in vitro and in silico approaches. PLoS One 9(1):e84531. https://doi.org/10.1371/journal.pone.0084531
10. Burchell JT, Panegyres PK (2016) Prion diseases: immunotargets and therapy. Immunotargets Ther 5:57–68. https://doi.org/10.2147/ITT.S64795
11. Martins PM (2013) True and apparent inhibition of amyloid fibril formation. Prion 7(2): 136–139. https://doi.org/10.4161/pri.23111
12. Nielsen L, Khurana R, Coats A, Frokjaer S, Brange J, Vyas S, Uversky VN, Fink AL (2001) Effect of environmental factors on the kinetics of insulin fibril formation: elucidation of the molecular mechanism. Biochemistry 40(20):6036–6046. https://doi.org/10.1021/bi002555c
13. Kamal Zaidi F, Bhat R (2018) Resveratrol interferes with an early step in the fibrillization pathway of human lysozyme and modulates it towards less-toxic, off-pathway aggregates. Chembiochem 19(2):159–170. https://doi.org/10.1002/cbic.201700207
14. Hyeon JW, Kim SY, Lee SM, Lee J, An SS, Lee MK, Lee YS (2017) Anti-prion screening for Acridine, dextran, and tannic acid using real time-quaking induced conversion: a comparison with PrPSc-infected cell screening. PLoS One 12(1):e0170266. https://doi.org/10.1371/journal.pone.0170266
15. Wilham JM, Orru CD, Bessen RA, Atarashi R, Sano K, Race B, Meade-White KD, Taubner LM, Timmes A, Caughey B (2010) Rapid end-point quantitation of prion seeding activity with sensitivity comparable to bioassays. PLoS Pathog 6(12):e1001217. https://doi.org/10.1371/journal.ppat.1001217
16. Schmitz M, Cramm M, Llorens F, Müller-Cramm D, Collins S, Atarashi R, Satoh K, Orrù CD, Groveman BR, Zafar S, Schulz-Schaeffer WJ, Caughey B, Zerr I (2016) The real-time quaking-induced conversion assay for detection of human prion disease and study of other protein misfolding diseases. Nat Protoc 11(11):2233–2242. https://doi.org/10.1038/nprot.2016.120
17. Baskakov IV, Bocharova OV (2005) In vitro conversion of mammalian prion protein into amyloid fibrils displays unusual features. Biochemistry 44(7):2339–2348. https://doi.org/10.1021/bi048322t43
18. Bocharova OV, Breydo L, Salnikov VV, Gill AC, Baskakov IV (2005) Synthetic prions generated in vitro are similar to a newly identified subpopulation of PrPSc from sporadic Creutzfeldt-Jakob disease. Protein Sci 14(5):1222–1232. https://doi.org/10.1110/ps.041186605
19. Naiki H (2017) Thioflavin T: not an all-rounder, but a trustworthy friend for over 27 years. Amyloid 24(sup1):9. https://doi.org/10.1080/13506129.2017.1278688
20. Serio TR, Cashikar AG, Kowal AS, Sawicki GJ, Moslehi JJ, Serpell L, Arnsdorf MF, Lindquist SL (2000) Nucleated conformational conversion and the replication of conformational information by a prion determinant. Science 289(5483):1317–1321. https://doi.org/10.1126/science.289.5483.1317
21. Lee J, Culyba EK, Powers ET, Kelly JW (2011) Amyloid-Œ≤ forms fibrils by nucleated conformational conversion of oligomers. Nat Chem Biol 7(9):602–609. https://doi.org/10.1038/nchembio.624
22. Nizynski B, Dzwolak W, Nieznanski K (2017) Amyloidogenesis of tau protein. Protein Sci 26(11):2126–2150. https://doi.org/10.1002/pro.3275

Chapter 21

SolubiS: Optimizing Protein Solubility by Minimal Point Mutations

Rob van der Kant, Joost van Durme, Frederic Rousseau, and Joost Schymkowitz

Abstract

Protein solubility is adapted to endogeneous protein abundance in the cell where protein folding is also assisted by multiple chaperones. During recombinant protein production, purification and storage proteins are frequently handled at concentrations that are several orders of magnitude above their physiological concentration, often resulting in protein aggregation. Here we describe SolubiS, a method allowing for (1) detection of aggregation prone linear segments within a protein sequence and (2) identification of mutations that abolish the aggregation propensity of these segments without affecting the thermodynamic stability of the protein. Provided the availability of structural information this method is applicable to all globular proteins including antibodies, resulting both in increased in vitro protein solubility and in better protein production yields.

Key words SolubiS, Protein aggregation, Protein design

1 Introduction

1.1 General Introduction

Protein aggregation remains a major factor limiting the production and application of many proteins, including monoclonal antibodies and enzymes. Protein aggregates are assemblies consisting of native and/or unfolded proteins [1] that have a negative effect on stability, safety, titers, immunogenicity, manufacturability, and solubility of biologics in general [2–6].

Currently, it is widely accepted that aggregation is mediated by the presence of aggregation- prone regions (APRs) , which assemble by intermolecular β-structured interactions, forming the core of the aggregate. Although these APRs are mostly hydrophobic in nature, they additionally require other key properties as a high β-sheet propensity and a low net charge. Methods of aggregation prediction are tailored to identify APRs in the primary sequence, which are extensively reviewed elsewhere [7–9]. These methods predict the theoretical aggregation potential of proteins in the

Cláudio M. Gomes (ed.), *Protein Misfolding Diseases: Methods and Protocols*, Methods in Molecular Biology, vol. 1873, https://doi.org/10.1007/978-1-4939-8820-4_21,

unfolded state, termed the "intrinsic aggregation propensity." Most of these methods are based on the aggregation of short stretches of amino acids in vitro, outside of the context of the protein.

To nucleate aggregation, an APR is required to be solvent exposed to allow the formation of stable interactions with other like sequences. Yet, in most globular proteins, these APRs are buried inside the hydrophobic core of the folded structure, where they are protected from nucleating aggregation by the thermodynamic stability of the protein [8, 10]. Significant exceptions are active sites and protein–protein interaction sites [11–14], where functional requirements appear to lead to the occurrence of APRs that can be detrimental when the protein is not engaged in functional interactions. For example, at formulation conditions of therapeutic proteins, high concentration and the absence of ligand or other interaction partners promote protein self-interaction and the formation of aggregates in the stock solutions [15].

Besides structural stabilization, a number of additional mechanisms can contribute to suppress aggregation [16]. One of these mechanisms is the presence of aggregation gatekeeper residues, which are generally charged residues or proline residues that decelerate the aggregation reaction [17–20]. In natural protein sequences, such gatekeeper residues are strongly enriched at the flanks of APRs. Moreover, molecular chaperones, such as Hsp70, bind to exposed APRs, therefore preventing intermolecular assembly of APRs to nucleate aggregation [21]. Finally, protein turnover rates [22] as well as protein expression levels [23] are evolutionary adapted to minimize protein aggregation under physiological conditions.

The SolubiS method allows for the introduction of gatekeeper residues in structurally unstable APRs, while maintaining the proteins structural integrity and function (Fig. 1).

1.2 The SolubiS Theory in Short

Aggregation is a stochastic process; this means that the effective aggregation of a protein is dependent on the population of aggregation-compatible conformations in which the APRs are exposed. The cooperation of physicochemical parameters such as protein and ion concentrations, pH, and temperature contributes to determine the concentration of aggregation-prone conformers in a protein solution. Therefore, the challenge for relatively thermodynamically stable proteins is identifying sequences that will aggregate under native conditions. The solution to this challenge lies in the distinction between APRs that are thermodynamically protected by folding and those that occur in aggregation-competent

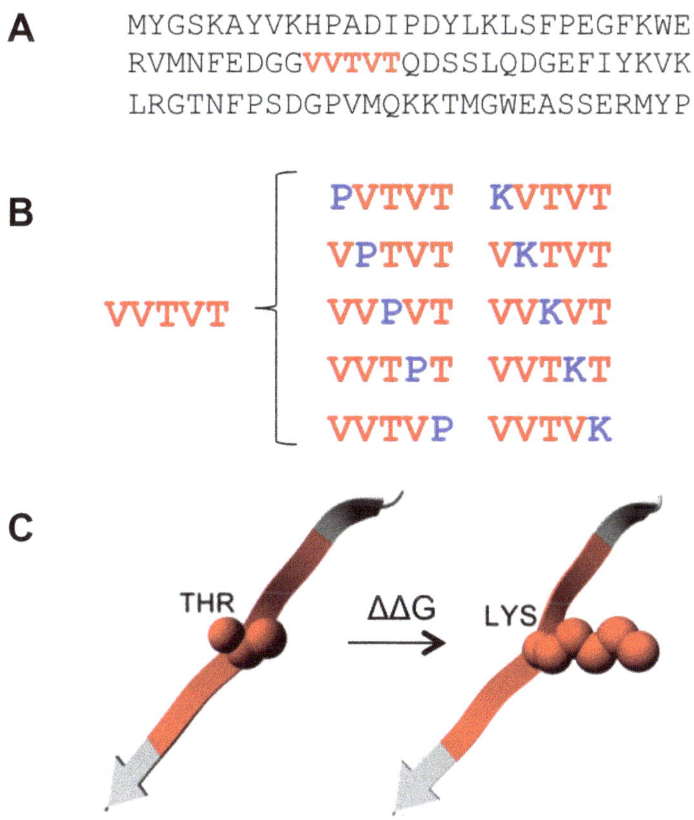

Fig. 1 The basic principle of the SolubiS method. (**a**) Locate APRs using TANGO. (**b**) Mutate every residue in an APR to the different gatekeeper residues (Arg, Lys, Asp, Glu, and Pro). (**c**) Calculate the effect of each mutation on the thermodynamic stability of the protein using FoldX

conformations that can form without major unfolding transitions (Fig. 2a).

2 Methods

The SolubiS method uses TANGO to predict APRs from the primary peptide sequence [24] and the thermodynamic contribution of these APRs to the stability of the protein using FoldX [25]. FoldX performs best on crystal structures, but can also perform on a reliable homology model when the crystal structure has not been elucidated. The prediction of the effect of a mutation on the thermodynamic stability of a structure is also performed using FoldX.

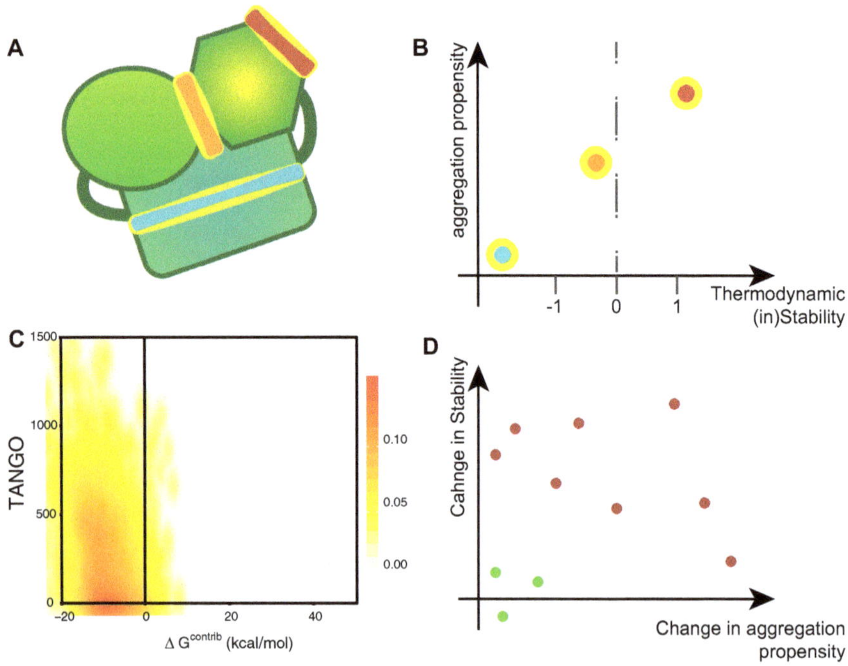

Fig. 2 Principles behind the SolubiS method. (**a**) Example schematic representation of possible locations of APRs in proteins. APRs can occur buried in the protein (Blue), at the functional sites like domain interfaces (Orange) or at an unstable region of the protein (Red). (**b**) Stretch-plots: representation of aggregation propensity and local stability of APRs. Problems increase toward the top right of the plot; ideally, APRs would be located in the bottom left. A single point in the stretch-plot represents an APR. In this theoretical representation, the colors represent the APRs in Fig. 2a. (**c**) Density plot of aggregation propensity and local stability of APRs in crystal structure of globular proteins. The analysis is based on a set of 2650 high quality structures (*R*-factor of <0.20 and resolution of <1.9, with 30% sequence identity) of globular proteins generated by the Whatif software suite [15, 30]. (**d**) MASS-plots: representation for the potential aggregation reducing mutations in an APR. Green dots represent mutations that decrease the aggregation propensity, while not destabilizing the structure. Red dots are mutations that either decrease the aggregation propensity insufficiently or are too destabilizing to the structure

2.1 APR Prediction from Primary Sequence

While there are many predictors for intrinsic aggregation propensity published, we prefer to use TANGO [24], as it predicts APRs with well-defined sequence boundaries, that is, with a definitive segregation between the APR and the surrounding residues. Additionally, and especially important for the purpose of identifying critical APRs, TANGO accomplishes high specificity and thus predicts few false-positive APRs [7]. Moreover, no other predictors have been tested in combination with FoldX, meaning that the success rates are unknown. The SolubiS method currently only uses TANGO to predict APRs in the primary sequence and subsequently uses it to predict the effect of gatekeeper mutations on these APRs. The added benefit is a clear metric for the effect of each mutation, which makes it more straightforward to select a mutant that decreases the aggregation potential the most.

2.2 Structural Context

Previous successful applications of the SolubiS methodology have been accomplished by starting with a high resolution crystal structure [15, 26]. In the absence of a high-resolution structure, the best way to go is building a reliable homology model. Many methods have been developed to obtain such a homology model [25, 27–29], with the I-TASSER suite ranking number one in the most recent community-wide protein structure prediction experiment, CASP12. In the SolubiS method, as FoldX is used to predict the effect of a mutation on the thermodynamic stability of a protein, a model built or optimized by FoldX is preferred. In order to achieve optimal results, it is evident that a model should resemble the natural occurrence of the protein, meaning that the oligomerization state of the protein should be respected. To generate a reliable structure, attention should also be paid to crystal contacts, that is, to artifactual protein interactions resulting from protein crystallization, or by fusion proteins, stabilizing nanobodies, molecular crowders, and others. Conversely, attention should be paid to leave natural ligands or cofactors in place. Once a reliable and clean structure is obtained, the structure should be "repaired" using FoldX. FoldX uses a rotamer database, which contains a large set of side-chain rotamers, extracted from high-resolution PDB structures. This process optimizes the conformation of the side chains, allowing for an accurate calculation of the effect of each mutation. It is always wise to use different structures to perform the mutation scan and to compare the consistency of the predictions afterward, thereby decreasing the risk of selecting destabilizing mutations.

2.3 Stretch-Plots

To easily visualize the aggregation susceptibility of a protein in function of both its intrinsic aggregation propensity and thermodynamic stability, we devised the "stretch-plot" (Fig. 2b). In a stretch-plot, each APR is represented by a single data point, with on the y-axis the local "intrinsic" aggregation propensity of the sequence segment (summed TANGO, 5–100 per residue) and on the x-axis its contribution to the free energy of folding ($\Delta G^{contrib}$). The stretch-plot thereby allows for a simple description of the aggregation potential of each APR within a protein. As mentioned before, for a protein to aggregate, an APR has to be solvent exposed in order for it to interact with other like sequences. Local stability has been shown to be a good variable to predict the contribution of APRs to the aggregation propensity of a protein [15]. The stretch-plot therefore allows for identifying and prioritizing potential problematic APRs by both their intrinsic aggregation propensity and their contribution to the stability of the native protein. In theory, APRs in the top right corner are "critical" since they occur in less stable regions of the protein and display higher intrinsic aggregation propensities (Red dot in Fig. 2b). A density plot of all stretch-plots based on a set of 2650 high resolution structures of globular proteins generated by the WhatIf software suite [15, 30], showed the distribution of APRs in globular proteins (Fig. 2c). There is a clear tendency for proteins to have APRs located in the bottom left part of the stretch-plot, although there are many exceptions.

2.4 Scoring Function

To extract critical APRs from the stretch-plot of each protein and represent these by a single metric, we developed a simple integrative scoring function. A corrected TANGO score for each APR was constructed by cutting off the $\Delta G^{contrib}$ at −5 and +5 kcal/mol and by normalizing this value between 0 and 1 (0 representing a thermodynamically stable APR, 1 representing a highly unstable APR). Multiplying this value with the TANGO score allows for the reduction of stable APRs in a way that the total score is dominated by the contribution of critical APRs. We call this scoring scheme the "SolubiS score." APRs with a low SolubiS score are deemed noncritical, that is, having a good local stability. Low scoring APRs are usually buried in the core of the protein, where they contribute to stability. As they are unlikely to become solvent exposed under native conditions, they are unlikely to be the drivers of aggregation. High scoring APRs have a high TANGO score, that is, aggregation propensity, and have an unfavorable $\Delta G^{contrib}$, that is, not contributing to the stability of the protein. These APRs might become solvent exposed and are more likely to drive aggregation. Each APR is scored and the scores of all APRs in the protein are summed. A low score means that the protein contains no "critical" APRs, while a high score represents a protein that might aggregate under native conditions. The score can be used to compare proteins that are relatively similar, as previously applied for monoclonal antibodies [15], where the SolubiS score was used to classify antibodies based on aggregation propensity.

2.5 Predicting SolubiS Mutations

Once a reliable structure has been chosen, all APRs in the primary sequence have been located and their structural context has been determined, the mutant prediction can start. In general, every APR is scanned for possible aggregation reducing mutations. Per APR each residue is mutated to a gatekeeper residue (Arg, Lys, Glu, Asp or Pro) (Fig. 1b). For all mutations, the effect on aggregation propensity (total TANGO score) and stability (FoldX output) should be assessed (Fig. 1c). To easily select mutations one can plot the $\Delta\Delta G$ ($\Delta G_{Mutant}-\Delta G_{WT}$) against the ΔTANGO ($TANGO_{Mutant}-TANGO_{WT}$), resulting in a MASS-plot (mutant aggregation and stability spectrum) for each APR (Fig. 2d) [15]. Three variables are important when considering which mutation to select, (1) the effect of each mutation on the aggregation propensity, (2) the effect of a mutation on the thermodynamic stability of the protein, (3) the effect of a mutation on the stability of the complex between the protein and its possible interaction partners.

2.6 Mutant Selection

As a MASS-plot is generally obtained for each APR, the MASS-plot of the most critical APRs should be carefully examined. The perfect SolubiS mutation maximally decreases the aggregation propensity (ΔTANGO) and does not destabilize the structure. This means the best candidate mutations are in the bottom left corner of the plot (Fig. 2d). Concerning ΔΔG, it would be optimal to find a stabilizing mutation, but most importantly it should not destabilize the protein. As this is not always possible, we prefer trying the least destabilizing mutation and in the subsequent round, try to find a mutation that counters the destabilizing effect of the SolubiS mutation.

2.7 Pipelines

There are several ways to obtain SolubiS mutations. The first is using the webserver [31] (http://solubis.switchlab.org/), which produces the full output and guarantees a proper use of the relevant software. It makes a report, including the stretch-plot and MASS-plots, which makes the selection of mutations straightforward. The second is using the SolubiS plugin for YASARA [32, 33], which is more interactive and has been thoroughly documented, guiding the user through the process step-by-step. The last possibility is running TANGO and FoldX manually, which allows for more control over all steps. As it requires some scripting, output analysis and plotting, we suggest this method should only be used by people with substantial experience in protein structure bioinformatics. A benefit of this method is that one can use different aggregation propensity predictors, as well as different in silico predictors for calculating the effect of point mutations on the thermodynamic stability of a protein.

2.8 Validating Promising SolubiS Mutations

After the in silico analysis is performed and the data is collected, one can start choosing promising mutations. In our experience, selecting about five of the best ranking mutations generally results in the identification of a mutant with significantly decreased aggregation propensity. This number is only an indication to increase the success rate of the effort. Obviously, these mutant proteins should be produced and purified, after which aggregation propensity, thermodynamic stability and functionality should be tested and compared to the wild type. In our experience, the largest jump in aggregation resistance is obtained when multiple SolubiS mutations are combined. As mentioned before, globular proteins have multiple aggregation prone regions, while one SolubiS mutation can only target one APR at the time. Generally, we combine successful SolubiS mutations, resulting in a synergy effect. As multiple APRs can be available to cause aggregation, knocking one out might reduce the aggregation propensity, but the largest effect can be seen when removing both APRs. An additional factor to keep in mind is the effect of a SolubiS mutation on the isoelectric point (pI). Depending on your buffer pH, a mutation can cause the pI to come closer to the pH, thereby reducing the absolute net charge of the protein. This decreases the electrostatic repulsion between proteins, allowing them to get closer together, increasing the stochastic chance of interaction.

3 Case Studies

3.1 α-Galactosidase (α-Gal)

Human α-galactosidase (α-Gal) is a lysosomal hydrolase whose deficiency results in Fabry disease, a metabolic X-linked inherited lysosomal storage disorder (LSD) [34, 35]. Enzyme replacement therapy, where recombinantly produced α-Gal is intravenously delivered to the patient to replace the endogenous protein, is currently used to treat Fabry disease. The structure of α-Galactosidase consists of a homodimer in which each monomer contains a (β/α) domain (Fig. 3a), harboring the active site and an antiparallel β domain. Since the wild type protein was aggregation prone, we used the SolubiS method to design mutations to suppress this aggregation [26]. α-Gal has three APRs, two of which are targetable by SolubiS. During the first round, single SolubiS mutations were selected, resulting in an increase in solubility, while retaining their enzymatic function. In the second round, double mutants and triple mutants were designed by combining the best mutations with additional mutations designed to increase the thermodynamic stability of the protein. As in the first round of mutations, the specific enzymatic activity remained comparable to WT. Interestingly the fraction of protein in the insoluble fraction was clearly reduced, and the accumulation of soluble aggregates in a size exclusion assay was clearly reduced as well.

3.2 Protective Antigen (PA)

PA is an element of the protein toxin secreted by *Bacillus anthracis*, the pathogen that causes anthrax [36]. PA is an important component of recombinant Anthrax vaccines, since the immune response against this virulence factor confers protection against Bacillus anthracis infection [37]. However, recombinant PA (rPA) stored in aqueous buffers is aggregation-prone [38]. The structure of PA reveals four domains (Fig. 3b): Domain 1 contains two calcium binding sites and the furin cleavage site required for proteolytic

Fig. 3 (continued) sites highlighted. The respective APRs are highlighted in both the structure and the stretch-plot. The large red circle shows the same area in the aggregation resistant mutant (A348R/A368R/S405L). Mutations are highlighted in green. (**b**) Crystal structure of PA wild type (PDB ID: 1T6B [43]). The small red circle shows the location of the "critical" APR in the stretch-plot (red dot). The large red circle shows the same area of the aggregation resistant mutant (T576E/S559 L), based on a crystal structure (PDB ID: 5FR3 [26]). Mutations are highlighted. Images of protein structures were generated using YASARA Structure [27]

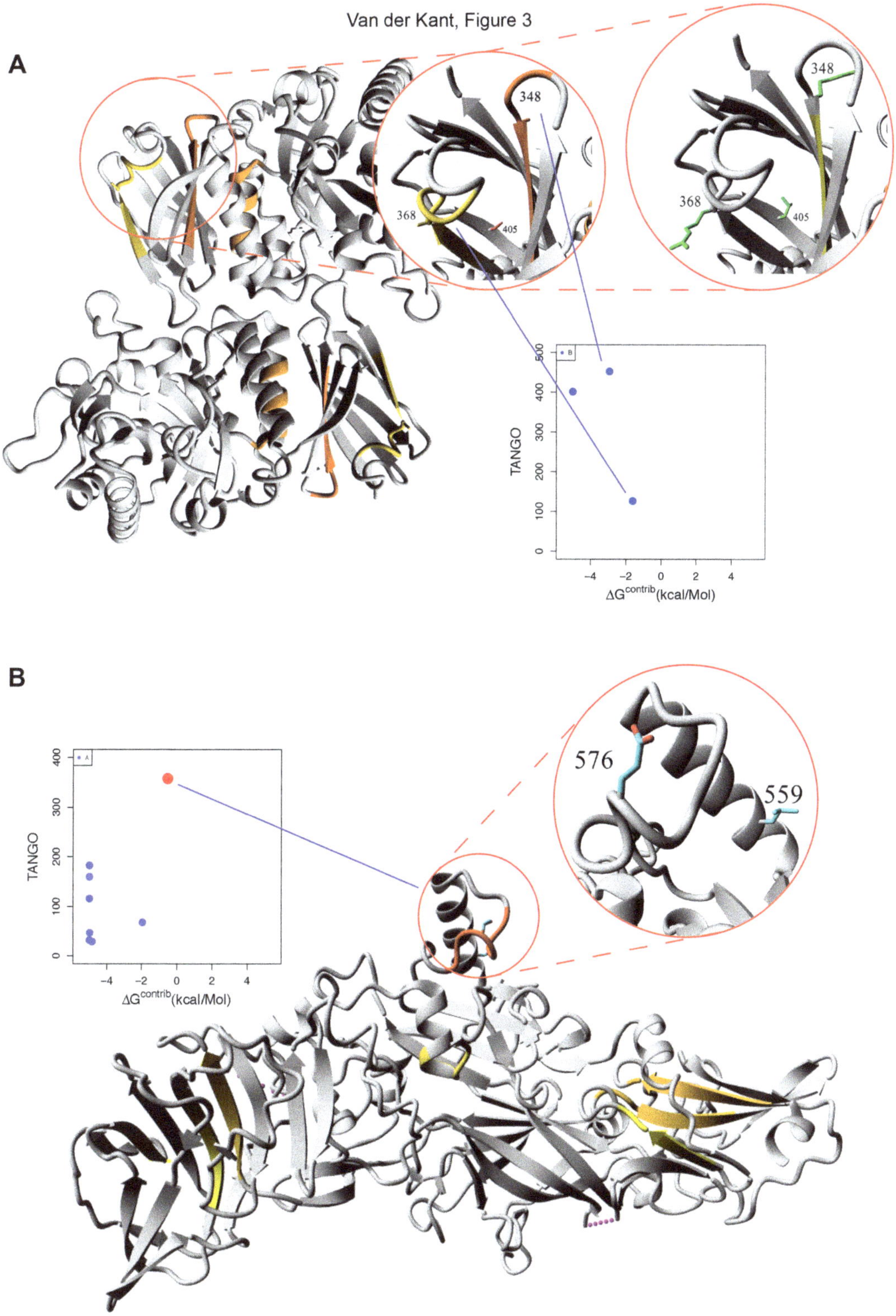

Fig. 3 Application of the SolubiS method on α-galactosidase (α-Gal) and protective antigen (PA). (**a**) Crystal structure of α-Gal homodimer (PDB ID: 3GXP [42]) with APRs highlighted. The small red circle shows the location of the two "critical" APRs, and the middle red circle shows the same region zoomed in with the mutation

activation of the molecule. Domain 2 is required for membrane insertion and together with domain 3 for heptamer formation. Domain 4 is essential for binding cell surface receptors [39]. TANGO detects several APRs in the amino acid sequence of the protein, of which one clearly stands out in the stretch plot (Fig. 3b). One APR is located toward the top right, indicating low stability and high intrinsic aggregation propensity. Incidentally, previous experimental work identified this region as the key determinant of rPA aggregation in vitro [40]. Using SolubiS, three mutations were designed to reduce the aggregation propensity of PA, resulting in mutants that were more resistant to heat-stress-induced aggregation. An additional mutation was designed using FoldX to increase the thermodynamic stability of mutant PA, resulting in an improved thermodynamic stability and increased aggregation resistance at 40 °C. The scoring function facilitates converting the stretch-plot to a single value, the SolubiS score (Table 1). From both the stretch-plots and the SolubiS scores for WT and mutant PA, we can see the mutant is predicted to be significantly less aggregation prone than the wild type. The "critical" APR is clearly visible in the WT stretch-plot, it has a high TANGO score and a low stability according to FoldX. This APR has disappeared in the stretch-plot of the aggregation resistant mutant, resulting in a much lower SolubiS score.

Table 1
SolubiS scores based on the stretch-plots of Figs. 3, 4, 5, and 6. Names, representative structures, references, and SolubiS scores are listed. For mutants where no structure has been elucidated, the mutation was done by FoldX on the representative wild-type structure and the SolubiS score was calculated based on the resulting mutant structure

Name	PDB ID	SolubiS score
α-Gal WT	3GXP [42]	166.8
α-Gal A348R/A368R/S405L	FoldX [25]	0
α-Gal A348R/A368P/S405L	FoldX [25]	0
Protective Antigen WT	1T6B [43]	181.2
Protective Antigen T576E/S559L	5FR3 [26]	21.7
Human β2-microglobulin	2XKS [44]	531.5
Mouse β2-microglobulin	1LK2 [45]	0
mAb2 WT	2FJF [47]	277.0
mAb2 SL50K/FH101P	FoldX [25]	30
Herceptin F_V WT	4HKZ [49]	234.0
Herceptin F_V SL52D/KH30D	4X4Y [50]	78.8

3.3 β2-Microglobulin (β2m)

Although the study by Radford and co-workers into the differences in aggregation between Human β2-microglobulin (hβ2m) and Mouse β2-microglobulin (mβ2m) was not designed with SolubiS in mind, it provides an interesting case for post-hoc analysis. hβ2m is an aggregation-prone protein involved in pathological amyloid deposition in haemodialysis patients. mβ2m on the other hand barely forms amyloid aggregates under physiological conditions despite having 100 times higher serum levels than hβ2m, which is more than five times higher than required to trigger dialysis-related hβ2m. Mouse and human β2-microglobulin orthologs are 70% identical in sequence and share 88% overall sequence similarity. Considering only the primary structure, both sequences are predicted to have similar aggregation and amyloid propensities by various algorithms. Differences in aggregation propensity have been attributed to differences in solubility of the unfolded proteins at pH 2.0 due to differences in (charge) composition of the FG-loop [41]. Analysing the structural context of each APR for the human (Fig. 4a) and the mouse variant (Fig. 4b) by means of stretch plots (Fig. 4c, d, respectively), there also appears to be a large difference in the local contribution of the APR to the stability of the folded protein under physiological conditions: the APR in hβ2m is less stable than the APR in mβ2m, explaining the ability of

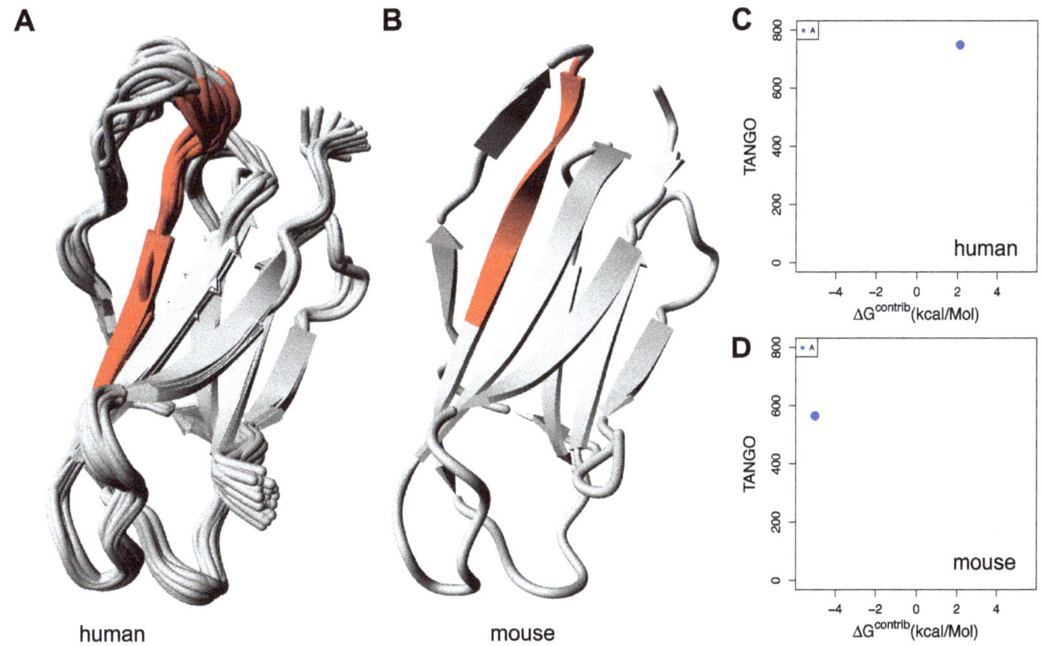

Fig. 4 Structural context of the APRs in Human and mouse β2-microglobulin. (**a**) Solution NMR structure ensemble of human β2-microglobulin (PDB ID: 2XKS [44]) showing the APR in red. (**b**) Crystal structure of mouse β2-microglobulin (PDB ID: 1LK2 [45]) showing the APR in red. (**c**) Stretch-plot of human β2-microglobulin. (**d**) Stretch-plot of mouse β2-microglobulin. Images of protein structures were generated using YASARA Structure [27]

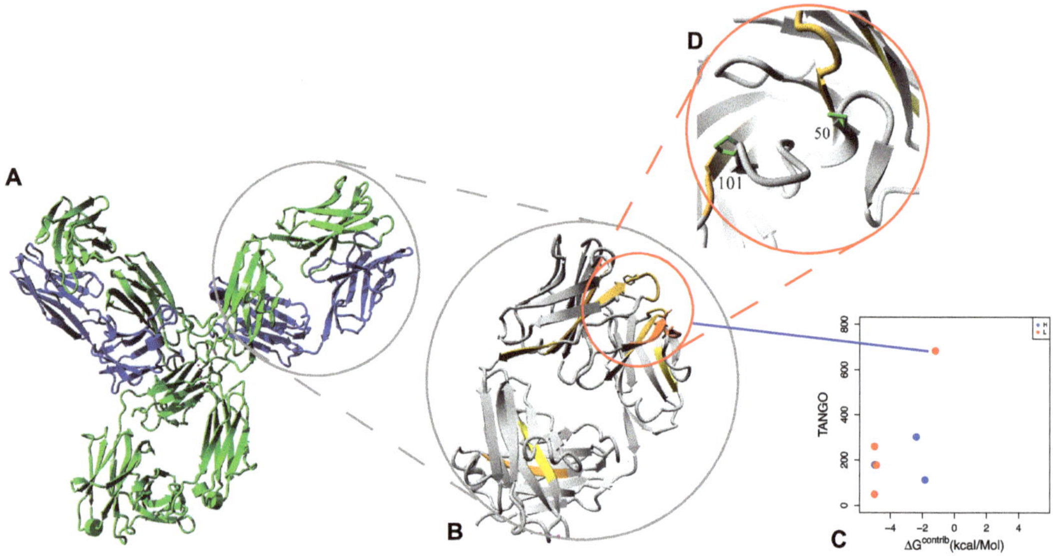

Fig. 5 Application of the SolubiS method on mAb2. (**a**) Crystal structure of a full monoclonal antibody (PDB ID: 5DK3 [46]). Blue: light chain, green: heavy chain. (**b**) Crystal structure of the Fab region of mAb2 (PDB ID: 2FJF [47]) with APRs highlighted. Small red circle shows the location of the "critical"APRs in the variable domain. (**c**) Stretch-plot of the Fab region of mAb2 with the light chain critical APR indicated. (**d**) Top view of the antigen-binding domain of an aggregation resistant mutant (SL50K/FH101P) with introduced mutations in green. Images of protein structures were generated using YASARA Structure [27]

mβ2m to achieve much higher serum levels without forming amyloid deposits. In line with the experiments performed with these two proteins, the SolubiS score is high for hβ2m and zero for mβ2m (Table 1).

3.4 Monoclonal Antibodies

Monoclonal antibodies are the largest class of protein therapeutics and continue to grow. Antibodies or immunoglobulins are large Y-shaped polypeptides composed of two heavy chains, containing four Ig-domains each, and two light chains, containing two Ig domains each (Fig. 5a). This backbone structure is relatively constant and evolutionarily optimized to be soluble at the concentrations required in the blood. However, the antigen binding loops, or complementarity determining regions (CDRs) are hypervariable sequences that make up the loops of the terminal Ig domains of both heavy and light chains. We recently analysed experimentally characterized antibody sequences and showed that the aggregation behavior of monoclonals under mild temperatures is largely determined by the APRs that occur in the CDRs, since they are selected ad hoc to provide good binding to a specific antigen and are not optimized for solubility (Fig. 5b). We demonstrated that the SolubiS score can be used to accurately distinguish between aggregation prone and not-aggregation prone monoclonal sequences, potentially providing a method to prioritize candidates for drug development [15]. Moreover, we analysed a monoclonal against the vascular endothelial growth factor VEGF in more detail. This antibody contained two APRs in the complementarity determining regions (CDRs) and stood out in the stretch-plot (Fig. 5b, c). We used SolubiS to design mutations targeting both APRs (Fig. 5d), which resulted in mutant antibodies with improved aggregation resistance under temperature stress as well as long-term storage conditions. However, perhaps equally important, the production yield of these mutants was up to five-fold higher than the wild type. Single mutants showed significant improvements in biophysical properties, but once more the combining of mutations resulted in the largest improvements.

3.5 Antibody Variable Domains

A final example can be found in mutational studies by Dudgeon et al. aimed at lowering the aggregation of antibody variable domains by combinatorial means [2]. Phage display was used to find mutations that decreased the aggregation propensity of human antibody variable domains. Variable domains were chosen that belong to the most common families in human repertoire (V_H3, $V_\kappa1$). Mutations that protected against aggregation clustered in the light chain variable domain CDR2 and in the heavy chain variable domain CDR1. Two mutations were found that improved the variable domain the most (52D in the light chain and 30D in the heavy chain). These findings were retrofitted to a single chain F_v (scF_v) of trastuzumab (Herceptin), a monoclonal that binds the extracellular domain of the HER2 receptor (Fig. 6a). The mutations reduced the aggregation propensity of Herceptin and the stretch-plots of WT Herceptin (Fig. 6b, c) and its double mutant can confirm that, as well as the SolubiS Score (Table 1). Interestingly these two mutations have been patented (StAbilize™ Technology) and gave rise to a company (Solvanix). Coincidentally the APR in Herceptin CDR L2 is identical to the APR in mAb2 in [15], where different mutations were chosen to increase aggregation resistance and yield.

4 Conclusion

Protein aggregation is a problematic phenomenon often observed when proteins are employed as therapeutic agents, thereby removing them from their native biological context and naturally occurring concentration. In other words, therapy often requires the production, storage and application of proteins above their natural solubility. And whereas buffer formulation can provide a quick fix to suppress the aggregation, this will not fundamentally eliminate the possibility. In contrast, altering the primary sequence has the potential to eliminate the aggregation propensity, increasing product safety.

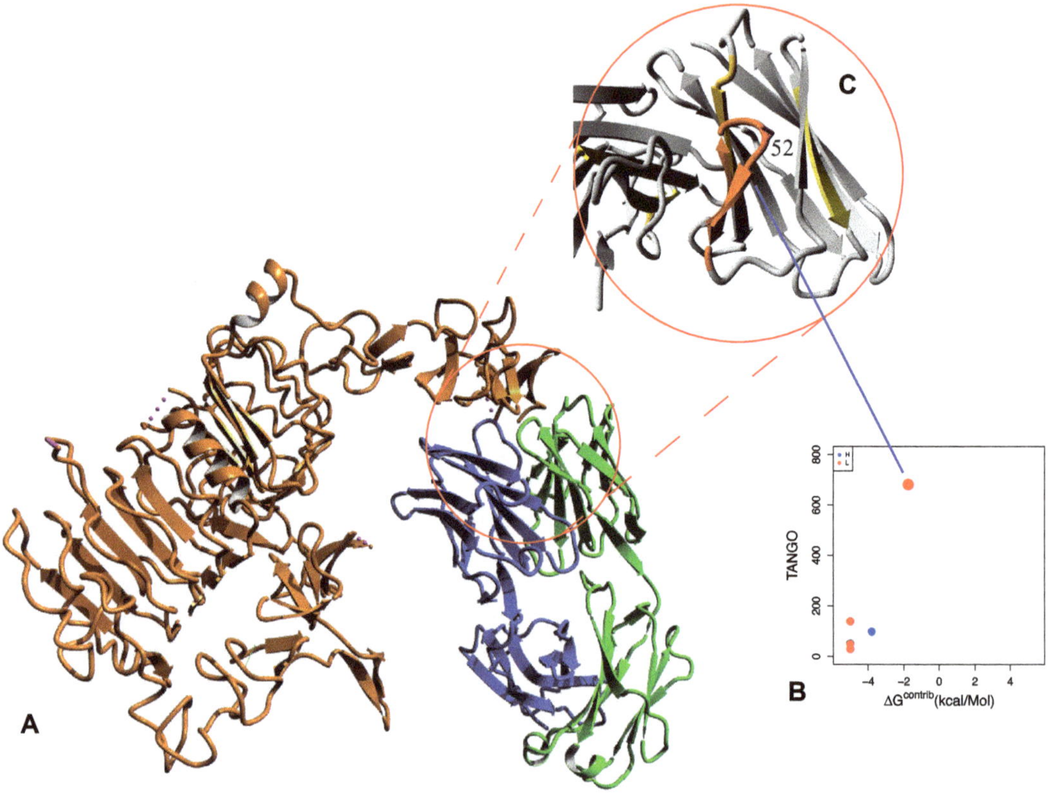

Fig. 6 Structural context of trastuzumab (Herceptin) APRs. (**a**) Crystal structure of Herceptin wild-type Fab bound to its target. (PDB ID: 1N8Z [48]). Orange: extracellular domain of human HER2 receptor, green: light chain, blue: heavy chain. (**b**) Stretch-plot of Herceptin wild-type F_V region based on the variable domains of a Herceptin crystal structure (PDB ID: 4HKZ [49]). (**c**) Location of "critical" APR in Herceptin wild-type F_V in the light chain. Images of protein structures were generated using YASARA Structure [27]

SolubiS is a rational protein design method to minimize protein aggregation that combines prediction of aggregation hot spots in the primary sequence of a protein with the calculation of the thermodynamic stability of the structure. By screening many mutations in silico, we were able to show for diverse protein molecules, including monoclonal antibodies, that it is possible to significantly improve their aggregation behavior by a few well-chosen mutations, without affecting their desired function.

The method is publicly available at http://solubis.switchlab.org.

Acknowledgments

R.V.D.K. was supported by Boehringer Ingelheim Pharma GmbH & Co. The VIB Switch Laboratory was supported by grants from VIB, University of Leuven, the Funds for Scientific Research

Flanders (FWO), the Flanders Institute for Science and Technology (IWT), and the Federal Office for Scientific Affairs of Belgium (Belspo), IUAP P7/16, and by the European Research Council under the European Union's Horizon 2020 Framework Programme, ERC Grant agreement 647458 (MANGO) to J.S.

References

1. Garidel P, Kebbel F (2010) Protein therapeutics and aggregates characterized by photon correlation spectroscopy. BioProcess Int
2. Dudgeon K, Rouet R, Kokmeijer I et al (2012) General strategy for the generation of human antibody variable domains with increased aggregation resistance. Proc Natl Acad Sci U S A 109:10879–10884. https://doi.org/10.1073/pnas.1202866109
3. Kumar S, Singh SK, Wang X et al (2011) Coupling of aggregation and immunogenicity in biotherapeutics: T- and B-cell immune epitopes may contain aggregation-prone regions. Pharm Res 28:949–961. https://doi.org/10.1007/s11095-011-0414-9
4. Vázquez-Rey M, Lang DA (2011) Aggregates in monoclonal antibody manufacturing processes. Biotechnol Bioeng 108:1494–1508. https://doi.org/10.1002/bit.23155
5. Roberts CJ (2014) Protein aggregation and its impact on product quality. Curr Opin Biotechnol 30C:211–217. https://doi.org/10.1016/j.copbio.2014.08.001
6. Wang W, Singh SK, Li N et al (2012) Immunogenicity of protein aggregates—concerns and realities. Int J Pharm 431:1–11. https://doi.org/10.1016/j.ijpharm.2012.04.040
7. Hamodrakas SJ (2011) Protein aggregation and amyloid fibril formation prediction software from primary sequence: towards controlling the formation of bacterial inclusion bodies. FEBS J 278:2428–2435. https://doi.org/10.1111/j.1742-4658.2011.08164.x
8. De Baets G, Schymkowitz J, Rousseau F (2014) Predicting aggregation-prone sequences in proteins. Essays Biochem 56:41–52. https://doi.org/10.1042/bse0560041
9. Belli M, Ramazzotti M, Chiti F (2011) Prediction of amyloid aggregation in vivo. EMBO Rep 12:657–663. https://doi.org/10.1038/embor.2011.116
10. Fitzpatrick AW, Knowles TPJ, Waudby CA et al (2011) Inversion of the balance between hydrophobic and hydrogen bonding interactions in protein folding and aggregation. PLoS Comput Biol. https://doi.org/10.1371/journal.pcbi.1002169
11. Castillo V, Ventura S (2009) Amyloidogenic regions and interaction surfaces overlap in globular proteins related to conformational diseases. PLoS Comput Biol. https://doi.org/10.1371/journal.pcbi.1000476
12. Rousseau F, Wilkinson H, Villanueva J et al (2006) Domain swapping in p13suc1 results in formation of native-like, cytotoxic aggregates. J Mol Biol 363:496–505. https://doi.org/10.1016/j.jmb.2006.07.061
13. Wang X, Singh SK, Kumar S (2010) Potential aggregation-prone regions in complementarity-determining regions of antibodies and their contribution towards antigen recognition: a computational analysis. Pharm Res 27:1512–1529. https://doi.org/10.1007/s11095-010-0143-5
14. Bennett MJ, Sawaya MR, Eisenberg D (2006) Deposition diseases and 3D domain swapping. Structure 14:811–824. https://doi.org/10.1016/j.str.2006.03.011
15. van der Kant R, Karow-Zwick AR, Van Durme J et al (2017) Prediction and reduction of the aggregation of monoclonal antibodies. J Mol Biol 429:1244–1261. https://doi.org/10.1016/j.jmb.2017.03.014
16. Balch WE, Morimoto RI, Dillin A, Kelly JW (2008) Adapting proteostasis for disease intervention. Science 319:916–919. https://doi.org/10.1126/science.1141448
17. Otzen DE, Kristensen O, Oliveberg M (2000) Designed protein tetramer zipped together with a hydrophobic Alzheimer homology: a structural clue to amyloid assembly. Proc Natl Acad Sci U S A 97:9907–9912. https://doi.org/10.1073/pnas.160086297
18. Richardson JS, Richardson DC (2002) Natural beta-sheet proteins use negative design to avoid edge-to-edge aggregation. Proc Natl Acad Sci U S A 99:2754–2759. https://doi.org/10.1073/pnas.052706099
19. Rousseau F, Schymkowitz J, Serrano L (2006) Protein aggregation and amyloidosis: confusion of the kinds? Curr Opin Struct Biol 16:118–126. https://doi.org/10.1016/j.sbi.2006.01.011
20. Monsellier E, Chiti F (2007) Prevention of amyloid-like aggregation as a driving force of

protein evolution. EMBO Rep 8:737–742. https://doi.org/10.1038/sj.embor.7401034
21. Van Durme J, Maurer-Stroh S, Gallardo R et al (2009) Accurate prediction of DnaK-peptide binding via homology modelling and experimental data. PLoS Comput Biol 5:e1000475. https://doi.org/10.1371/journal.pcbi.1000475
22. De Baets G, Reumers J, Delgado Blanco J et al (2011) An evolutionary trade-off between protein turnover rate and protein aggregation favors a higher aggregation propensity in fast degrading proteins. PLoS Comput Biol 7:e1002090. https://doi.org/10.1371/journal.pcbi.1002090
23. Tartaglia GG, Pechmann S, Dobson CM, Vendruscolo M (2009) A relationship between mRNA expression levels and protein solubility in E. coli. J Mol Biol 388:381–389. https://doi.org/10.1016/j.jmb.2009.03.002
24. Fernandez-Escamilla A-M, Rousseau F, Schymkowitz J, Serrano L (2004) Prediction of sequence-dependent and mutational effects on the aggregation of peptides and proteins. Nat Biotechnol 22:1302–1306. https://doi.org/10.1038/nbt1012
25. Schymkowitz J, Borg J, Stricher F et al (2005) The FoldX web server: an online force field. Nucleic Acids Res 33:W382–W388. https://doi.org/10.1093/nar/gki387
26. Ganesan A, Siekierska A, Beerten J et al (2016) Structural hot spots for the solubility of globular proteins. Nat Commun 7:10816. https://doi.org/10.1038/ncomms10816
27. Krieger E, Vriend G (2002) Models@home: distributed computing in bioinformatics using a screensaver based approach. Bioinformatics 18:315–318
28. Arnold K, Bordoli L, Kopp J, Schwede T (2006) The SWISS-MODEL workspace: a web-based environment for protein structure homology modelling. Bioinformatics 22:195–201. https://doi.org/10.1093/bioinformatics/bti770
29. Yang J, Yan R, Roy A et al (2015) The I-TASSER suite: protein structure and function prediction. Nat Methods 12:7–8. https://doi.org/10.1038/nmeth.3213
30. Hooft R, Sander C, Vriend G (1996) Verification of protein structures: side-chain planarity. J Appl Crystallogr 29:714–716
31. Van Durme J, De Baets G, van der Kant R et al (2016) Solubis: a webserver to reduce protein aggregation through mutation. Protein Eng Des Sel 29:285–289. https://doi.org/10.1093/protein/gzw019
32. De Baets G, Van Durme J, van der Kant R et al (2015) Solubis: optimize your protein. Bioinformatics 31:2580–2582. https://doi.org/10.1093/bioinformatics/btv162
33. Van Durme J, Delgado J, Stricher F et al (2011) A graphical interface for the FoldX forcefield. Bioinformatics 27:1711–1712. https://doi.org/10.1093/bioinformatics/btr254
34. Brady RO, Gal AE, Bradley RM et al (1967) Enzymatic defect in Fabry's disease. Ceramidetrihexosidase deficiency. N Engl J Med 276:1163–1167. https://doi.org/10.1056/NEJM196705252762101
35. Eng CM, Desnick RJ (1994) Molecular basis of Fabry disease: mutations and polymorphisms in the human alpha-galactosidase a gene. Hum Mutat 3:103–111. https://doi.org/10.1002/humu.1380030204
36. Collier RJ (2009) Membrane translocation by anthrax toxin. Mol Asp Med 30:413–422. https://doi.org/10.1016/j.mam.2009.06.003
37. Brey RN (2005) Molecular basis for improved anthrax vaccines. Adv Drug Deliv Rev 57:1266–1292. https://doi.org/10.1016/j.addr.2005.01.028
38. Singh S, Singh A, Aziz MA et al (2004) Thermal inactivation of protective antigen of bacillus anthracis and its prevention by polyol osmolytes. Biochem Biophys Res Commun 322:1029–1037. https://doi.org/10.1016/j.bbrc.2004.08.020
39. Petosa C, Collier RJ, Klimpel KR et al (1997) Crystal structure of the anthrax toxin protective antigen. Nature 385:833–838. https://doi.org/10.1038/385833a0
40. Ganesan A, Watkinson A, Moore BD (2012) Biophysical characterisation of thermal-induced precipitates of recombinant anthrax protective antigen: evidence for kinetically trapped unfolding domains in solid-state. Eur J Pharm Biopharm 82:475–484. https://doi.org/10.1016/j.ejpb.2012.05.019
41. Pashley CL, Hewitt EW, Radford SE (2016) Comparison of the aggregation of homologous β2-microglobulin variants reveals protein solubility as a key determinant of amyloid formation. J Mol Biol 428:631–643. https://doi.org/10.1016/j.jmb.2016.01.009
42. Lieberman RL, D'aquino JA, Ringe D, Petsko GA (2009) Effects of pH and iminosugar pharmacological chaperones on lysosomal glycosi-

dase structure and stability. Biochemistry 48:4816–4827. https://doi.org/10.1021/bi9002265
43. Santelli E, Bankston LA, Leppla SH, Liddington RC (2004) Crystal structure of a complex between anthrax toxin and its host cell receptor. Nature 430:905–908. https://doi.org/10.1038/nature02763
44. Eichner T, Kalverda AP, Thompson GS et al (2011) Conformational conversion during amyloid formation at atomic resolution. Mol Cell 41:161–172. https://doi.org/10.1016/j.molcel.2010.11.028
45. Rudolph MG, Shen LQ, Lamontagne SA et al (2004) A peptide that antagonizes TCR-mediated reactions with both syngeneic and allogeneic agonists: functional and structural aspects. J Immunol 172:2994–3002
46. Scapin G, Yang X, Prosise WW et al (2015) Structure of full-length human anti-PD1 therapeutic IgG4 antibody pembrolizumab. Nat Struct Mol Biol 22:953–958. https://doi.org/10.1038/nsmb.3129
47. Fuh G, Wu P, Liang W-C et al (2006) Structure-function studies of two synthetic anti-vascular endothelial growth factor Fabs and comparison with the Avastin fab. J Biol Chem 281:6625–6631. https://doi.org/10.1074/jbc.M507783200
48. Cho H-S, Mason K, Ramyar KX et al (2003) Structure of the extracellular region of HER2 alone and in complex with the Herceptin fab. Nature 421:756–760. https://doi.org/10.1038/nature01392
49. Donaldson JM, Zer C, Avery KN et al (2013) Identification and grafting of a unique peptide-binding site in the fab framework of monoclonal antibodies. Proc Natl Acad Sci U S A 110: 17456–17461. https://doi.org/10.1073/pnas.1307309110
50. Langley DB, Roome B, Christ D (2014) Retrofitting antibodies with stabilizing mutations. Herceptin scFv mutant K30D/S52D. RCSB PDB. To be published. doi: https://doi.org/10.2210/pdb4x4y/pdb

Index

Cláudio M. Gomes (ed.), *Protein Misfolding Diseases: Methods and Protocols*, Methods in Molecular Biology, vol. 1873, https://doi.org/10.1007/978-1-4939-8820-4, © Springer Science+Business Media, LLC, part of Springer Nature 2019

U

V

W

X

Y

Lightning Source UK Ltd.
Milton Keynes UK
UKHW05n1614231018
331042UK00004BB/124/P

9 781493 988198